Catalysis –
Science and Technology

Volume 6

EDITORS:

PROF. DR. J. R. ANDERSON (Melbourne/Australien)
PROF. DR. M. BOUDART (Stanford/USA)

CONTRIBUTORS:

J. BUTT, K. FOGER, F. G. GARIN,
V. GIANNINI, G. MAIRE,
J. PASQUOUN

CATALYSIS –
Science and Technology

Volume 6

With 111 Figures and 42 Tables

Akademie-Verlag · Berlin
1985

Die
Originalausgabe erscheint
im Springer-Verlag Berlin·Heidelberg
New York

Vertrieb
ausschließlich für alle Staaten mit Ausnahme
der sozialistischen Länder:
Springer-Verlag Berlin·Heidelberg
New York

Vertrieb
für die sozialistischen Länder:
Akademie-Verlag Berlin

Erschienen im
Akademie-Verlag, 1086 Berlin, Leipziger Straße 3–4

Lizenznummer: 200 · 100/498/84
Printed in the German Democratic Republic
Gesamtherstellung: VEB Druckerei „Thomas Müntzer", 5820 Bad Langensalza
Umschlaggestaltung: Eckhard Steiner
LSV 1215
Bestellnummer: 763 398 0 (6850)
13200

General Preface to Series

In one form or another catalytic science reaches across almost the entire field of reaction chemistry, while catalytic technology is a cornerstone of much of modern chemical industry. The field of catalysis is now so wide and detailed, and its ramifications are so numerous, that the production of a thorough treatment of the entire subject is well beyond the capability of any single author. Nevertheless, the need is obvious for a comprehensive reference work on catalysis which is thoroughly up-to-date, and which covers the subject in depth at both a scientific and at a technological level. In these circumstances, a multi-author approach, despite its well-known drawbacks, seems to be the only one available.

In general terms, the scope of *Catalysis: Science and Technology* is limited to topics which are, to some extent at least, relevant to industrial processes. The whole of heterogeneous catalysis falls within its scope, but only biocatalytic process which have significance outside of biology are included. Ancillary subjects such as surface science, materials properties, and other fields of catalysis are given adequate treatment, but not to the extent of obscuring the central theme.

Catalysis: Science and Technology thus has a rather different emphasis from normal review publications in the field of catalysis: here we concentrate more on important established material, although at the same time providing a systematic presentation of relevant data. The opportunity is also taken, where possible, to relate specific details of a particular topic in catalysis to established principles in chemistry, physics, and engineering, and to place some of the more important features into a historical perspective.

Because the field of catalysis is one where current activity is enormous and because various topics in catalysis reach a degree of maturity at different points in time, it is not expedient to impose a preconceived ordered structure upon *Catalysis: Science and Technology* with each volume devoted to a particular subject area. Instead, each topic is dealt with when it is most appropriate to do so. It will be sufficient if the entire subject has been properly covered by the time the last volume in the series appears. Nevertheless, the Editors will try to organize the subject matter so as to minimize unnecessary duplication between chapters, and to impose a reasonable uniformity of style and approach. Ultimately, these aspects of the presentation of this work must remain the responsibility of the Editors, rather than of individual authors.

The Editors would like to take this opportunity to give their sincere thanks to all the authors whose labors make this reference work possible. However, we all stand in debt to the numerous scientists and engineers whose efforts have built the discipline of catalysts into what it is today: we can do no more than dedicate these volumes to them.

Preface

For catalytic practitioners who are concerned with laboratory studies of reaction mechanisms, as often as not catalyst deactivation is treated as a nuisance to be ignored or factored out of the experimental results. However, the engineer concerned with the design and operation of real catalysts and processes cannot afford this luxury: for him deactivation and the need for regeneration are inevitable facts of life which need to be treated as quantified design parameters. The first chapter in this volume by Prof. J. B. Butt deals with catalyst deactivation and regeneration as processes in their own right, and shows how they are to be approached from kinetic and design points of view.

Catalytic olefin polymerization spans a very wide field in catalytic process chemistry and technology. Processes of this sort range from the generation of high volume products such as polyethylene and polypropylene, through more specialized commercial products, to conversions that still remain laboratory curiosities. The reaction chemistry is, in detail, often very complex. However, because of the insight provided by organometallic reaction chemistry, many of the polymerization mechanisms are reasonably well understood, and the way in which product stereospecificity may be obtained is also understood in considerable detail. This highly complex subject is reviewed in detail in the second chapter of this volume by Prof. I. Pasquon and Dr. G. Giannini.

Skeletal reactions of hydrocarbons, skeletal isomerization and hydrogenolysis, occur over a number of metallic catalysts. These reactions have been studied rather intensively for their own intrinsic interest, and because they provide very good examples of the way in which reaction selectivity is related to catalyst structure. From work which has involved the use of both dispersed metal catalysts and single crystal catalysts, together with carbon isotopic tracer studies, some of the main mechanistic features of these processes are now reasonably well understood, and this is the subject of the third chapter in this volume by Dr. G. Maire and Dr. F. G. Garin.

The generation of catalysts consisting of small metallic particles dispersed over a support has a long history in catalysis as a means of maximizing the available metallic surface area. Although widely practiced, this concept was developed to a high level of sophistication with the introduction several decades ago of dispersed platinum catalysts for catalytic reforming in the petroleum refining industry. Catalytic scientists and materials scientists have long been fascinated by the problem of characterizing the very small metal particles which such catalysts contain, and assessing whether such particles have properties which differ from those of the bulk metal. These problems become even more acute if one has to consider multimetallic rather than monometallic systems. The chapter by Dr. K. Foger provides a general review of this subject in which emphasis is particularly placed upon characterization and properties.

Contents

List of Contributors

Professor John B. Butt
Department of Chemical Engineering
Northwestern University
Evanston, IL 60201, USA

Dr. K. Foger
CSIRO Div. of Materials Science
Catalysis and Surface Science Lab.
University of Melbourne
Parkville, Victoria 3052, Australia

Dr. F. G. Garin
Institut de Chimie
Université Louis Pasteur de Strasbourg
4, rue Blaise Pascal
F-67008 Strasbourg, France

Dr. Umberto Giannini
Istituto Guido Donegani
Montedison
V. G. Fauser, 4
I-28100 Novara, Italy

Dr. G. L. C. Maire
Institut de Chimie
Université Louis Pasteur de Strasbourg
4, rue Blaise Pascal
F-67008 Strasbourg, France

Professor Italo Pasquon
Dipartimento di Chimica Industriale
e Ingegneria Chimica "Giulio Natta"
del Politecnico Piazza Leonardo da Vinci 32
I-20133 Milano, Italy

Chapter 1

Catalyst Deactivation and Regeneration

John B. Butt

Department of Chemical Engineering, Northwestern University,
Evanston, Il 60201, USA

Contents

1. Introduction

If one considers the entire realm of chemical kinetics, one distinguishing feature of reaction on surfaces is that the rates of such reactions tend to decrease as time increases under otherwise steady state conditions. Such changes are seldom known in homogeneous reactions in which catalytic surfaces are not involved. From the point of view of experimental observation, such changes may or may not be correlated with the kinetic variables of importance for the main reaction such as concentration or temperature since as will be seen they may be the result of parallel, independent chemical processes, or they may be associated with the chemical behavior of the reactants or products of the main reaction, or they may be the result of some degradation of the catalytic surface.

Any process, physical or chemical, that decreases the intrinsic activity of a catalytic surface can be termed deactivation. However, in this chapter we shall be concerned primarily with deactivation arising from chemical processes. The foremost of these are chemical poisoning of the surface, in which reaction mixture impurities are strongly chemisorbed on active sites and block access of reactant molecules to the surface, and coke deposition in hydrocarbon reactions in which hydrogen-deficient carbonaceous residues accrue on the surface, similarly blocking access of reactant molecules. Other chemical mechanisms of deactivation that are not so widely discussed include deposition of materials other than coke on the surface — notably various metals in certain applications — and transformations associated with the solid state chemistry of the catalyst. Physical processes of deactivation are associated with sintering or agglomeration of metal crystallites in supported metal catalysis, collapse of the pore structure in oxide catalysts and support materials, and volatilization of active catalytic components. In some cases chemical and physical processes of deactivation are interrelated as for example in the case of a solid state transformation that leads to sintering. Life is always grey.

Life is also rendered complicated in the analysis of deactivation phenomena, either chemical or physical, by the wide variety of ways various workers have chosen to define "activity". Thus one may encounter one definition as the temperature required for set conversion and fixed space velocity or, just the other way around, conversion attained for set temperature and space velocity. Or, why not set the conversion and temperature and use the space velocity as an activity measure? All these, as well as more fundamental comparisons (*i.e.*, turnover frequency) have been reported in the literature and in attempting to make comparisons of any latitude one is often faced with the uncertainties associated with interconversion of these various measures. Hence, in what follows concerning the discussion of literature data it should be kept in mind that a certain amount of caution should be exercised; for the most part comparisons are qualitative rather than quantitative.

In most cases of practical catalysis, the reaction selectivity is as important, if not more so, than activity. Accordingly, deactivation effects on selectivity can be more important than those on activity and well documented cases have been reported in the literature in which relatively minor deactivation (on the basis of activity change) has been accompanied by changes in selectivity of an order of magnitude. Alterations in selectivity are always a factor to be kept in mind when trying to cope with catalyst deactivation.

At the opposite, hopeful, end of the scale from deactivation is regeneration. In realistic terms, the practicality of regeneration depends upon the mechanism by which the catalyst has been deactivated. The classical cases of regeneration processes and the literature dealing with them have almost all been concerned with the removal of coke deposits by oxidation (burning). The only other significant literature on regeneration deals with the redispersion of supported metals, which is excluded from consideration here. The regeneration of catalysts that have been subjected to chemical poisoning seems not to admit of any generalized treatment; in fact a good working rule has it that the best way to cope with poisoning is to avoid it in the first place. Unfortunately, such noble sentiments tend to camouflage reality.

The organization of this chapter is based on successive magnitudes of scale. We shall first consider some of the basic mechanisms of chemical deactivation via poisoning and coke formation. At the same time, we will wish to keep in mind subsequent implications in catalytic reaction engineering applications, so the development of appropriate kinetic models in light of the mechanisms will be important also at this stage. At the second level of scale we shall investigate several problems that arise when one considers deactivation within the context of the individual catalyst particle. Here there appears a significant complication to the classical Thiele problem of intraparticle diffusion and reaction, since one must now deal with the interrelation of three characteristic rates: chemical, transport, and deactivation. Then we graduate to the level of the chemical reactor and larger scale process implications. In both intraparticle and reactor applications, the effect of deactivation is to change a large number of steady state problems into unsteady state problems. This has very important practical implications for, as will be seen, consideration of the unsteady state nature of the problem addressed can sometimes lead one into strategies of catalyst selection or reactor operation that never would be considered were the catalyst immortal. Catalyst deactivation also leads to a number of well posed problems in reactor design and operation, and there exists a flourishing literature on reactor optimization in the face of catalyst decay. Unfortunately, these become very complicated mathematically and we must consider such problems outside the scope of the present offering.[1]

Finally, the chapter is not a review. Literature references will be rather sharply focused to the problem areas discussed. A general review, constructed

---

[1] Indeed, most of the problems of intraparticle deactivation or deactivation in fixed bed reactors involved coupled systems of equations that require numerical solution. Since this is a chapter on deactivation and not on mathematics, we will provide the analytical formulation of such problems and examples of final results, but will not attempt detailed description of the intervening steps.

somewhat along the lines of this chapter was published some years ago by the writer [1], with subsequent treatments more directed toward the interaction of poisoning and coking on chemical process dynamics [2, 3]. Recent reviews of catalyst deactivation by coke formation have been given by Froment [4], of deactivation by poisoning by Hegedus and McCabe [5], and of deactivation by solid state transformation by Delmon and Grange [6] and Wanke and Flynn [7].

2. Poisoning

A. General Description

In the most general sense it is convenient to think of poisoning as the removal of sites from the active catalyst surface by the strong, competitive chemisorption of reaction mixture impurities. Such impurities are normally present in small amounts, and surface coverages of posions causing significant changes in activity or selectivity may be as low as small fractions of a monolayer.

Poisons can be characterized as to a number of properties; here we shall consider the primary distinctions to be between "temporary" and "permanent", and between "selective" and "nonselective". That between temporary and permanent is fairly evident from the terminology, resting upon the degree of reversibility of chemisorption on the surface. Since equilibrium chemisorption is a dynamic process and therefore strongly dependent upon temperature, it is possible for a poison to be permanent at a low temperature level and temporary at a higher temperature. For example, sulfur is a permanent poison for a metal such as platinum in lower temperature hydrogenation reactions but is not an important factor in poisoning Pt at the high temperatures encountered in automobile exhaust converters. We should also make here a distinction between temporary poisons and inhibitors of reaction rate that are either reactants or products of the main reaction. In conventional Langmuir-Hinshelwood rate correlations, inhibition terms often appear in the denominator of the correlation expressing depression of the reaction rate via reactant or product adsorption. Similarly, negative order dependence on reactant concentration may be obtained in power-law correlations. Some authors refer to this as self poisoning, but this is excluded from the definition here.

Observation of differences in the degree of deactivation with surface coverage of poison have led to the selective-nonselective terminology. This is best illustrated in Figure 1. For a nonselective poison each increment of poison on the surface results in an identical decrement in the catalyst activity and hence the activity-poison relationship is linear. This is often viewed as the result of essentially uniform sites on the catalyst and no interaction between poison molecules, the poisoning analogy of the Langmurian surface. For selective poisoning, as shown, the initial increment of poison on the surface results in a disproportionately large deactivation with subsequent uptake giving a diminishing rate of change of activity. It can be seen that at high uptakes rather large changes in poison on the surface have relatively small effects on activity. Explanations often suggested for selective poisoning are nonuniform distribution of surface site strengths, interaction between poison molecules, multiple site poisoning at low coverage with corresponding steric hindrance at high coverage, or all of the above. Note that in a reaction engineering context one might be able to take advantage of the flat response at high loadings by designing a reaction system to operate in this region, albeit at greatly diminished activity levels.

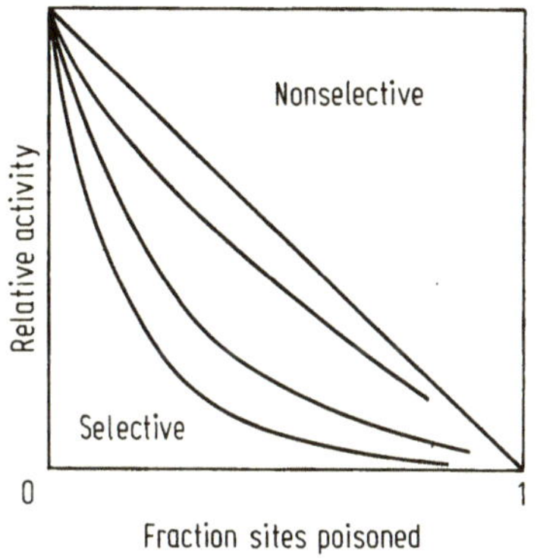

Figure 1. Qualitative activity-poison content relationships for selective and nonselective poisoning

One modification should be made at this point to the definition of poisons given in the first paragraph of this section. For more generality one should also include as potential poisons elements or compounds that may be present in the catalyst itself and that are not associated with the reaction mixture. For example, acidic catalysts such as alumina or silica-alumina are deactivated by small amounts of sodium or potassium residual from the method of preparation. Metal impurities found in small amounts in materials such as silica gel or alumina may also modify the catalytic properties of metals deposited on these supports.

B. Early Work on Metals and Oxides

Systematic investigation of catalyst poisoning has had its origins rather recently if one considers the time span from the present to the day of Berzelius. However, some observations of poisoning, if we interpret them in modern terms, are indeed very old.[1] One of the first observations of this sort was reported by Faraday [8], who among other things was interested in the reaction of hydrogen and oxygen in the presence of platinum of various forms. Such experiments were conducted in a glass apparatus and one would presume that in general the success of the experiment must have been measured by the force of the explosion destroying the apparatus. In one

[1] The author is indebted to Professor R. L. Burwell, Jr. for several stimulating discussions on the history of catalysis.

experiment, some ethylene was added to the hydrogen-oxygen mixture, and nothing happened for a long time. Finally the apparatus exploded in its customary fashion, presumably much to Faraday's satisfaction. One would say now that the ethylene was preferentially adsorbed on the platinum surface and only after it had been oxidized and removed could the violent coupling between hydrogen and the remaining oxygen occur. In this case, then, ethylene was acting as a temporary poison for the hydrogen-oxygen reaction. A number of examples of this sort can be found in the chemical literature of the 19th century, but the rationalization of catalysis itself, much less poisoning, was severely hampered by the absence of periodic classification of elements prior to 1860 and yet afterwards until the development of active site theory in the 1920–1930's.

In view of the above, it is not surprising that one does not find much specific reference to catalyst poisoning until the literature of the 1940–1950 period. One of the landmark publications of that period was the review of the poisoning of metals published by Maxted in 1951 [9]. Some earlier instances can be cited, including the poisoning by CO of the copper-catalyzed ethylene hydrogenation [10] and oxygen poisoning of ortho-para hydrogen conversion on tungsten [11], but Maxted was really the first to have sufficient data to attempt some systemization. Available to him was a large amount of information concerning hydrogenations, primarily on platinum and palladium, in the presence of a number of different poisons. Perhaps it was fortunate that the number of catalysts was somewhat limited, since this shifted the focus from the susceptibility of an individual metal to poisoning to the potential of various materials to be poisons. From this, a rather simple picture emerges. Since all metals active in chemisorption and catalysis must have available bonding orbitals (otherwise they would not be effective catalysts), one is then forced to consider the electronic configuration of candidate poison molecules. Potential nonmetallic poisons, for example, would be molecules containing the Group VB or VIB elements of proper electronic configuration. By the latter, in the most simple sense, is meant that unshared electron pairs or unoccupied orbitals must exist for chemisorption (poisoning) to occur. These were termed "unshielded" compounds by Maxted as, for example, arsine which is a strong poison for hydrogenation catalysts

$$\begin{array}{c} \mathrm{H}:\underset{\cdot\cdot}{\ddot{\mathrm{As}}}:\mathrm{H} \\ \mathrm{H} \end{array}$$

such as platinum and palladium. The reasonableness of this simple hypothesis can be further supported by the fact that when the unshielded compound is rendered into a shielded form it becomes non-toxic. In the case of arsine this was demonstrated by Maxted by examining the behavior of platinum in the decomposition of hydrogen peroxide. Here, under strong oxidizing conditions arsine is transformed to the arsenate, which is a shielded structure:

$$\left[\begin{array}{c} \mathrm{O} \\ \mathrm{O}:\underset{\cdot\cdot}{\ddot{\mathrm{As}}}:\mathrm{O} \\ \mathrm{O} \end{array}\right]^{-3}$$

and is non-toxic for the decomposition reaction on platinum. From experimental information such as this, Maxted developed the classifications shown in Table 1.

Table 1. Relationship of electronic structure to toxicity, after Maxted [9]

Toxic Compounds — Unshielded	Nontoxic Compounds — Shielded
Hydrogen Sulfide, Phosphine	Phosphate Ion
Sulfite Ion, Selenite, Tellurite	Sulfate Ion, Selenate, Tellurate
Organic Thiols	Sulfonic Acid
Organic Sulfides	Sulfone
Pyridine	Pyridinium Ion
Piperidine	Piperidinium Ion

What is not included or intimated in such classification schemes is the effectiveness of a given poison. In one view, this would be translated as to which of the curves of Figures 1 represents the behavior of a given poison on a given catalyst, and this cannot be predicted *a priori*. However, the basis for a general framework of rationalization of poisoning can be established even if we go hardly beyond the work of Maxted, which in the end has as its main value the clear establishment of competitive chemisorption as a mechanism. *Plus ça change, plus c'est la même chose.* The "strength" of a poison is thus directly related to the strength of a chemisorptive bond, and

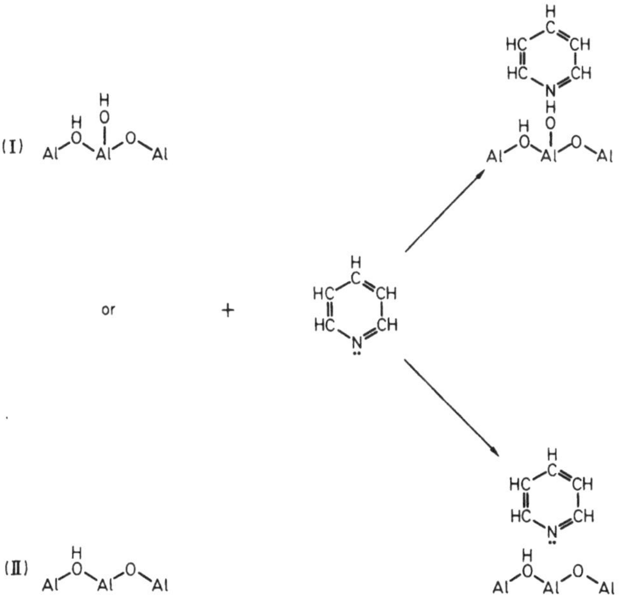

Figure 2. The titration of acid sites on alumina with pyridine. Chemical reactions. (Reproduced with permission from Butt, J. B., AIChE Jl., *22*, 1 (1976); 1976; American Institute of Chemical Engineers)

in turn the covalent chemisorptive bond should be analyzed in terms similar to those for bonds in covalent molecules. This in turn, however, suggests another level of complication depending upon the level of complexity involved [5]. The simplest category then would be poisoning of monofunctional catalysts with uniform sites (nonselective), next monofunctional catalysts with nonuniform sites (selective), next multifunctional catalysts, and so on. In the more detailed discussion of poisoning given later in this chapter we shall follow this classification.

Turning now to earlier studies of oxides, one quickly finds that much of that work was concerned with determining the interrelation between the acidity of oxide surfaces and their activity in acid-catalyzed reactions such as hydrocarbon isomerization or cracking. Again following the concept of competitive chemisorption, any basic compound should be strongly bonded to the acidic sites and serve as a poison for these acid-catalyzed reactions. The chemistry of such chemisorptions is by now well understood and is illustrated in Figure 2 for the chemisorption of the organic base pyridine on either the Brønsted (I) or Lewis (II) acid sites of alumina, and some results of an experiment [12] following up this idea are given in Figure 3. Here are presented the results of a series of experiments in which a silica/alumina catalyst was pre-poisoned by contact with varying amounts of a number of different organic bases and then examined for activity in that workhorse of model compound reactions, cumene cracking.[1] As suggested by the results of Maxted, it is fruitful to examine the potential of various materials

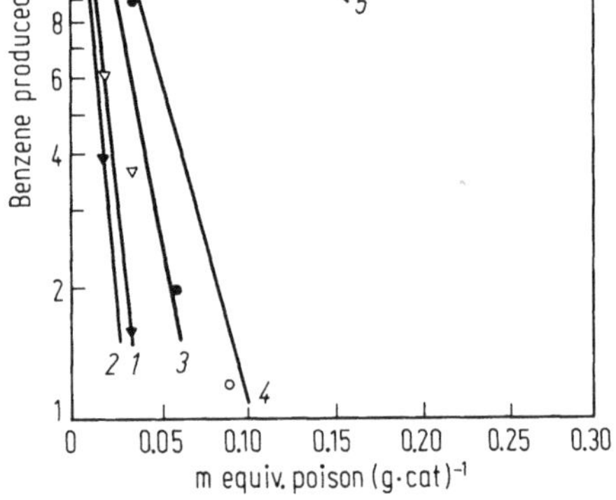

Figure 3. Poisoning by organic nitrogen compounds of cumene dealkylation on silica-alumina. 1, quinoline; 2, quinaldine; 3, pyrrole; 4, piperidine; 5, decylamine; 6, aniline. Temperature, 698 K, LHSV = 1.5. (Reproduced with permission from ref. [12])

[1] Note here that we are examining a particular type of activity correlation, as discussed before. In this case, relative evaluation is based on extent of conversion at fixed temperature and space velocity.

to be poisons; here the order is quinaldine > quinoline > pyrrole > piperidine > decylamine > aniline. One would wish for a direct correlation in terms of basic strength but this is not to be so directly since these compounds differ somewhat in their lability to the acidic surface. However, when correction is made for the extent of cracking of the individual poison molecules, the correlation with basic strength is direct. It is worth noting that extension of this work lead to a direct correlation between cracking activity for a number of different catalysts and acidity, as determined by quinoline chemisorption.

The studies of metals and oxides by workers such as Maxted and Mills, *et al.*, have set a pattern for much that has been done in the ensuing three decades. One caveat applies. Correlations of activity (however measured) *vs.* amount of poison of the form in Figure 1 has been popular in the literature and, as discussed before, have been used to imply selectivity or nonselectivity in terms of interaction with surface sites. In technological catalysts such as supported metals this implies that the intrinsic properties of the catalytic surface have been observed. We shall see later that mass transport effects can in some instances completely disguise the intrinsic poisoning properties of a catalyst in ways similar to kinetic disguises, so one should avoid hasty interpretation of such poisoning curves.

C. Homogeneous Surfaces

About the closest we can come to a homogeneous surface is a single crystal surface corresponding to a low Miller index. The last two decades have seen great strides in our ability to examine such surfaces, primarily via various types of electron spectroscopy under high vacuum conditions. The relationship of such studies to the catalysts of commerce have been questioned; however in the examination of direct poison-catalytic surface interactions, with none of the complications mentioned above, they cannot be surpassed. The most useful of these has probably been low energy electron diffraction (LEED) and Auger electron spectroscopy (AES), although a number of other methods have been used.

Platinum is perhaps the most extensively studied metal, and sulfur from a number of origins, the most investigated poison species. An excellent review is available concerning sulfur-poisoning of well characterized metal surfaces [13]. A good illustration is the results that have been reported for sulfur bonding on the platinum (100) surface, obtained by LEED [14, 15]. Two structures are observed. In the first, five sulfur atoms are arranged in cross-fashion:

each coordinated with four platinum atoms in the surface layer. This is called a centered c (2 × 2) structure and corresponds to half a monolayer of sulfur at full coverage. The second involves only the four corner sulfur atoms:

This is called a primitive p (2 × 2) structure and corresponds to one-fourth monolayer coverage at saturation.

While we wish to avoid overly general extrapolation from one set of results, the above are not atypical of the chemisorption of many types of potential poison molecules on homogeneous metal surfaces. The process appears to be geometrically complex, the nature of the geometry may vary with surface coverage (*i.e.*, reconstruction) and in many cases ensembles of surface metal atoms can be interacting with a single poison atom or molecule. It is of interest that the latter possibility was envisioned by Herington and Rideal many years ago [16].

Now, how does such chemisorption behavior of the poison atom correlate with the activity of the catalytic surface? To continue with the (100) platinum surface, we illustrate the results of Fischer and Kelemen [17] for sulfur poisoning of the reduction of carbon monoxide by nitric oxide in Figure 4. It is seen that the initial portion of the activity-surface coverage

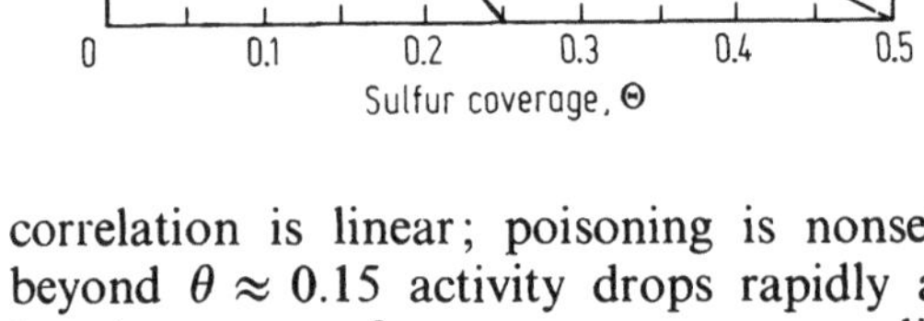

Figure 4. Relative rate of CO + NO reaction as a function of sulfur coverage on Pt (100). (Reproduced with permission from ref. [17])

correlation is linear; poisoning is nonselective in this region. However beyond $\theta \approx 0.15$ activity drops rapidly and the surface becomes totally inactive at a surface coverage corresponding to saturation via the (p 2 × 2) structure. Even though there is a platinum atom available at the center of the p (2 × 2) ensemble, there would be obvious hindrance of the bimolecular surface reaction. Thus, these poisoning results give a nice corroboration of the chemisorption studies and an understanding of the deactivation of the surface in rather simple geometric terms. Strict geometric interpretation, however, is probably oversimplified. One might suspect that the presence of sulfur on the surface could have some effect on the strength of bonding of the reactant molecules, even at relatively low coverages and, in fact, this has been demonstrated for carbon monoxide on (100) platinum [18]. Hence, even poisoning by simple molecules of homogeneous surfaces involves the entangled contributions of geometric and electronic factors that appear in so many aspects of catalysis.

D. Heterogeneous and Bifunctional Surfaces

It is probably safe to say that the moment one turns to any other than well defined single crystal surfaces, heterogeneity of some sort has entered the picture. For simplicity here we will consider heterogeneity to be a state in which there exist sites of different strengths for chemisorption of poison molecules on the catalytic surface. In this sense a bifunctional catalyst surface represents the ultimate in heterogeneity since there are two sets of sites that are involved with different catalytic reactions.

A major effect of heterogeneity is that the characterization of poisoning behavior becomes dependent not only on the poison involved but also on the reaction. A classical example of this is the work of Pines and Haag [19] on some isomerization and dehydration reactions on a series of modified aluminas. This has been discussed in detail in an earlier review [1], but remains one of the best documented studies of its sort. Central to the discussion here are the conversion results shown in Figure 5 for cyclohexene (CH) isomerization and butanol dehydration on a series of alkali-impregnated aluminas. Now the CH isomerization requires a 1° carbonium ion intermediate that in turn requires strongly acidic sites, whereas the alcohol dehydration should proceed on both strong and weak acidic sites. One would thus suspect that the activity of fresh catalysts would be higher for the latter reaction, and a given amount of alkali poison would have less effect than for CH isomerization. Both of these effects are supported by the

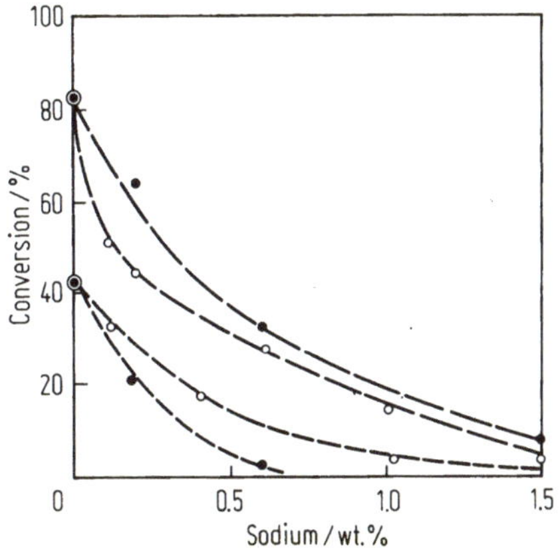

Figure 5. Isomerization of cyclohexane (dotted line) and dehydration of 1-butanol (dashed line) over a series of impregnated aluminas. —○—, NaOH impregnated; —●—, NaCl impregnated. Temperature, 623 K, LHSV = 2.0. (Reproduced with permission from ref. [19])

data in Figure 5. What happens if we now attempt to develop a correlation between activity and acidity similar to that of Mills *et al.* [12]? In the case of heterogeneous site distributions, this turns out to be catalyst and reaction-specific. Figure 6 demonstrates the result for CH isomerization on several

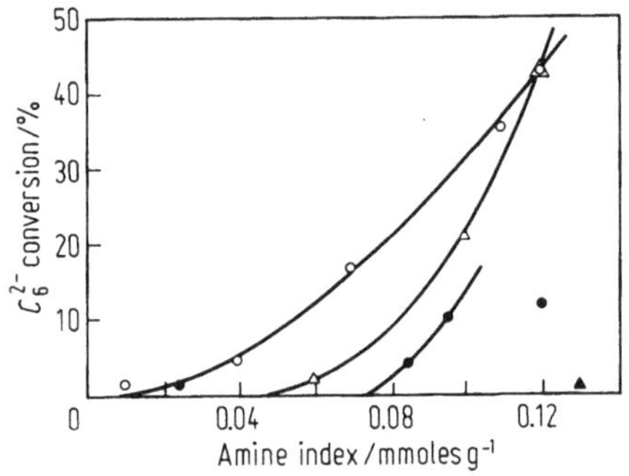

Figure 6. Cyclohexene conversion *vs.* acidity. ○ — NaOH impregnated; ● — from potassium aluminate; △ — NaCl impregnated; ▲ — from sodium aluminate. Temperature, 623 K, LHSV = 2.0. (Reproduced with permission from ref. [19])

series of catalysts of different preparation (poisoning) histories. For any given series there appears to be a relationship. However, generality is lost in the correlation perhaps because amine index is not a very good measure of strong acidity. By contrast, if the same correlation is attempted for butanol dehydration, less demanding in its acidity requirement, a good single relationship is obtained for all catalysts.

As mentioned above, bifunctional or polyfunctional catalysts represent in one sense the limit of heterogeneous surfaces. The Pt/Al_2O_3 family of reforming catalysts (and related alloys) serves as a basic example of multifunctionality, in which the metal plays the role of hydrogenation-dehydrogenation catalyst, and the support also plays the role of an acidic function catalyst. It is with such bifunctional catalysts that changes in selectivity upon poisoning can become particularly pronounced, since reaction sequences on such catalysts normally involve parallel or series pathways (or both) and if one function is deactivated more severely than the other, selectivity alteration is inevitable. Accounts of poisoning in supported metal catalysts have been given by Sterba and Haensel [20] and in multifunctional oxide catalysts by Knözinger [21].

As an example of the influence of series-parallel reaction structure on selectivity in the poisoning of bifunctional catalysts, consider the reaction of 1-butene in the presence of hydrogen on Rh/SiO_2 reported by Webb and Macnab [22]. In this case the reaction structure is parallel

$$n\text{-}C_4^{=} + H_2 \rightleftarrows n\text{-}C_4 \qquad (Rh)$$
$$n\text{-}C_4^{=} \rightleftarrows i\text{-}C_4 \qquad (SiO_2)$$

They investigated the individual rates of isomerization and hydrogenation after poisoning by mercury. The results as shown in Figure 7 indicate clearly the preferential poisoning of the metallic function for the hydrogenation reaction. More interestingly, let us follow the course of poisoning here from

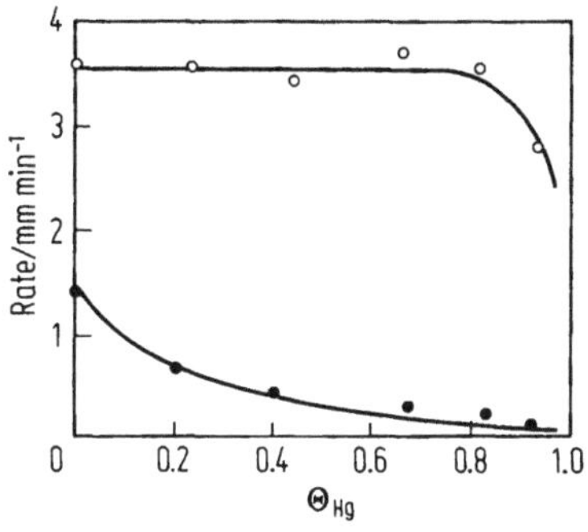

Figure 7. 1-butene isomerization and hydrogenation rates as a function of Hg coverage over Rh. —○—, isomerization; —●—, hydrogenation. (Reproduced with permission from ref. [22])

$\theta = 0$ to $\theta = 0.8$ from two points of view, (i) change in the total rate of reaction of *n*-butene (hydrogenation plus isomerization), and (ii) change in the selectivity of the reaction, defined as the ratio of the rate of hydrogenation to the rate of isomerization. The net change in activity given by (i) shows a decline of 25% upon mercury poisoning, however, the selectivity defined by (ii) changes by a factor of 8! Thus our concern about activity *vs.* selectivity changes on poisoning. Series mechanisms can be more complex. As a example here let us consider the detailed mechanism of isomerization of *n*-butane on Pt/Al_2O_3:

$$(n\text{-}C_4)_g \rightleftarrows (n\text{-}C_4^=)_a + H_2 \qquad (Pt)$$
$$(n\text{-}C_4)_a \rightarrow (i\text{-}C_4^=)_a \qquad (Al_2O_3)$$
$$H_2 + (i\text{-}C_4)_a \rightleftarrows (i\text{-}C_4)_g \qquad (Pt)$$

Now a poison selective for the platinum function has, in the limit, the ability to suppress all reaction of n-C_4 by intercepting the sequence in the first step. Conversely, one selective only for Al_2O_3 could in the limit suppress isomerization, but one might see some reaction of n-C_4 to n-$C_4^=$. It is apparent that as one goes to more complicated reactions than these simple hydrogenation-isomerization models, complex selectivity patterns can develop upon poisoning, and it well be important to understand poisoning chemistry in some detail.

3. Coking

A. General Description

It is informative at the outset to make some general comparisons between deactivation by coking and deactivation by poisoning. If we were to summarize as concisely as possible what has been said in the foregoing to describe poisoning, it would probably be to say that the poisoning process is understandable in terms of competitive chemisorption for the most part, and as a result is a fairly precise event from a chemical point of view. Understanding the nature of the active site and the nature of the poison provides a means for quantitative analysis of the overall deactivation process, although it is recognized that such understanding may only be difficultly achieved. Owing to this precision, it is possible to make rather careful definitions of the term "poisoning" and to worry about details of homogeneous and heterogeneous surfaces, etc. Coke formation and the deactivation resulting is almost an entirely different matter. The chemistry associated with coke formation is complex and not well understood, dependent upon the reaction and the catalyst, and the coke itself may alter its nature with age.

Indeed, we can start with two basic questions, neither of which has a general answer. These are (i) "... where did the coke come from ..." and, (ii) "... what do we mean by coke anyway ..."? One of the problems with coking is that people have not yet decided what to call it, since the phenomenon is referred to otherwise as fouling, self-poisoning or blocking. Let us give these terms a gentle burial. Basically what we mean by coking is the formation of hydrogen-deficient carbonaceous residues on the catalytic surface of empirical formula approximately CH. The chemical nature of such residues is at issue, as well as is the nature of the deactivation resulting from their formation. As far as chemical nature is concerned, the process has been visualized as a type of condensation polymerization on the surface resulting in large, polymeric structures that overlay active sites and physically block access of reactants; alternatively, the formation of polycyclic and heterocyclic aromatics has been described in which the mechanism of deactivation may be a combination of site blockage and specific chemical interaction akin to poisoning. Further to complicate matters is that in some cases the extent of coke formation is sufficient that the deposits can become extremely large macromolecules of dimension sufficient to fill and block access of reactant to the interior of catalyst pores, much less the active surface.

Since we will not address the question of pore blockage via coke formation specifically in later discussion, it may be appropriate to make a few comments on this here. Particularly to the point are the results of Levinter *et al.* [23] for coke formation on $SiO_2-Al_2O_3$ with styrene-benzene mixtures. In these experiments, all conducted at 773 K and LHSV $= 5\ cm^3\ h^{-1}$ (g catalyst)$^{-1}$, a limiting degree of coke formation was observed that varied with reaction conditions. A functional dependence of coke formation on styrene concentration was indicated, since benzene was inert for coke formation under the conditions employed; the extent of coking was also dependent upon catalyst average pore radius and particle size. It was shown that the limiting amount of coke formation, corresponding to the catalyst pore volume, was never attained, but was limited by catalyst pore blockage and consequent elmination of accessible surface area. In general this blockage was associated with preferential build-up of coke near the catalyst particle surface; some further experimental studies of this have been reported by Muno *et al.* [24] for cumene cracking on H-mordenite.

Although we have heavily qualified any precise description of coking to this point, there are some specific reactions that lie outside of the definition of coke we have adopted. In these, the element carbon itself, rather than hydrogen-deficient polymers or aromatics, is involved. Such cases include Fischer-Tropsch synthesis on iron-based catalysts, where the formation of metal carbide phases can occur[1], and steam reforming of methane with Ni catalysts in which the deposited material is a form of carbon. Even more striking is the high temperature decomposition of hydrocarbons on supported nickel. Here the solubility and diffusivity of carbon in nickel is sufficient that the carbon can diffuse through the supported metal crystallite, grow to a large agglomerate between the crystallite and support, and effect deactivation by detaching the metal from the support. Indeed there is no wonder in the fact that one can describe catalysis and deactivation as an adversary proceding.

B. The Origins of Coke Formation

Since coke formation has less well definable chemical bases than poisoning, literature accounts of deactivation by coking also often use somewhat different approaches than in the case of poisoning. Most notable is the extensive use of time-on-stream correlations relating the weight of coke on catalyst to time of catalyst utilization. It is important to note that the variable correlated is weight of coke, not activity, so that in general it is necessary to develop a separate relationship between weight of coke on catalyst and catalyst activity. Most such correlations derive from the early work of Voorhies [26] that proposed a power-law correlation of the form

$$C_c = a t_c^b \tag{1}$$

This is, of course, a linear relationship in logarithmic coordinates. Some typical data are shown in Figure 8. Often the catalyst activity factor, defined

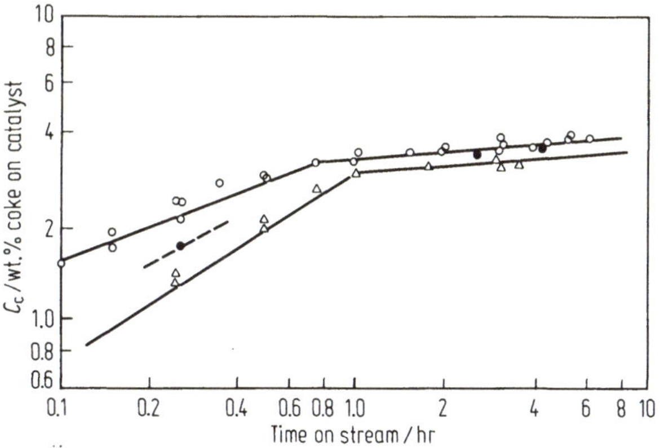

Figure 8. Voorhies Correlation for coke formation: cumene cracking on H-Mordenite at space velocity of 0.33 hr^{-1}. —○—, 503 K, 533 K; —●—, 513 K; —△—, 623 K

[1] Interestingly, the metal carbide phases often demonstrate enhanced activity for the synthesis compared to the reduced metal itself [25].

as fraction of original activity, s is inversely related to C_c by a form such as

$$s = 1/C_c^m \tag{2}$$

such that a corresponding activity-time relationship is

$$s \approx 1/t_c^n \tag{3}$$

with $n = mb$. The power law form of equation (3) is more conveniently approximated in some applications by an exponential function:

$$s = e^{-\alpha t_c} \tag{4}$$

in which the constant α assumes the significance of a characteristic decay constant. It is thus a characterization factor specific to reaction, catalyst and conditions. A single measurement of α has little value in discerning the origins of coke formation, but systematic investigation for a series of related reactions on a series of related catalysts can provide useful information, and we shall return to a discussion of this later on. A number of different forms for time-on-stream correlations have been proposed and used successfully in individual instances, however from a qualitative point of view they yield results in reaction/reactor analysis that are similar to those presented above, so we shall stay with equations (1)–(4) in subsequent illustrations.

Useful as the Voorhies correlation has been over the years, two incorrect inferences made early on have crept their way into the history and lore of coke deposition. These were that coke deposition is diffusion-limited and not dependent upon space velocity. The former was arrived at by the fact that the exponent b of equation (1) is often near 0.5, and the latter because early experiments did not cover a very large range of space velocities. One counterexample is all that is required to refute the concept of a universally diffusion-limited reaction; for example Ozawa and Bischoff [27] report values of b varying from 0.55 to 0.92 for ethylene cracking on a commercial silica/alumina catalyst in the temperature range 620–770 K. The question of space velocity dependence was examined in detail by Eberly *et al.* [28] with an affirmative result and subsequently substantiated by numerous reaction and reactor simulation studies. In spite of all this there persists the idea that coke formation can be treated as a conversion (concentration) independent process. To the extent that deactivation by coking can be correlated by time-on-stream, to be sure, concentration-dependency is camouflaged, but that does not mean that it is absent.

The question was asked earlier "... where did the coke come from ..."? In surveying a large amount of literature dealing with coke formation in various reaction systems (but mostly for catalytic cracking) one can make a general association in noncyclic materials with the degree of unsaturation, with reactivity generally increasing in the direction of olefins, dienenes, acetylenes. However in most reactions involving mixed feedstocks, the evidence points mostly to polycyclic aromatics as primarily coke precursors — materials such as naphthalene, phenanthrene, chrysene and pyrene with various alkyl substituents. In cracking reactions the side chains are first removed, so the bare rings may be the active intermediates in coking. An interesting study of this was reported by Appelby *et al.* [29], in which a series of aromatic, naphthenic, heterocyclic and paraffinic hydrocarbons were cracked on either silica-alumina or silica-zirconia-alumina catalysts. Reactions were generally conducted at 773 K, atmospheric pressure and flow rates on the order of 10 mol (liter catalyst)$^{-1}$ h^{-1}. After 15 min on stream the catalysts were removed and analyzed for coke content. Experiments with *n*-butene and butadiene showed formation of about 15 wt-% coke in the catalyst over the period and, interestingly, about half of this was of definite aromatic structure. Experiments with the series of cyclic compounds are summarized in Figure 9. If we confine our attention for simplicity to those

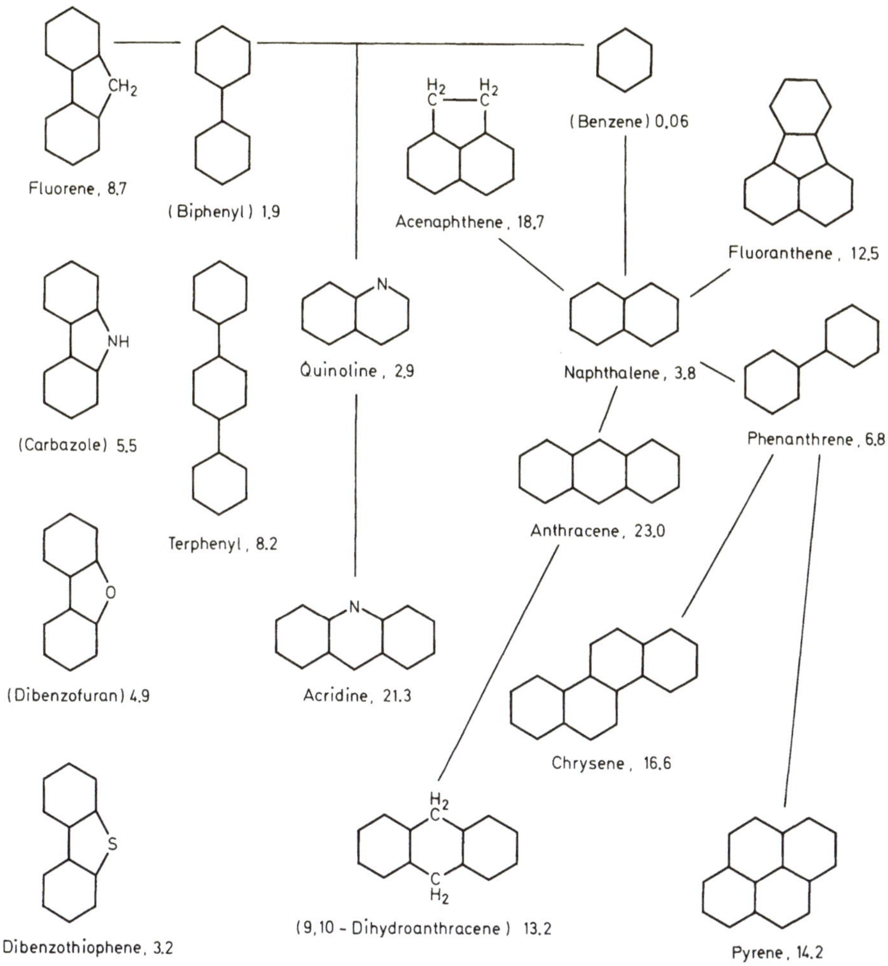

Figure 9. Ranking of coke formation from polycyclic and heterocyclic aromatics: weight percent coke on silica-alumina after 15 min at 773 K. (Reproduced with permission from ref. [29])

compounds not containing heteroatoms, it can be seen that as one proceeds in a general southeasterly direction on the figure, the amount of coke formed during the process period increases dramatically. X-ray diffraction analysis of the deposits indicated a turbostratic layer structure suggestive of condensed aromatic nuclei of high molecular weight. These results, together with the

Table 2. A carbonium ion mechanism for formation of higner aromatics from benzene and naphthalene. (Original H atoms and double bonds eliminated for simplicity)

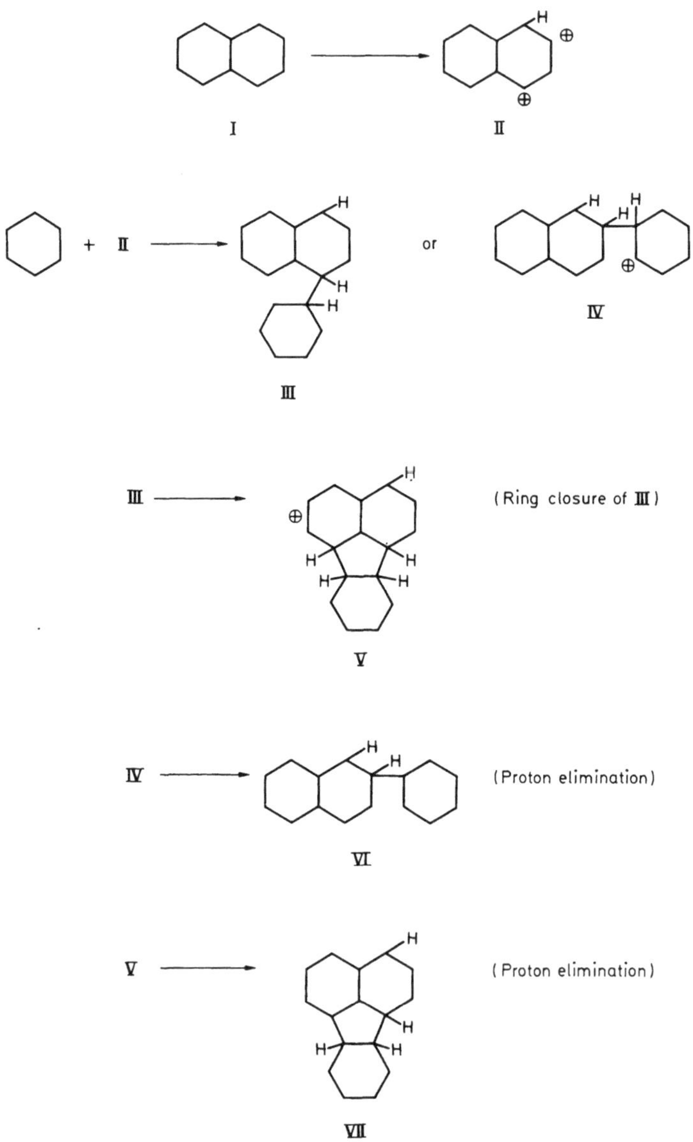

aromatic nature of coke formed from linear compounds such as butadiene point strongly to aromatic ring structures as important coke intermediates. Similar inferences were arrived at by Eberley *et al.* [28] *via* infra-red study of coke deposits on silica alumina resulting from cracking of a number of compounds, including hexadecane and 1-methylnaphthalene.

If we accept the circumstantial evidence pointing to the importance of aromatic ring structures, then it is possible to understand some of the chemistry involved through carbonium ion reactions. In cracking the catalysts are acidic and the aromatics basic. Hence, starting with benzene and naphthalene, we could envision the creation of high molecular weight aromatics as shown in Table 2.

Detailed structural factors may also play a role. Even though side chains are first removed in cracking this is not to say that alkyl subtituted aromatics, for example, cannot form coke precursors. One possible way to incorporate structural factors is simply to define the basicity of the compound with reference to the acidic catalyst. This approach was also investigated by Appleby et al. [29], with fruitful results.

The foregoing has dealt with studies involving model compounds; what of more realistic cases involving industrial processes such as hydrotreating or catalytic cracking where feeds are complex mixtures of many (thousands) of individual components? Understandably, such data are difficult to come by, but there exists an excellent case history in the literature for catalytic cracking [30–36]. The problem is rendered even more complicated because in addition to accounting for coke formation it is also necessary to determine how the kinetics of the main cracking reactions should be handled. In the case of Weekman and coworkers, the follwing reaction sequence was employed

$$C_1 \rightarrow a_1C_2 + a_2C_3 \qquad (k_0)$$
$$C_2 \rightarrow C_3 \qquad (k_1)$$

where C_1 is gas-oil feedstock, C_2 is the gasoline fraction and C_3 low molecular weight gas and coke. The time-on-stream correlation of equation (4) was used to correlate coke deposition effects on activity of a commercial silica alumina cracking catalyst. Using the sequence above for kinetic modeling of a large amount of data on cracking activity for numerous feedstocks, they were able to establish the correlations for kinetic constants, k_0 and k_1, and decay constant, α, shown in Figure 10 simply in terms of the aromatic to naphthene weight ratio in the feed. The correlation is based on the results of evaluation of α for a large number of feedstocks characterized by relative content of paraffins, naphthenes and aromatics. It was found that coke formation was relatively insensitive to paraffin content, but that the decay constant was increased significantly upon addition of aromatic components to a naphthenic feed. Conversely, the intrinsic rate constants were decreased by this procedure. Hence, there is the reduction to practice of Figure 10. The two variants designated as PC-32 and PA-37 have an interesting history. Of all the feeds investigated, they were the only ones containing

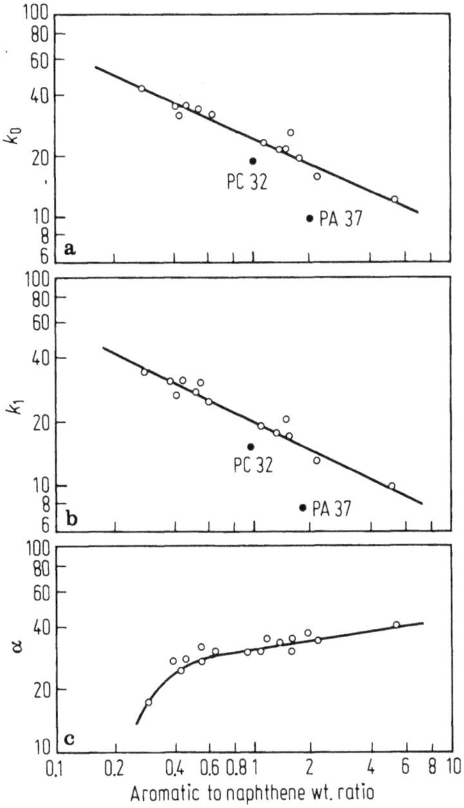

Figure 10. Correlations of the catalytic cracking model parameters:
a Overall gas-oil cracking rate constant (k_0); relative values
b Gasoline formation rate constant (k_1); relative values
c Catalyst decay constant (α); relative values
Temperature 755 K. (PC32 and PA37 contain recycled feed. Other data are for virgin feedstocks). (Reproduced with permission from ref. [33])

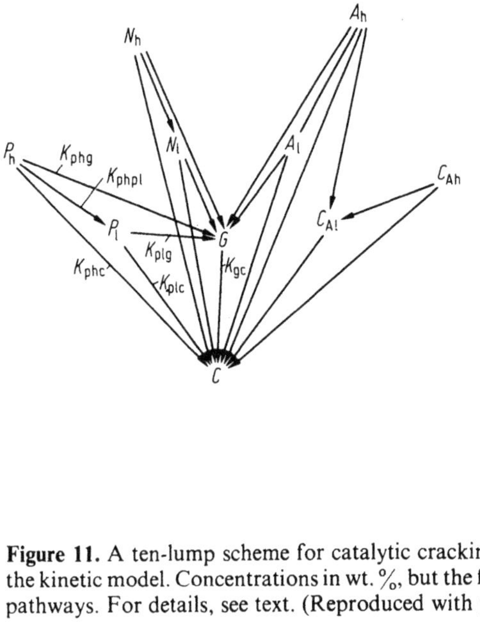

P_l = wt.% paraffinic molecules, (mass spec analysis), 430° - 650°F
N_l = wt.% naphthenic molecules, (mass spec analysis), 430°- 650 °F
C_{Al} = wt.% carbon atoms among aromatic rings, (*n-d-M* method), 430° - 650 °F
A_l = wt.% aromatic substituent groups, (430° - 650 °F)
P_h = wt.% paraffinic molecules, (mass spec analysis), 650 °F$^+$
N_h = wt.% naphthenic molecules, (mass spec analysis), 650 °F$^+$
C_{Ah} = wt.% carbon atoms among aromatic rings, (*n-d-M* method), 650 °F$^+$
A_h = wt.% aromatic substituent groups, (650 °F$^+$)
G = G lump (C_5 - 430 °F)
C = C lump (C_1 to C_4 + coke)
$C_{Al} + P_l + N_l + A_l$ = LFO (430°-650 °F)
$C_{Ah} + P_h + N_h + A_h$ = HFO (650 °F$^+$)

Figure 11. A ten-lump scheme for catalytic cracking. The lumps are identified as contained in the kinetic model. Concentrations in wt. %, but the figure is intended primarily to depict reaction pathways. For details, see text. (Reproduced with permission from ref. [36])

recycle feed, *i.e.*, previously treated material. Obviously the chemistry of these materials has been altered in such a way that the aromatic/naphthene correlation is not obeyed. The resolution of this problem involved re-examination of the kinetic sequence given above, since obviously the coke formation chemistry is not as simple as given there. It was found that division of the feedstock lump, C_1, into paraffinic, naphthenic and aromatic constituents, further subdivided into high boiling and low boiling fractions, gave a representation of the overall cracking reaction that could be represented by a group of first-order reactions in parallel with much more structure than the three-lump scheme. The elements of this are shown in Figure 11. For example, consider the reactions of P_h, the lump consisting of high boiling paraffinic molecules (616 K^+). There are three reaction pathways, one to the gasoline lump, G (rate constant K_{phg}), one to lower boiling paraffins, P_l (rate constant K_{phpl}), and one directly to dry gas plus coke, C (rate constant K_{phc}). In turn, the lower boiling paraffins can go to G with rate constant K_{plc}, and so on. Interpretation of the other reaction pathways shown on the figure is similar (rate constants have not been lettered in to avoid clutter). Most important are the lumps C_{Ah} and C_{Al} that represent aromatic rings, high and low boiling, with no substituent groups. While there is a reaction pathway from C_{Ah} to C_{Al}, and from the high boiling aromatics (with substituents) A_h to C_{Al}, these lumps are otherwise isolated from the rest of the reaction network. They can crack only to coke and do not contribute to the gasoline fraction G. Confidence in this lumping scheme is established by extensive correlations with experimental data given by Jacob et al. [36].

The discussion and interpretations above have largely dealt with coke formation in catalytic cracking which is fairly specific to silica-alumina or zeolitic catalysts and which deals with time scales of deactivation less than one second. Characteristic time scales for deactivation by coking may be much greater than this — at the opposite extreme are some hydrotreating processes in which the useful life of the catalyst may extend over a period of one to two years. In such cases the nature of the coke itself may change. We have identified the nature of coke as relatable to polycyclic condensed aromatic rings, however the feedstocks in hydrotreating are heavier and more complex than those in cracking, particularly in regard to asphaltene content. These materials are layered agglomerates in which individual sheets are formed from a composite of condensed aromatic rings together with saturated carbon chains or a net of naphthenic rings. The sheets are separated by *ca.* 0.4 nm. The composite asphaltene "molecule" is very large, of the order of several nm. Beuther *et al.* [37] have associated coke formation from these asphaltene molecules with the formation of a mesophase in which the naphthenes are either cracked or dehydrogenated and the aromatic concentrates orient themselves in specific clusters. In this view, then, the coke production is almost a macroscopic phenomenon — on the order of pore dimensions — and the result a very large cluster molecule that can continue to undergo chemical change with time. Some confirmation of this has been provided by Riley *et al.* [38], who showed a decrease in carbon free radical density with age in coke from heavy feed hydrotreating, indicative of a trend towards graphitization of an initially rather reactive hydrogen-containing material. Thus the definition of coke may depend upon age as well as origin.

4. Kinetics and Reaction Networks

For the purposes of convenience in subsequent modeling work, it is convenient to write some simple schemes for the reaction pathways involved in deactivation by poisoning and coking. In the discussion of poisoning we emphasized the parallel, competitive nature of the process, hence we may write the following reaction network to represent the essential features of poisoning

$$\begin{aligned} &A + L \rightarrow B + L \\ &P + L \rightarrow P \cdot L \end{aligned} \tag{I}$$

where the overall transformation A → B occurs on a surface with vacant sites L and poison P removes these irreversibly. Coke formation we have associated with a number of different kinds of molecules that in a given case could either be reactants or products (generally not considered to be impurities). Thus two limiting forms of coking schemes may be envisioned in which either reactant or products are the coke precursors:

$$\begin{aligned} &A + L \rightarrow B + L \\ &A + L \rightarrow A \cdot L \end{aligned} \tag{II}$$

or

$$\begin{aligned} &A + L \rightarrow B + L \\ &B + L \rightarrow B \cdot L \end{aligned} \tag{III}$$

Obviously, there will be instances where both reactants and products act as coke precursors.

The reaction schemes above are overly simplified, but deliberately so. One could easily include additional factors such as reversibility of reactions and poisoning, more complicated transformations than A → B (Academic Reaction I), reaction intermediate pathways, and so on. However, by keeping these networks simple we are able to focus clearly on specific effects of deactivation without detailed kinetics getting in the way; as will be seen there is plenty of time to get more complicated later on. A second argument for simplicity at this point is the profound influence that the structure — series or parallel — of the reaction network has on the practical analysis of deactivation effects within the context of catalytic reaction engineering, either intraparticle deactivation or reactor design.

On the other hand, analysis of the kinetics of deactivation would appear to be a complicated and formidable task, not susceptible to significant simplification, since the activity of a catalyst is a function of its entire history including preparation, handling, storage, pretreatment and reaction conditions. However, this does suggest that one might be able to approach the problem by proper definition of a history variable such that the rate of a reaction undergoing deactivation would be described in terms of the product of kinetic dependences that are time-invariant and the activity dependence that is not [39]. Such a formulation is termed "separable" deactivation kinetics and in general the rate of the main reaction at anytime is given by:

$$(-r)_T = r_1(C)\, r_2(T)\, r_3(s) \tag{5}$$

where the first two rate factors refer to the time-independent kinetic correlation and $r_3(s)$ is the history function and s the history variable. Since $r_3(s)$ is treated separately, we may write another equation describing its kinetic behavior as:

$$(-r)_s = r_4(C)\, r_5(T)\, r_6(s) \tag{6}$$

in which the individual rate factors express the dependence of the activity variable on concentrations, temperature and activity. Now, the utility of kinetic correlations is in engineering application, and engineers are fond of nondimensional numbers, so most applications have taken $r_3(s)$ to be the history variable, s, scaled such that $0 \leqq s \leqq 1$ with $s = 0$ representing a totally deactivated catalyst and $s = 1$ the fresh surface.

Let us now investigate the implications of equations (5) and (6) for a more detailed example. Take reaction scheme (I) as it pertains to a catalyst surface of activity s; we can write the elementary steps of this reaction as

$$\begin{aligned} &A_{(g)} + L \rightleftarrows A_{(ads)} \qquad (k_1, k_2)\\ &A_{(ads)} \rightarrow B_{(ads)} \qquad (k_r)\\ &B_{(ads)} \rightleftarrows B_{(g)} + L \qquad (k_3, k_4) \end{aligned} \tag{IV}$$

The kinetics of these elementary steps, for an ideal Langmuir surface, are:

$$k_1 s p_A (1 - \theta_A - \theta_B) = k_2 s \theta_A \tag{7}$$

$$(-r)_T = k_r s \theta_A \tag{8}$$

$$k_3 s \theta_B = k_4 s p_B (1 - \theta_A - \theta_B) \tag{9}$$

and

$$\theta_A + \theta_B + \theta_L = 1 \tag{10}$$

Following the normal procedure for Langmuir-Hinshelwood analysis, we eliminate θ_L *via* the equilibria of (7) and (9). The net adsorption-desorption constants for the deactivated surface, $k_i s$, always appear in ratio and divide out, so the activity variable appears only as a factor in k_r

$$(-r)_T = \frac{(k_r s)\, K_A p_A}{(1 + K_A p_A + K_B p_B)} \tag{11}$$

with $K_A = k_2/k_1$, *etc.* Thus

$$r_1(C)\, r_2(T) = \frac{k_r K_A p_A}{(1 + K_A p_A + K_B p_B)} \tag{12}$$

and

$$r_3(s) = s \tag{13}$$

One might question the approach of kinetic separability on the basis of the above derivation; since it works for ideal surfaces in the Langmuir sense, might not one expect it to fail for the non-ideal surfaces of technological catalysts? From a theoretical base the answer is a rather straightforward yes. The argument can be developed along the same principles used to derive adsorption isotherms for non-ideal surfaces [40], employing the concept that the total surface consists of the subassembly of ideal surfaces distributed according to the heat of chemisorption. Again returning to the isomerization example A → B, for a subunit of surface characterized by heat of chemisorption q the rate of reaction is:

$$(-r_{\mathrm{T}})_{\mathrm{q}} = \frac{s_{\mathrm{q}}k_{\mathrm{q}}K_{\mathrm{q}}p}{1 + K_{\mathrm{q}}p} \qquad (14)^1$$

and the overall reaction is obtained by summation over the n individual surfaces

$$(-r)_{\mathrm{T}} = \sum_{n} n_q(-r_{\mathrm{T}})_q = \int_0^{q_{\mathrm{m}}} n_q(-r_{\mathrm{T}})_q \, \mathrm{d}q \tag{15}$$

where q_{m} is the maximum heat of chemisorption. Expressing $(-r_{\mathrm{T}})_q$ directly gives

$$(-r)_{\mathrm{T}} = \int_0^{q_{\mathrm{m}}} \frac{n_q s_q k_q K_q p}{1 + K_q p} \, \mathrm{d}q \tag{16}$$

where $K_{\mathrm{q}} = K_0 \exp(q/RT)$ and $k_{\mathrm{q}} = k_0 \exp(-E/RT)$. Normally the activation energy can be approximated as some function of q; a form often employed is $E = -\beta q$. In addition, the dependencies of n_q and s_q must be specified. A corresponding derivation for a separable deactivation function for the non-ideal surface gives

$$(-r)_{\mathrm{T}} = \langle s \rangle \int_0^{q_{\mathrm{m}}} \frac{n_q k_q K_q p}{1 + K_q p} \, \mathrm{d}q \tag{17}$$

with

$$\langle s \rangle = \frac{1}{q_{\mathrm{m}}} \int_0^{q_{\mathrm{m}}} s_q \, \mathrm{d}q \tag{18}$$

[1] To simplify notation here, we will consider $K_q p$ in the denominator of equation (14) to include all inhibition effects.

One can say in general that the results of equations (16) and (17) are not the same because in general the sum of the averages is not necessarily equal to the average of the sum.

In practice, distinction between separable and nonseparable kinetics of deactivation seems to be dependent to some extent upon the mechanism of deactivation. Well defined examples of the failure of separable kinetics have been reported when deactivation is *via* poisoning [41, 42], yet broad experience with coke formation in many applications has shown that separable formulations are quite workable [4]. The vast majority of literature in poisoning as well as coking, however, has employed separable deactivation kinetics and we shall also in the following.

If we review some of the examples of poisoning and coke formation given earlier in this chapter with an eye to developing models for the rates of catalyst deactivation — a related problem of how to incorporate activity history in the kinetics of the main reaction — a simple picture emerges. Consider that in an operating situation the abscissae of Figures 1 and 3, for example, that give amount of poison on catalyst, are an indirect measure of time of operation. The log-linear correlation between conversion and time represented in Figure 3 is then the characteristic thumb-print of a first order reaction. Similarly, the very selective cases of poisoning in Figure 1 indicate exponential activity decay with time and are first order, while the non-selective case is an example of zero order decay. Finally, the exponential coking time-on-stream correlation of equation (4) is indicative of first order decay. These suggest that the kinetics of the deactivation steps in reaction schemes (I), (II) and (III) can be written in terms of simple power law forms. Thus, for poisoning in (I)

$$(-r)_D = k_D C_P^q s^v = -\frac{ds}{dt} \tag{19}$$

with the activity variable s scaled between zero and one. In this representation s is a measure of fraction remaining available sites with respect to those on a fresh surface. There is no reason why exactly the same formulation could not be employed for coking as per schemes (II) and (III). However, as stated before, historically most analysis of coke formation has been via a primary correlation for weight of coke on catalyst and a secondary one for activity as a function of weight of coke. Most workers who have elected to work with more fundamental kinetics for coke formation have nonetheless chosen weight coke formation as the primary kinetic statement; thus a typical kinetic expression for scheme II might be

$$(-r)_c = -k_c C_A^q s^v = -\frac{dC_c}{dt} \tag{20}$$

and

$$s = f(C_c) \tag{21}$$

Numerous relationships have been proposed and explored for the format of equation (20), and we shall discuss some of them later in the chapter.

Some rationalization for the Voorhies correlation can also be obtained from a more detailed examination of fundamental kinetic models [43]. Let us write a more detailed version of the coke formation reaction in scheme (II) as

$$\begin{aligned} &A + L \rightarrow B + L \\ &A + L \rightleftarrows A \cdot L \\ &A \cdot L + (A \cdot L)_n \rightarrow (A \cdot L)_{n+1} \end{aligned} \tag{IV}$$

in which the first step represents the formation of a coke precursor on the surface, $A \cdot L$, and the second the growth of an existing coke "molecule" $(A \cdot L)_n$. This second step is rate-determining. Now from a conventional Langmuir-Hinshelwood kinetic analysis for a rate-determining surface reaction between adjacent sites we obtain

$$(-r)_c = \frac{-K' C_{(AL)_n} p_A}{(1 + K_A p_A + K_B p_B + C_{(AL)_n})^2} \tag{22}$$

where $C_{(AL)_n}$ is the concentration of coke on the surface. Viewing the coking effect as a very strong adsorption of $(A \cdot L)_n$ on the surface, then the last term in the denominator above is predominant, and

$$(-r)_c = \frac{-K' p_A}{C_{(AL)_n}} = \frac{-d[C_{(AL)_n}]}{dt} \tag{23}$$

This integrates directly to

$$C_{(AL)_n} = C_c = K_c t^{1/2} \tag{24}$$

where $K_c = (K' p_A)^{1/2}$.

In summary, then, there appears to be a good case on the basis of published information to retain rather simple models for the kinetics of deactivation by poisoning or coke formation. Some complications are introduced by the problem of kinetic separability and this has important implications in the understanding of mechanisms. However, the information required for such precise models is not available in most cases.

5. Intraparticle Deactivation

We are now in a position to step up one level on the scale of the magnitude of events. Looking at mechanisms as microscopic analysis and reactors as macroscopic analysis, we are thus left at this point with the mesoscopic analysis.

It is important to keep in mind, as mentioned at the beginning of this chapter, that in such analysis we are in general concerned with the interactions of three characteristic rates: rates of reaction, rates of transport and rates of deactivation. The existence of deactivation turns a number of familiar steady state catalytic reaction engineering problems into not-so-familiar unsteady state ones, and the parallel or series nature of the reaction schemes involved will lead to spatial variation of gradients not encountered in non-deactivating systems. While these remarks are offered here in conjunction with the intraparticle deactivation problem, much the same pertains to later analysis of reactors.

A. Deactivation Disguises

In Section 2 considerable discussion was given concerning the nature of selective or nonselective poisoning in terms of typical poisoning curves such as those of Figure 1. In fact, such representations of experimental data may not be representative of intrinsic surface events, as shown by Wheeler [44] many years ago. His concern was with the effect of diffusional limitation on the apparent nature of deactivation, that is, the effect of transport interactions with poisoning rates on correlations such as Figure 1.

Let us consider first the case of intrinsic uniform poisoning (*i.e.*, nonselective) coupled with a diffusion-limited first order reaction. The theory of diffusion and reaction in porous catalysts tells us that in such a case the net or observable rate of reaction is given by

$$(-r)_{T_o} = \frac{\tanh \varphi_o}{\varphi_o} k_o C_o \tag{25}$$

where φ_o is the Thiele modulus which in slab geometry is defined for the fresh catalyst as $\varphi_o = W\sqrt{k_o/D_e}$. Now, in uniform poisoning the rate constant can be taken to decrease linearly with the fraction of surface poisoned, α, so that at any time

$$k = k_o(1 - \alpha) \tag{26}$$

The corresponding rate of reaction on the poisoned catalyst is

$$(-r)_T = \frac{\tanh \varphi}{\varphi} k_o(1 - \alpha)\, C_o \tag{27}$$

with φ defined as $W\sqrt{k_o(1 - \alpha)/D_e}$. If we define the relative activity factor F as the ratio of rates on the poisoned catalyst to the fresh catalyst, the following is readily obtained

$$F = \frac{(-r)_T}{(-r)_{T_o}} = \frac{\tanh[\varphi_o(1-\alpha)^{1/2}]}{\tanh \varphi_o} (1 - \alpha)^{1/2} \tag{28}$$

For large values of φ_o, F approaches $(1 - \alpha)^{1/2}$, while for small values of φ_o, F approaches $(1 - \alpha)$.

Making the association with Figure 1, in terms of the nomenclature above one would plot F *vs.* α. For large φ_0 one would have a curve, concave down, lying above the nonselective line. In this case, which one might term "antiselective", the existence of diffusion actually retards the rate of deactivation with respect to the intrinsic, non-selective behavior. As φ_o becomes very small the influence of diffusion disappears and the variation of F with α approaches the intrinsic poisoning relationship.

A second case may be envisioned that is probably more important. Here the affinity of poison for the surface is very high, so through the active catalyst dimension W for a fraction poisoned of α, length αW will be inactive and $(1 - \alpha) W$ active. Of course, owing to the physics of access of poison to the active surface, the inactive portion αW will lie adjacent to the surface and transport of reactant through this portion is by diffusion with a linear concentration gradient. A schematic is shown in Figure 12. In the steady state,

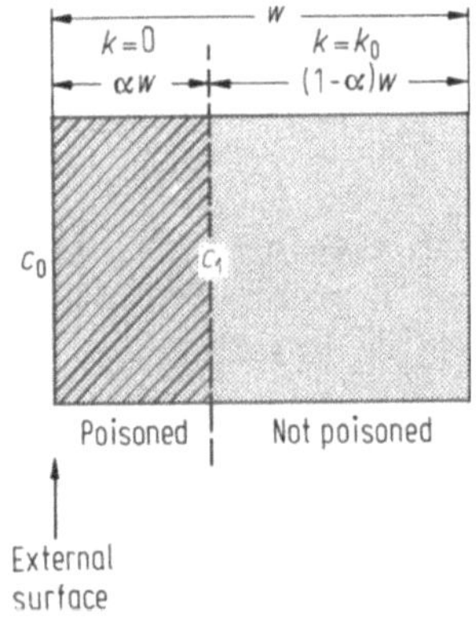

Figure 12. Schematic representation of diffusion and reaction with rapid, non-uniform catalyst poisoning

the diffusional flux will be balanced by the rate of reaction, so we may write the mass balance as

$$D_e \frac{(C_o - C_1)}{\alpha W} = \frac{\tanh \varphi}{\varphi} k_o(1 - \alpha) W C_o \tag{29}$$

where φ for the poisoned catalyst is $W(1 - \alpha)\sqrt{k_o/D_e}$. Solving for the unknown intermediate concentration C_1 gives

$$C_1 = \frac{C_o}{1 + \alpha \tanh [\varphi_o(1 - \alpha)] \varphi_o} \tag{30}$$

Resubstituting equation (30) into the equation for the diffusional flux gives for the rate on the poisoned catalyst

$$(-r)_T = \frac{D_o C_o}{\alpha W} \left[\frac{\varphi_o \alpha \tanh[\varphi_o(1 - \alpha)]}{1 + \varphi_o \alpha \tanh[\varphi_o(1 - \alpha)]} \right] \tag{31}$$

The corresponding rate on the unpoisoned catalyst is

$$(-r)_{T_o} = \frac{\tanh \varphi_o}{\varphi_o} k_o C_o W \tag{32}$$

and F is

$$F = \frac{\varphi_o \alpha \tanh[\varphi_o(1 - \alpha)]}{1 + \varphi_o \alpha \tanh[\varphi_o(1 - \alpha)]} \left(\frac{1}{\alpha \varphi_o \tanh \varphi_o} \right) \tag{33}$$

Again limiting values are of interest. For large φ_o, $\tanh \varphi_o$ approaches unity and

$$F = \frac{1}{1 + \alpha\varphi_o}; \qquad \varphi_o \text{ large} \tag{34}$$

This generates the characteristic selective poisoning curves of Figure 1. For small φ_o, $\tanh \varphi_o$ approaches φ_o and

$$F \approx (1 - \alpha); \qquad \varphi_o \text{ small} \tag{35}$$

Here we are back to the intrinsic poisoning behavior of the surface again.

We note that apparent nonselective poisoning represented by equation (34) is associated with a limiting case of poison "gradient" within the catalyst — discontinuous — with the poisoned zone located next to the external surface of the particle. Wheeler termed this "pore mouth poisoning" (his original derivation was done within the context of a single pore rather than the rectangular geometry used here); other terms have also been used, most notably "shell-progressive poisoning". The practical lesson here is very similar to that gleaned from diffusional disguises of kinetics and selectivity in unpoisoned catalysts: if intrinsic properties are to be measured experimentally it is important that transport limitations be absent [45].

The analysis here focuses on a specific relationship between F and α. However, in reality one is concerned with the trajectory of F with α in which time of operation is an implicit variable. Stated another way, in engineering problems associated with catalyst deactivation we are not at a fixed point on one of the curves of Figure 1, but rather are traversing them from the initial condition $F = 1$, $\alpha = 0$ to whatever final condition may be involved. In order to analyze this, then, it is necessary to include detail concerning poisoning or coking dynamics.

B. Dynamics of Intraparticle Deactivation

The essential features of the dynamics of intraparticle deactivation have been set forth by Smith and coworkers [46, 47] and Murakami *et al.* [48]. The analysis requires that we leave the realm of analytical mathematics and enter the world of numerical solution; in addition, the number of parameters involved becomes rather large, so we can only give some specific examples here.

Let us consider first coke formation within a particle under conditions of diffusional limitation according to the parallel scheme of (II) [46]. The mass balance on reactant A is

$$D_e \nabla^2 C_A - \varrho(k_A + k_C)\, C_A s = \varepsilon \frac{\partial C_A}{\partial t} \tag{36}$$

where separable deactivation kinetics are assumed. The rate constant k_A is in volume/weight-time units, ϱ is the particle density in weight/volume, and ε particle porosity. The effect of coke on activity is given by

$$\frac{dC_c}{dt} = -k_c C_A s \tag{37}$$

and

$$s = 1 - \frac{C_c}{C_{c_\infty}} \tag{38}$$

where C_{c_∞} is the limiting value for coke deposition corresponding to $s = 0$. The initial conditions required for equation (36) are complex, since they require specification of the concentration profile of A within the fresh catalyst particle. This amounts to requiring the solution of the familiar steady state effectiveness factor problem, *i.e.*

$$\begin{aligned} &D_e \nabla^2 C_A - \varrho k_A C_A = 0 \quad \text{at} \quad t = 0 \quad \text{and} \quad 0 \leqq r \leqq R_p \\ &s = 1 \quad \text{at} \quad t = 0 \quad \text{and} \quad 0 \leqq r \leqq R_p \end{aligned} \tag{39}$$

Boundary conditions required are the normal ones specifying concentrations at the particle surface ($r = R_p$), and symmetry at the particle center. Two simplifications are possible in the often encountered event that catalyst coke formation is slow compared to the rate of the main reaction, *i.e.*, $k_c \ll k_A$. The problem assumes a quasi-steady form and the time derivative becomes small, so equation (36) reduces to

$$D_e \nabla^2 C_A - \varrho k_A C_A s = 0 \tag{40}$$

The overall solution remains an unsteady state one, however, since the time derivative of equation (37) is retained.

Figure 13 gives a qualitative picture of what is obtained upon numerical solution of these equations. For the very active fresh catalyst sharp gradients in reactant concentration are initially present, while the activity profile is uniform at $s = 1$. Now as the reaction proceeds and coke begins to build

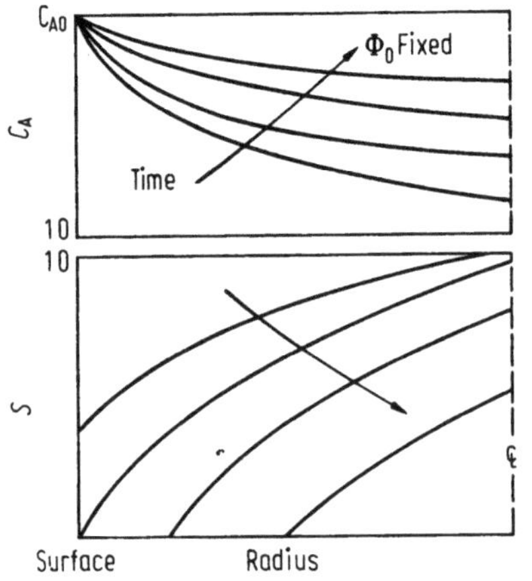

Figure 13. Evolution of concentration and activity profiles within a catalyst particle for parallel coking mechanism

up on the internal surface one sees a decrease in the sharpness of the reactant concentration gradient and the development of nonuniformity in the activity profile. In effect, the net reaction is becoming less diffusion-limited since the effective rate constant $\varrho k_A s$ is decreasing while the transport coefficient D_e remains unchanged.[1] Hence the associated Thiele modulus, $\varphi = R_p\sqrt{\varrho k_A s/D_e}$, also decreases with time and the overall effectiveness factor increases.

An important lesson is to be learned from the evolution of the activity profiles shown in Figure 13. We have stated several times that the parallel or series nature of the kinetic scheme describing deactivation has an important impact on the overall process; this is shown directly here. One may resolve the point by a series of questions and answers. (1) Where will the rate of deactivation be the greatest? Where the rate of coke formation is the greatest. (2) Where will the rate of coke formation be the greatest? Where the concentration of coke precursor A is the greatest. (3) Where is the concentration of A the greatest? At the external surface of the particle. The inescapable result of this is simply that the overall particle is deactivated from the outside to the inside. In extreme cases, when D_e is very small, the concentration and activity profiles become very steep, essentially discontinuous, and the deactivated zone will assume the shell-progressive nature discussed in the previous section.

The parallel nature of the coking scheme of (II) also leads us to consider some strategies in catalyst formulation that would never be considered if only the steady state, nondeactivating catalyst performance were analyzed. Since there is a common precursor for both product and coke, it is apparent that the influence of a diffusion limitation will be seen in both steps of (II) and this is confirmed in Figure 13. Carrying the thought one step further, one could ask the question whether there might be any advantage in deliberately inducing a certain amount of diffusional restriction into a catalyst employed for scheme (II). The thought is that while some sacrifice in initial activity would be required, there might be long range benefit since one is at

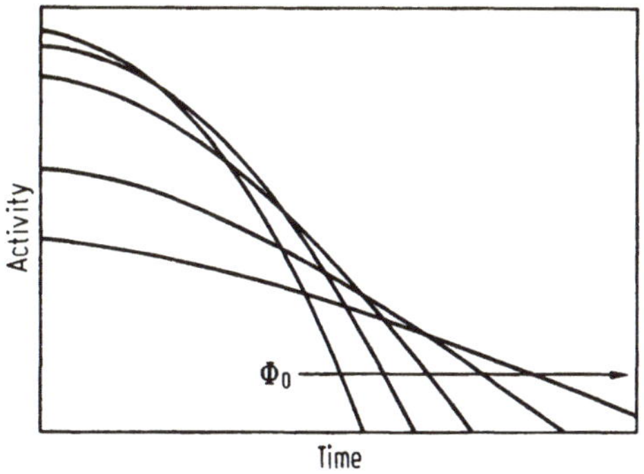

Figure 14. Relative activity as a function of time for catalysts of differing initial Thiele modulus; parallel coking mechanism

[1] This is an assumption that probably has good basis for poisoning mechanisms, however coke formation — particularly on fine pore catalysts — may severely affect the effective diffusivity. This problem has been reviewed elsewhere [2, 24].

the same time retarding the rate of coke formation. The question then is one of a tradeoff between activity and activity maintenance. The result of such a strategy is illustrated in Figure 14; we see a crossover in the relative activity curves occurring at differing times of operation, dependent upon the initial value of the Thiele modulus, φ_0. On the basis of this crossover, one would attempt to engineer a compromise between activity and total life in terms of φ_0 to maximize the area under the curve to some limiting final effectiveness.

The foregoing discussion and results, illustrated in Figures 13 and 14, have been set forth in the context of parallel coking *via* scheme (II). Obviously the poisoning of scheme (I) is also parallel and, in general terms, similar behavior would be obtained for this situation. One important distinction between the two should be kept in mind, however. The effective diffusivity for reaction and deactivation is the same for scheme (II), but not necessarily for scheme (I); whether the strategy of diffusion limitation for increasing activity maintenance would be effective in parallel poisoning is very dependent upon relative values of the diffusion coefficients for reactant and poison.

What of the series deactivation according to scheme (III)? Here the precursor for coke formation is the reaction product and if the reaction on the fresh catalyst is severely diffusion limited one will find the maximum concentration of this precursor in the center of the pellet. Following the same type of dialogue used for scheme (II) one concludes that deactivation occurs from the inside to the outside in series coking. Further reflection also yields the conclusion that deliberate diffusional limitation in this case increases rather than decreases the rate of deactivation; as one increases the Thiele modulus one increases the net time a product molecule of B must spend within the internal pore structure of the catalyst and hence increases the chance that coke can be formed.

A nice experimental confirmation of these results has been provided by Murakami *et al.* [48] on two reactions representative of schemes (II) and (III). These were the disproportionation of toluene in which toluene reacts to form benzene and xylene while simultaneously forming coke directly (scheme II), and the dehydrogenation of *n*-butyl alcohol in which the aldehyde product of dehydration further reacts to form coke (scheme III). Catalysts were a 10% boria-alumina for the disproportionation reaction and commercial alumina with 10% sodium for the dehydrogenation. Catalysts pellets were sectioned after a fixed amount of time-on-stream and photographed in cross-section. These revealed clearly the outside-inside coke formation in the disproportionation reaction and the inside-outside formation in the dehydrogenation reaction. Lest one become too complacent with the theory, however, evidence was also presented for the dehydrogenation at higher temperatures showing an apparent reversal of the coking pattern back to outside-inside. The interpretation of this seems reasonably clear: the higher temperature leads, of course, to greatly enhanced diffusional limitation and it is likely that the reactant never penetrated into the central core of the particle. Coke formed where it could, which was in the outer shell where reactants and products were present. It is likely also in this case that the kinetics of coke formation were not slow compared to the main reaction and the quasi-steady state assumption incorporated in equation (40) was no longer valid. Solutions are presented in this paper for the more general model of equation (36) and corresponding forms for the other deactivation schemes.

Intraparticle deactivation under nonisothermal conditions was first investigated by Sagara, *et al.* [47] in an extension of the work of Masamune and Smith. The general trends observed were qualitatively similar to those for isothermal conditions, however very complex activity profiles can be developed under certain conditions. An important added complication in nonisothermal systems is that the activation energy of the deactivation process enters as a sensitive parameter and the overall behavior of the intraparticle system is extremely dependent upon the relative magnitudes of the activation energies of the main and deactivation reactions. The nature of thermal gradients is also affected by intraparticle deactivation. Conventional wisdom in reaction engineering has it that if concentration and temperature gradients are present, the concentration gradients will be located within the particle, while the temperature gradients will be found in the external boundary layer surrounding the particle. This is based on sound reasoning concerning the relative magnitudes of effective diffusivities vs bulk diffusivities and effective thermal conductivities vs molecular thermal conductivities. However, shell-progressive deactivation can change all this. Figure 15 gives an example of intraparticle and interphase temperature gradients for a Ni/

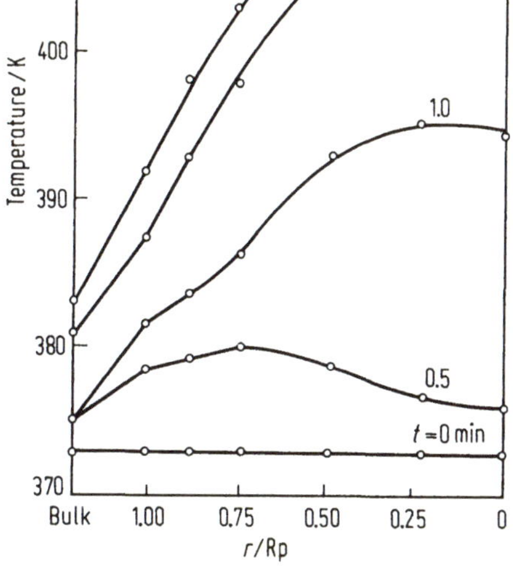

Figure 15. An example of the alteration of intraphase gradients by deactivation. Temperature profiles for the ignition to steady state (3.0 min) of thiophene-poisoned benzene hydrogenation on Ni/kieselguhr. The catalyst had been poisoned to 10% of original activity before the experiment. Inlet feed conditions: 373 K, 18% benzene, 82% hydrogen. (Reproduced with permission from ref. [49])

kieselguhr particle poisoned by thiophene for benzene hydrogenation. Depicted is the ignition of reaction in a particle that has been prepoisoned to $s = 0.10$ (based on global activity) to final steady state [49]. Thiophene poisoning in these experiments was shown to be shell-progressive and a dead zone in the particle extended to a relative radius of about 0.75. This zone is reflected in the shape of the transient gradients, particularly that shown at 1 min. The final steady state temperature gradient is here seen to be located almost entirely within the pellet. This is the result of "pushing" the zone of active catalyst away from the pellet external surface; heat generation occurs in the central core of the particle and the large intraparticle gradient is the result of thermal conduction through the poisoned portion.

C. Distributions and Composites

The rather pronounced effects of shell-progressive poisoning illustrated above have keyed a number of studies on what effect various distributions of active ingredient might have on activity maintenance under deactivating conditions. For example, if active ingredient (*i.e.*, the metal in a supported metal catalyst) was preferentially located near the surface of the particle one might be able to improve the tolerance to this type of poisoning. The thought amounts simply to putting the active sites where they will do the most good. Again, this is a strategy of catalyst design that might not be considered in the absence of deactivation, just as the idea of trading off diffusion limitations with deactivation rates in scheme (I) or (II). It should also be recognized that these approaches are not doing anything toward eliminating the basic causes of deactivation, rather they are efforts to cope with the effects.

Shell-progressive poisoning has been investigated in detail by Corbett and Luss [50] for a sequential reaction scheme $A \rightarrow B \rightarrow C$ with parallel deactivation. Various distributions were considered ranging from those with preferential loading at the particle center to those with preferential loading at the surface. In general, the latter resulted in superior activity maintenance, however there is a selectivity factor also involved, rate of A/rate of B, that can show rather different behavior from activity alone. In several instances explored computationally, there was demonstrated a minimum in selectivity with time-on-stream, either sharp and early or diffuse and later, dependent upon the nature of the distribution and the extent of diffusion limitation. It is important to note that in the opposite extreme of shell-progressive deactivation, that is when diffusion limitation is absent, preferential loading near the surface becomes a questionable procedure, since one has set the largest concentration of sites out front where they are just as accesible to poison or coke as to reactants.

The interactions between differing extents of diffusion limitation for poisoning and reaction are well illustrated by the shell models of Becker and Wei [51]. Four model systems were considered: active ingredient in an outer shell, in a annular ring, in an inner core, or uniformly distributed, and performance evaluated for reaction and poisoning according to a modification of scheme (I) in terms of the corresponding Thiele moduli, φ_A and φ_p. The modification refers specifically to the fact that poison was considered to be adsorbed on all internal surfaces of the particle, not just in the zone containing the active ingredient. This serves to emphasize those aspects of the particle dynamics that are dictated by the access of poison to the reactive zone. In the original work the results of a large number of parametric calculations were presented for a particular example in which a figure of merit was defined as the length of time required for the catalyst to be poisoned to a limiting final value referred to a basis of initial performance of the fresh catalyst in the absence of diffusion. These were intercompared for the four configurations and assembled in a kind of performance chart indicating the ranges of φ_A and φ_p preferential for a particular configuration. A generalization of this is shown in Figure 16. In the event, for example, that the

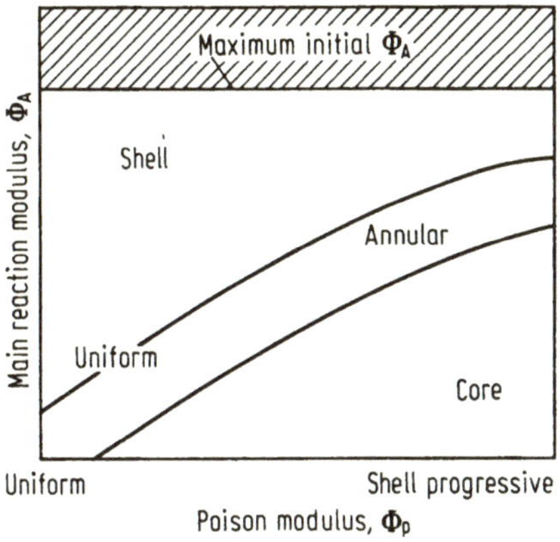

Figure 16. Performance chart for catalyst poisoning with differing intraparticle distributions of active component (log-log scale)

poisoning is shell-progressive but the main reaction is not diffusion-limited, high φ_p and low φ_A, one sees the preferential configuration marked as core, while in the opposite extreme of low φ_p and high φ_A the preferred configuration is shell. These are based on the relative rates of delivery of reactant and poison to the active zone. Core configuration allows access of the reactant to the active core while trapping the poison in the inactive outer portion of the particle. Shell configuration does not restrict access of poison to the active sites (*i.e.*, low φ_p), however does concentrate activity in the region accessible to reactant.

Composite support structures have also been explored as a means for controlling intraparticle deactivation [52]. Here the idea is to control access of the poison to the active zone, not by manipulation of the location of that zone, but by establishing a diffusional barrier within the particle prior to the locus of active ingredient. Of course, the loading of active ingredient may also be varied as a function of position as well and, as in that case, the strategy seems best suited for shell-poisoning. Thus, increasing diffusivity of poison in the inert zone relative to that in the active zone yields an enhancement in activity maintenance provided the poison is adsorbed to a sufficient degree in the inert zone.

D. Summary

The coupling of deactivation to intraparticle diffusion and reaction is a fundamental fact of life in most technological applications of catalysis. In the discussion above we have attempted to focus on a few examples that provide a broad perspective of what is involved in this coupling. There are many more specific and detailed aspects that have been left out, however, and for those interested it would be profitable to peruse the reviews cited before. One problem of considerable importance not treated here is deactivation via simultaneous coke deposition and poisoning. In hydrotreating, for example, this takes the form of metals deposition linked with coke deposition; the situation is exacarbated by the fact that metals deposition is likely shell-progressive and the amounts of materials involved are sufficient to cause decrease in effective pore diameter at the external surface, eventually with complete pore plugging. Many other equally evil scenarios can be envisioned for poisoning and coking in combination.

As far as what has been considered in this section, it is hoped that a strong case has been made for the influence of the kinetic scheme for deactivation, as well as the particular mechanism, in the dynamic behavior of individual particles. Also in the examples we have tried to provide some insight into strategies of catalyst design to cope with deactivation that derive largely from the kinetic structure of the process. As pointed out, in general these strategies are ones that would never be considered within the context of the steady state effectiveness factor problem. Finally, the examples have unfortunately been long on theory and short on experiment. Informative results have been obtained using single pellet diffusion reactors; the reader is referred to the review by Hegedus and Petersen [53] and the work of Lee *et al.* [49].

6. Chemical Reactor Analysis

The final level of scale is that of chemical reactor analysis under conditions of poison or coke deposition. Most of the remarks in this section will be devoted to fixed beds because of the very interesting parallels that exist between this class of problems and corresponding intraparticle problems, although some detail on moving and fluidized beds is provided in one of the examples.

It will again be seen that there exists a strong influence of the kinetic scheme of deactivation, by whatever mechanism, on the performance of a fixed bed. Particularly in the case of parallel deactivation, the problem becomes one of wave propagation involving the motion of reaction zones through the bed. Nonisothermal operation is much more important in the analysis of reactor behavior than for individual particles, so to a large extent one must rely upon numerical simulations. It should be mentioned that the mathematical problems involved in such simulations are rather challenging in themselves, since sharp gradients that move in time and position are often involved.

Strategies of operation here replace strategies for formulation discussed in the previous section. An important class of processes require operation at constant conversion (or selectivity) for reasons of downstream processing. Normally this is attained by increasing the temperature of the reactor to compensate for decline in catalyst activity and results in a very interesting example of positive feedback if the deactivation mechanism is thermally activated. In addition there is a class of deactivation disguises of activity and selectivity that bear a distant familial resemblance to the Wheeler interpretation of Figure 1 for single particles.

A. Superimposition Models

Earlier work on the analysis of deactivation in fixed bed reactors did not contemplate the coupled solution of reactor balance equations together with deactivation kinetics. Rather, they posed the problem in the form of given: i) poison or coke distributions within the reactor and, ii) relationship between the amount of coke or poison and catalyst activity; what is the resultant overall performance? The effect of such an inquiry is much that of taking a snapshot of the reactor at various periods in its history.

Consider, then, writing two balance equations, one for poison distribution through the bed and the second for activity as a function of poison concentration [54]. For the first, for example, we may have

$$C_p/C_p{}^{o} = (1 - bz) \;...\; \text{or} \;...\; \exp(-bz)\,, \qquad \text{etc.} \tag{41}$$

and for the second

$$s = 1 - aC_p \;...\; \text{or} \; \exp(-aC_p)\,, \qquad \text{etc.} \tag{42}$$

Average bed activity is obtained from

$$s = \frac{1}{L}\int_0^L (s)\,dz \tag{43}$$

and corresponding average poison concentration is

$$\langle C_p/C_p^o \rangle = \frac{1}{L}\int_0^L (C_p/C_p^o)\,dz \tag{44}$$

Combining a linear dependence of s upon C_p and an exponential distribution of poison in the bed, we arrive at a function such as

$$\langle s \rangle = \frac{1}{L}\int_0^L [1 - aC_p^o \exp(-bz)]\,dz \tag{45}$$

to evaluate for average activity. These values can then be associated with corresponding $\langle C_p/C_p^o \rangle$ to give a gross measure of reactor performance. Various combinations of such poison distributions and activity-poison correlations were explored by Anderson and Whitehouse [54] with an interesting result. Over a wide range of parametric values, log-log plots of $\langle s \rangle$ *vs.* $\langle C_p \rangle$ were linear; if one considers $\langle C_p \rangle$ to be an indirect measure of time-on-stream, then this turns out to be a kind of Voorhies correlation for fixed bed reactor poisoning. Obviously, similar conclusions would pertain to distributions written for coke deposition.

A somewhat more fundamental approach, but still amounting to a superimposition model, was taken by Wheeler and Robell [55]. They noted that particularly in the case of strong, irreversible poisoning the progression of poison profiles — that is, concentration of poison on the catalyst — strongly resembled the profiles of adsorbate or ionic concentration encountered in fixed bed adsorption or ion exchange. A classical theory for this is that of Bohart and Adams [56], which, if we adopt the notation to that for adsorption of poison, states that the kinetics are given at any point by

$$\frac{\partial C_p}{\partial t} = k_p (C_p)_g \, (1 - C_p/C_{p\infty}) \tag{46}$$

where $(C_p)_g$ is poison concentration in the gas (fluid) phase, and C_p corresponding concentration on the catalyst. Solving the fixed bed mass balance on poison concentration with the kinetics of equation (46) gives the following distribution

$$(C_p/C_{p\infty}) = \frac{1 - \exp(-N_t t/t_\infty)}{1 + \exp(-N_t t/t_\infty)\,[\exp(N_t z/L) - 1]} \tag{47}$$

where $N_t = k_p L/v$, v = velocity, and t_∞ is the ratio of the total capacity of the catalyst for poison adsorption to the rate at which poison is introduced into the reactor. The characteristic wave-front nature of poison profiles as predicted by equation (47) is shown in Figure 17. The steepness of the gradients involved would of course depend upon the magnitude of the parameters involved, particularly k_p. Characteristic of these profiles is that, after an initial transient, they assume a constant slope and pass through the

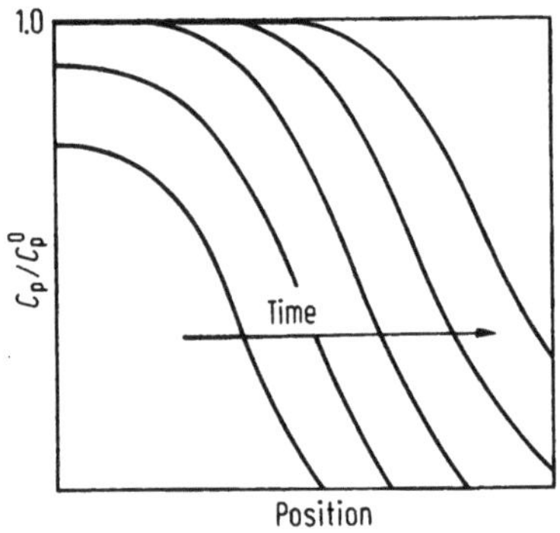

Figure 17. Poison wave profiles as obtained from Bohart-Adams theory

bed at a relatively constant velocity. This will be seen later, in fact, for much more complicated examples of poisoning in which the analysis does not involve any kind of superimposition model.

The overall bed performance is obtained by combining equation (47) with an activity-poison correlation and then averaging in the manner of Anderson and Whitehouse. Assuming a linear relationship for a first order reaction

$$s = \frac{k}{k_o} = 1 - C_p/C_{p\infty} \tag{48}$$

and

$$\ln (C_L/C_o) = \frac{t_R}{L}\int_0^L k \, dz \tag{49}$$

one obtains for conversion as a function of time

$$x = 1 - (C_L/C_o) \tag{50}$$

$$\ln (C_L/C_o) = -(k_o/k_p) \ln \{1 - \exp(-N_t t/t_\infty) + \exp[N_t(1 - t/t_\infty)]\} \tag{51}$$

Extensions of this analysis to cases in which intraparticle diffusion is important are given both by Wheeler and Robell and by Haynes [57].

B. Coupled Balances

As stated above, the superimposition models do not attempt to explain the process of deactivation *via* separate kinetic models but rather express the existence of a particular activity or coke profile and then attempt to see what the implications are. The work of Froment and Bischoff [58] was the first to explore analysis of the coupled balances for reaction and deactivation and remains topical at present. It is of particular interest for presentation here, since we may cast the analysis within the context of the deactivation schemes (I)–(III). As an example consider the parallel coking of scheme (II), with a time scale for deactivation similar to that for the main reaction. For a fixed bed in plug flow operation, the mass balance for the reactant is

$$\varepsilon \frac{\partial C_A}{\partial t} = -\varepsilon v \frac{\partial C_A}{\partial z} + s\varrho_c(-r)_A \tag{52}$$

The rate of coke formation is

$$(-r)_c = -\varrho_c k_c C_A s = -\frac{dC_c}{dt} \tag{53}$$

and the net rate of reaction of a is

$$(-r)_A = (k_A + k_c)\, C_A s \tag{54}$$

In the original work [58], several expressions for the relation between s and C_c were explored, *i.e.*, exponential, hyperbolic, etc.; however the qualitative nature of the results was unaffected.

Figure 18a gives a schematic of the general nature of the coke profiles developed within the reactor for parallel reactions where $k_c \ll k_A$. The situation is in a sense analogous to the corresponding intraparticle problem where coke was formed from the outside to the inside of the particle. Here the coke is also perferentially deposited at the location where the concentration of coke precursor is the highest, namely at the entrance to the bed.

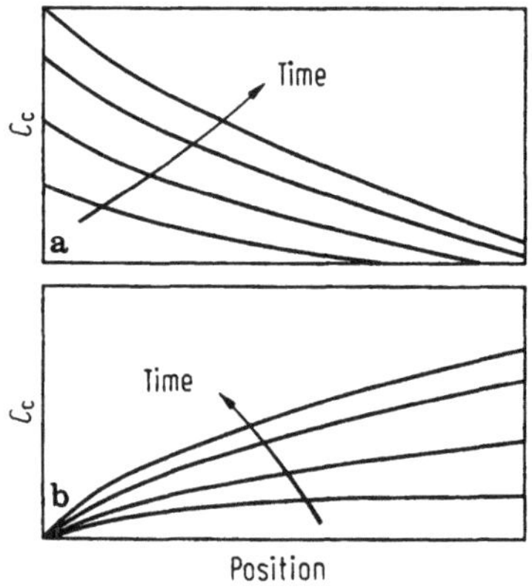

Figure 18. Development of coke profiles in a fixed bed reactor; **a** parallel scheme (II); **b** series scheme (III)

As the concentration of reactant decreases down the length of the bed the rate of coke formation decreases proportionately and one observes a coke profile decreasing with bed length. A similar formulation and solution of the problem for the series reactions of scheme (III) gives the general results shown in Figure 18b. Again the profile reflects the nature of the kinetic network; coke is formed only from product, and if this is absent in the feed, none can be formed at the entrance to the bed. Hence, one obtains coke profiles that increase with bed length. Each scheme thus has its characteristic fingerprint, be it either with respect to intraparticle or intrareactor profiles.

The nature of reactor dynamics arising from such deactivation is rather complex, particularly for the parallel schemes of (I) and (II). Let us consider in more detail the effects of the decreasing profiles of Figure 18a on the rate of reaction per unit volume in the reactor. A two-dimensional representation in time and distance is given in Figure 19. For the fresh bed this rate would decline exponentially with length in accord with first order kinetics. However as time goes on the entrance to the bed is preferentially deactivated and the rate declines faster at that point than in the rest of the bed. One can see from the figure that this leads to the development of a local maximum in activity that travels through the bed and eventually out the exit. At this end-point, the reaction profile, now increasing with bed length, is qualitatively the opposite of the initial conditions. This wave-like behavior

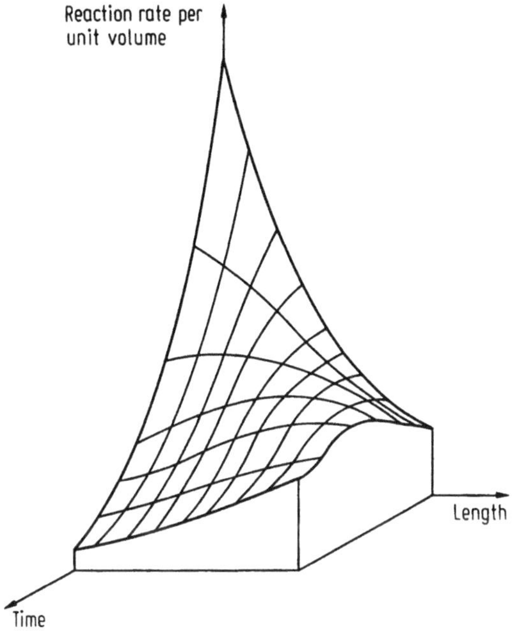

Figure 19. Activity waves in an isothermal reactor. (Reproduced with permission from ref. [53])

is characteristic of parallel mechanisms of deactivation as shown here and previously via the Bohart-Adams approach. For additional information on isothermal fixed bed deactivation the reader is referred to further studies of Bischoff [59], of Ozawa [60], and Wentrcek *et al* [61].

C. Deactivation Disguises

Just as in the case of intraparticle deactivation, the existence of coke or poison gradients within a fixed bed can disguise kinetics and/or selectivity. Often these assume the form of time-averaging disguises, a topic that has been treated extensively by Weekman and co-workers [30–36]. At a very basic level, the problem has familial similarity to that of separable kinetics. Consider the familiar plug flow reactor design equation for a first order reaction:

$$(W/F) = -(1/k_A p_{A_0}) \int_0^x dx/x \tag{55}$$

For slow deactivation, then, a possible experimental strategy might be to obtain k_A via measurement of conversion at fixed space velocity at different times and extrapolate back to zero time. However, if there are gradients of activity present in the bed, the rate constant cannot be independent of position, and the quantity measured at any given time is an average:

$$\langle k_A \rangle = \frac{1}{L} \int_0^L k_A(z)\, dz \tag{56}$$

We also in general could consider this average rate constant to be a function of the average poison or coke concentration, say $\langle C_c \rangle$, expressed by the form of equation (56). However, it is incorrect to imply

$$\langle k_A \rangle = k_A[\langle C_c \rangle] \tag{57}$$

That is, the average rate constant is not the same as the rate constant at the average coke content. The reader may quickly prove this by carrying out the indicated integrations for the case, for example, where the activity-coke (poison) content correlation is given by equation (42). The only sure procedure to employ when making kinetic measurements in a deactivating fixed bed is to solve the basic equations for the coupled model of poisoning or coking, as discussed by Wentrcek *et al.*

D. Non-isothermal Operation

Most reactions of commercial importance have significant heat effects associated, and commercial reactor operation is often conducted under adiabatic or near-adiabatic conditions. Some early experimental studies of deactivation in nonisothermal fixed beds were reported by Menon and Sreeramamurthy [62] for air oxidation of hydrogen sulfide on carbon (deactivation by product sulfur) and by Pexidr et al. [63] for benzene hydrogenation in Ni/Al_2O_3 (bed prepoisoned by carbon disulfide). Both of these investigations provided direct evidence of the motion of thermal waves through the reactor as witnessed by the displacement of the hot spot down the bed upon progressive poisoning.

As one might expect, the coupled heat and mass balances, together with poisoning or coking kinetics, admit of no simple solution, and a number of studies have been carried out in terms of parametric investigations *via* numerical methods. The ratio of activation energies for the main reaction and the deactivation process is a very important quantity in the behavior of deactivating nonisothermal reactors, and a very detailed theoretical study of this has been given by Blaum [64], based on a two phase reactor model using parameters for carbon monoxide oxidation on NiO with either parallel

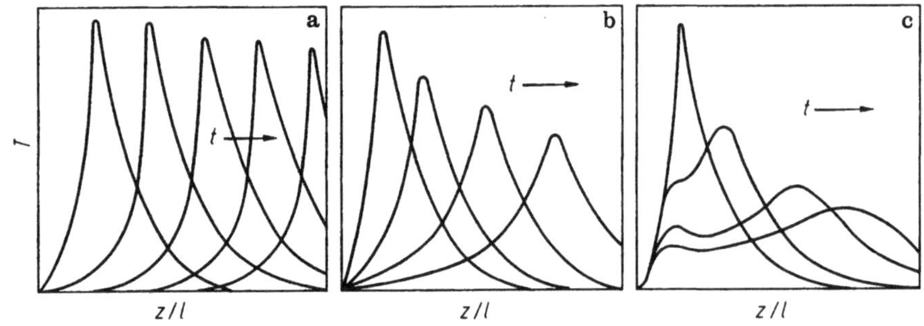

Figure 20. Thermal waves for slow deactivation in a fixed bed as a function of (fractional distance along the bed, z/l). **a** $E_c = 0$; **b** $E_c < E_A$; **c** $E_c > E_A$. (Reproduced with permission from ref. [64])

or series coke formation. As an example we consider the results obtained for the specific case of slow deactivation (rate of decay insignificant over the time scale required to establish the steady state in the reactor) for the three cases in which $E_c = 0$, $E_c < E_A$ and $E_c > E_A$. In Figure 20 are shown qualitatively the progression with time of the temperature profiles associated with each of these activation energy ratios plotted as a function of fractional distance along the bed, z/L. When $E_c = 0$ obviously the magnitude of the hot spot has no effect on the deactivation process and the thermal wave assumes a fixed shape and passes through the bed at essentially constant velocity. For $E_c < E_A$ the hot spot becomes more diffuse and decreases in magnitude with increasing time, while for $E_c > E_A$ the profiles become very broad and can assume very complex shapes, as shown. In the latter two cases, it turns out that the bed activity profiles pass through a minimum near the bed entrance which remains at that position throughout the bed life. The minimum is quite sharp for $E_c > E_A$.

A number of other examples give correspondingly complex results. When the time scale of deactivation is rapid and $E_c < E_A$, then the magnitude of the hot spot grows as it passes down the bed. The point here is not to submerge the reader in a vast amount of detail concerning one numerical simulation example, but rather to point out the complex coupling between thermal parameters, temperature waves and activity profiles. One's intuition is often a false friend in such situations.

An experimental study corresponding rather well to Blaum's case of slow deactivation, $E_c < E_A$, has been reported by Weng et al. [65] for fixed bed deactivation of Ni/kieselguhr by thiophene poisoning in benzene hydrogenation. A rather complex mathematic model is required to describe the deactivation transients, involving mass balances for reactant and poison, reactor energy balance, and kinetic expressions for the main and poisoning reactions. In generalized form, these are

$$\varepsilon \frac{\partial C_A}{\partial t} = -\varepsilon \frac{\partial v C_A}{\partial z} + (-r)_A \, s \tag{58}$$

$$\varepsilon \frac{\partial C_p}{\partial t} = -\varepsilon \frac{\partial v C_p}{\partial z} + (-r)_p \, C_{p\infty} \tag{59}$$

and

$$\frac{\partial T}{\partial t} = \frac{\lambda_{eff}}{\varrho c_p} \frac{\partial^2 T}{\partial z^2} - \frac{\varepsilon \varrho_g c_{pg}}{\varrho c_p} \frac{\partial v T}{\partial z} + \mathrm{h}(T) + \frac{(-\Delta H)}{\varrho c_p} (-r)_A \, s \tag{60}$$

In the specific study $(-r_A)$, the rate of benzene hydrogenation, was correlated by a Langmuir-Hinshelwood form rate equation, the kinetics of poisoning by a power law expression similar to equation (19) with $q = v = 1$, and the separate model employed as shown above.

Such simulations are not without their ugly surprises. In this case there was an extreme parametric sensitivity to the saturation poison capacity, C_{p_∞}.

This parameter serves to define the rate of passage of the poisoning front through the bed and therefore dictates the time scale of the entire process. The sensitivity is amplified by the fact that rapid, irreversible poisoning such as that in the thiophene-nickel system is characterized by very low values of

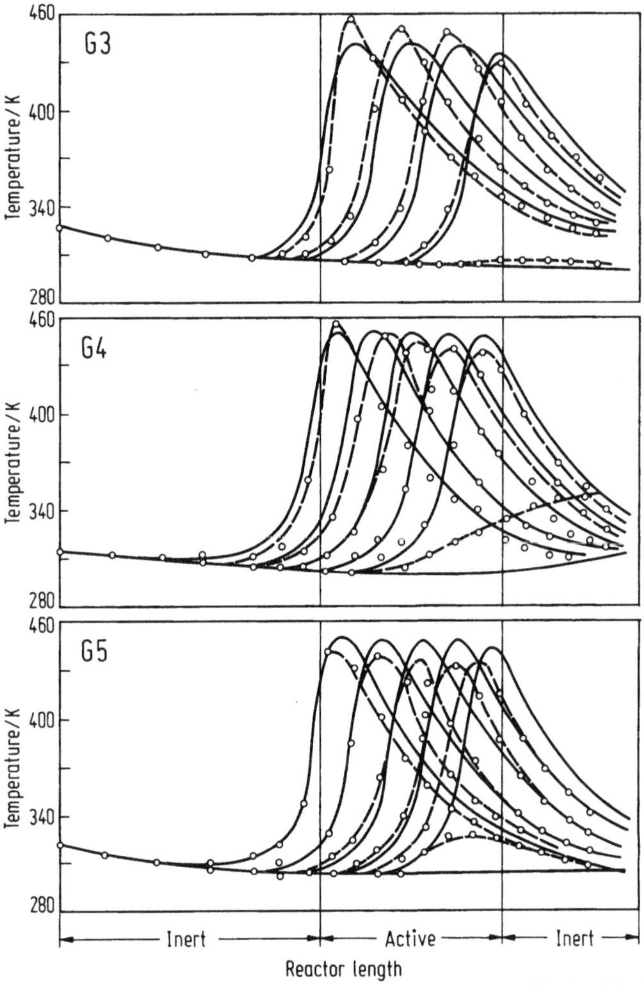

Figure 21. Experimental and computed temperature profiles in nickel-catalyzed benzene hydrogeneration with thiophene poisoning.

Experiment	Space velocity/ min^{-1}	Inlet compositions/Mol %		
		Thiophene	Benzene	H_2
G3	193	0.04	3.30	96.7
G4	130	0.04	4.30	95.7
G5	130	0.02	4.30	95.7

—O—, measured; ———, calculated. (Reproduced with permission from ref. [65])

the activation energy for deactivation[1], hence there is no compensating temperature effect. A second sensitive parameter is the wall heat transfer coefficient contained within the heat removal term, h(T), of equation (60). As one might expect, this is important in determining the magnitude of the hot spot. Nonetheless, if one spends sufficient time, the relative importance of the various parameters can be sorted out, and then they can be measured to the precision required for a useful simulation.

Figure 21 gives some indication of the results that can be obtained after this painful process has been completed. These represent temperature profile simulations in the experiments of Weng et al. over approximately a two-fold range of space velocity and inlet poison concentration. The three experiments represent conditions of high and low levels of inlet poison concentration in combination with high and low space velocities. It can be seen that the results fall rather well within what would be expected from the computational results of Blaum. Although strictly speaking this falls within his case of $E_c < E_A$, the value for E_c is so small here (*ca.* 4.1 kJ mol^{-1}) that one observes only a slight decrease in the magnitude of the hot spot as it progresses down the bed. Note, however, the fixed shape of the profiles, essentially until they are ready to emerge from the active section of the bed, and the constant velocity of progression down the bed. A corresponding study of this system has also been reported for adiabatic reactor operation [66]; here the impact of the wall heat transfer coefficient is eliminated but sensitivity to C_{p_∞} remains.

Not all of the interesting behavior associated with thermal waves in nonisothermal operation belongs to exothermic reactions. Dumez and Froment [67] have reported a detailed study of the dehydrogenation of 1-butene to butadiene in a fixed bed of chromia-alumina catalyst. This is an endothermic reaction, and deactivation occurs by coke formation from both reactant and product precursors. The correlation of coking kinetics reported probably represents the state of the art in detail at least, although a rate

Table 3. Characteristics of an industrial reactor for butene dehydrogenation

Length	0.8 m
Cross section	1 m^2
Catalyst and inert diameter	0.0046 m
Catalyst bulk density	400 kg of cat m^{-3} diluted bed
Inert bulk density	900 kg of inert m^{-3} diluted bed
Catalyst external surface area	274 m^2 m^{-3} diluted bed
Inert surface area	411 m^2 m^{-3} diluted bed
Inlet total pressure	0.25 atm
Inlet butene pressure	0.25 atm
Molar flowrate	15 kmol m^2 h^{-1}
Feed temperature	873 K
Initial bed temperature	873 K

[1] In effect, this is just the way the Arrhenius equation tells us that the uptake will be rapid and irreversible regardless of temperature level.

equation derived from first principles did not fit the experimental data well. A best-fit form was given by

$$(-r)_c = -\frac{dC_c}{dt} = \frac{-k^o_{cB}p^q_B - k^o_{cD}p^v_D}{\left(1 + K_H\sqrt{p_H}\right)^2} \times s \tag{61}$$

with

$$s = \exp(-aC_c) \tag{62}$$

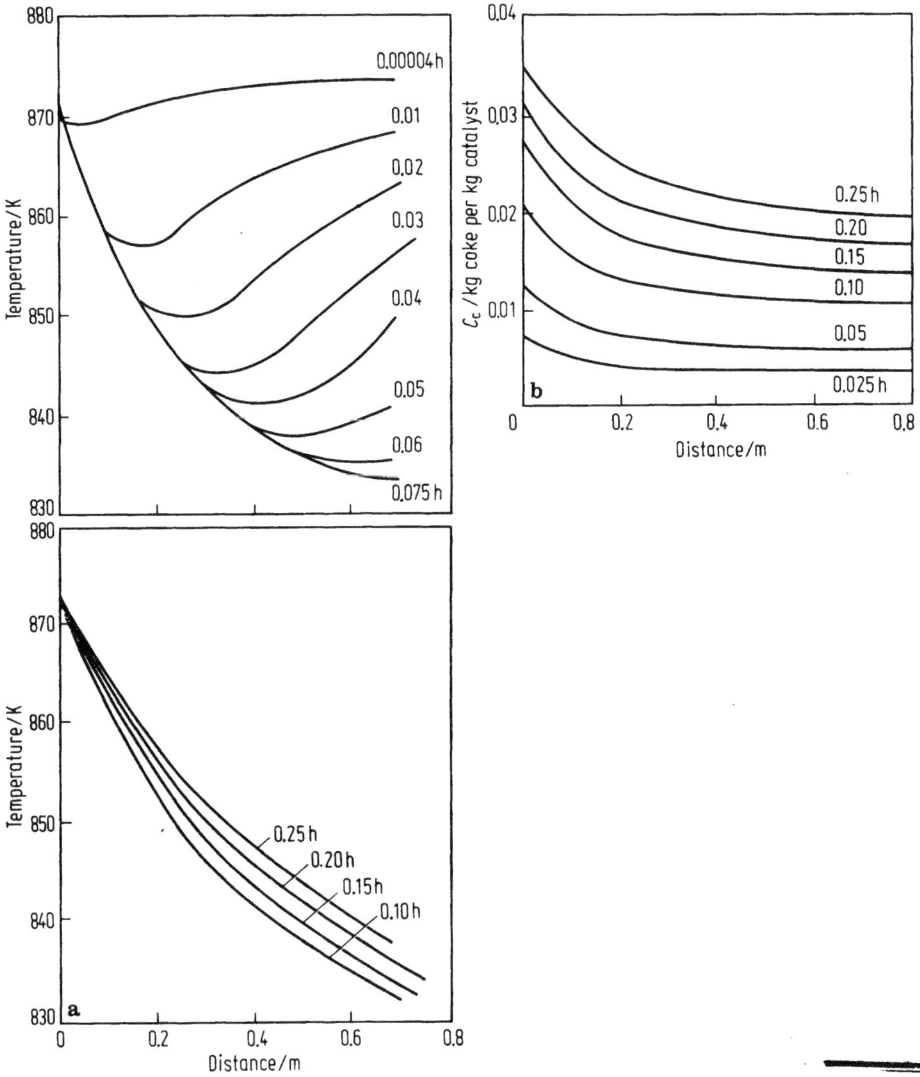

Figure 22. **a** Development of temperature profiles in the dehydrogenation reactor; **b** Development of coke profiles in the dehydrogenation reactor. (Reproduced with permission from ref. [67])

The reactor model included two-phase heat and mass balances with intraparticle diffusion limitations in the solid phase. In addition to the extensive experimental investigation, there is reported an interesting simulation for an industrial reactor based on experimental kinetics and the parameters of industrial operation given in Table 3. Typical reactor behavior is shown in Figures 22a and b. The large initial decrease in reactor temperature is the result of the initial high activity catalyst and the substantial endotherm associated with the dehydrogenation reaction. This travels down the reactor for some time until the apparent minimum passes out the end; subsequently the temperature assumes a quasi-steady state behavior. Although the coke profile decreases with bed length, an interpretation that butene is the predominant coke precursor is probably simplistic. One could argue that contributing factors are also the higher inlet temperature of the reactor and inhibition of coke formation by hydrogen (see equation (61)) near the reactor exit.

E. Constant Conversion Operation

As stated earlier, a number of important commercial processes, including hydrocracking and hydrodesulfurization, require reactor operation at constant conversion or selectivity for reasons of downstream processing. Commonly this involves increasing temperature at constant space velocity to compensate for deactivation, although the inverse is theoretically also possible. We shall consider here the case in which at any instant the reactor may be assumed to be isothermal; the time scale of deactivation is slow, so the (isothermal) temperature level gradually is increased with time of operation. In the case that deactivation rates are functions of activity level alone

$$(-r)_D = k_D s^n = -\frac{ds}{dt} \tag{63}$$

it has been shown [68] that the relationship between temperature and time-on-stream is given by

$$t = \frac{\exp(E_D/RT_0)}{k_D^0(E_D/E_A - n + 1)}\{1 - \exp[(E_D - nE_A + E_A)/R] \times (1/T - 1/T_0)\} \tag{64}$$

Expressions such as equation (64) are generally able to reproduce typical time-temperature trajectories obtained in industrial operation. Equation (64) also has the pleasing property that it can be separated into constant intervals of reciprocal temperature, *i.e.* $\left(\frac{1}{T_0} - \frac{1}{T_1}\right)$, $\left(\frac{1}{T_1} - \frac{1}{T_2}\right)$, etc., which leads to a direct method of evaluation of deactivation constants [68].

Unfortunately, as many of the examples presented here have shown, deactivation rates are most often not concentration-independent functions. More detailed analysis requires considerably more mathematical effort. If the chemical reactor is modeled as a series of perfect mixing cells then it is possible to obtain parametric solutions with a relatively small amount of computation [69]. For parallel poisoning according to scheme (I) we may write

$$x_A = 1 - \prod^{N}(1 + k_{A_i}s_i\theta)^{-1} \tag{65}$$

$$x_p = 1 - \prod^{N}(1 + k_{p_i}s_i\theta)^{-1} \tag{66}$$

where θ is the holding time in each of the N cells used to represent the reactor. It is possible to show from combination of equations (65) and (66) that a required temperature schedule is of the form

$$\frac{dT}{dt} = \frac{\prod^{N} k_A(ds_i/dt)}{\prod^{N} s_i(k_A/RT^2)} \tag{67}$$

where (ds_i/dt) represents the deactivation rate $(-r)_D$ in the ith cell.

Whether one employs the approach of equation (64) or (67), it should be kept in mind that both contain the ratio of the activation energies for reaction and deactivation. We have seen the profound influence that this has on reactor operation earlier in the calculations of Blaum. In constant conversion operation this ratio assumes the role of a forcing function in the following sense. The analysis above does not concern itself with the practicalities of temperature manipulation: a continuous trajectory of temperature with time is assumed. In practical operation, however, some type of control system would undoubtedly be used for adjustment of temperature and it is likely that this would be done at discrete time intervals, not continuously. Now the existence of a non-zero value for E_D ensures thermal activation of the catalyst decay process so we have a classical case here of positive feedback. The control policy employed can have a critical effect on decay rate and hence total run length. High gains associated with longer time intervals between control action can greatly shorten cycle length [70]. Primarily for this reason the use of constant conversion operation in the laboratory for determination of catalyst activity maintenance is a risky business. Confusion of selectivity is also a possible result of this positive feedback, particularly for bi- or polyfunctional catalysts. Consider a two-step reaction A → B → C in which the first step occurs on one function and the second on another, with both functions being independently poisoned. Schematically

$$\begin{aligned} &A + L_1 \rightarrow B + L_1 \\ &B + L_2 \rightarrow C + L_2 \\ &P_1 + L_1 \rightarrow P_1 \cdot L_1 \\ &P_2 + L_2 \rightarrow P_2 \cdot L_2 \end{aligned} \tag{IV}$$

Selectivity in this case is determined by the relative magnitudes of the rate constants, in turn determined by the relative values of the activation energies, as well as catalyst composition. Unless the poisoning activation energies are equal for the two steps, one can see that progressive increase of the temperature will drive the selectivity to an ultimate limit involving only B or only C [71].

7. Catalyst Regeneration

Because poisoning is a very specific type of chemical interaction, at least in the way we have defined it, it is very difficult to discuss in any general way the regeneration of poisoned catalysts. Specific detoxification procedures are given in the patent literature for various materials, or efforts may be made to render potential poisons nontoxic via variation of process conditions as per the example of the shielded compound concept of Maxted.

It seems reasonable, then, to focus on those aspects of catalyst regeneration that yield to generalization — in specific this means the regeneration of coked catalysts. Here the carbon burning reaction provides the thread of continuity. As before we shall proceed in sections according to the scale of events being considered.

A. Kinetics of Coke Oxidation

The oxidation of coke is a rapid reaction, for essentially we are talking about a combustion process, that one might reasonably assume to bear some relationship to the kinetics of carbon oxidation. In fact comparison of coke burning kinetics to those for various forms of pure carbon may assist in understanding better the nature of coke itself.

A number of studies of carbon or coke burning kinetics have been reported. The work of Bondi et al. [72] asserts the following reasonable form

$$(-r)_o = k_o C_{O_2} C_c \tag{68}$$

While no detailed data were presented in that source to support the form of equation (68), it would perhaps be in accord with intuition, prior experience, and pure carbon burning kinetics. There may be one reservation to this, however, given the nature of the coke deposit within the pores of a typical catalyst. If the amount of coke deposited is sufficiently large, there could well be multilayer coverage of coke molecules on the surface and under this circumstance it is not clear that there necessarily would be any dependence of $(-r)_0$ on C_c. Reaction under these conditions would be zero order in C_c, and only at coverages less than monolayer would the first order kinetics appear. A second reservation would have to do with what exactly is meant by C_c. In general this value is obtained from gross measurements that determine carbon plus hydrogen and make no distinction between stoichiometry or chemical nature. Presumably this reservation is not sufficiently serious to affect the form of the kinetic correlation, but would be reflected in differing values of k_c for individual cases.

It has been reported [73] that on a typical cracking catalyst (amorphous SiO_2/Al_2O_3) of surface area *ca.* 250 m^2 g^{-1} a monolayer of coke corresponds to about 12.5 wt.-% carbon (based on the carbon atom area in graphite). Since the coke levels encountered in catalytic cracking, at least, are generally below this level, then the assumption of first order in C_c seems justified. This has been confirmed in part by Weisz and Goodwin [74]. For a given temperature and oxygen partial pressure they report

$$\frac{dC_c}{dt} = (-r)_o = k_o C_c \tag{69}$$

and report data illustrating the kinetic transition from zero to first order for catalysts above ca. 10% C_c. If we define θ_{85} as the time required to remove 85% of the coke initially present (a typical figure for commercial operation), then based on equation (69)

$$\theta_{85} = -\frac{1}{k_o} \ln 0.15 \tag{70}$$

or

$$k_o = \frac{1.89}{\theta_{85}} \tag{71}$$

Comparison of this analysis with experimental data indicated that θ_{85} was independent of initial coke level, independent of the morphology of the catalyst (outside the region of diffusion limitation), independent of the origin of the coke, and not catalyzed by the SiO_2—Al_2O_3 catalysts in

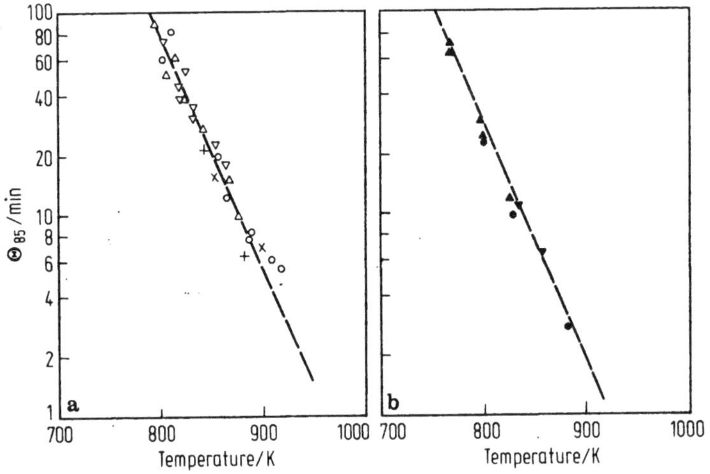

Figure 23. a Burning rate independence from origin of coke. (Differing symbols represent composite of experiments with different feeds, laboratory *vs.* commercial operation, etc.). **b** Intrinsic combustion rates on noncatalyzing oxide bases. ▲, Filtrol 110; ▼, silica-magnesia; ●, Fuller's earth. Dashed line is standard non-catalyzed kinetics. (Reproduced with permission from ref. [74])

use at the time. All these are implicit in the simple model of equation (69) if k_0 is to be considered an intrinsic constant for coke burning kinetics. The last two points are illustrated in Figure 23, where the activation energy corresponding to the indicated lines is 157.3 ± 6.7 kJ mol^{-1}, in good agreement with the 144.4 kJ mol^{-1} previously reported [73]. Further, these numbers are in agreement with the value of 153.6 kJ mol^{-1} reported by Gulbransen and Andrew [75] for the oxidation of graphite. It would then appear that coke oxidation is relatively independent of carbon structure and is not catalyzed by typical SiO_2—Al_2O_3 structures. Oxidation is catalyzed, however, by addition of transition metal oxides to the catalyst [74].

Having established the validity of first order burning with respect to C_c, we should now return to the more general form of equation (68). Kinetics with respect to oxygen have been established as first order by Walker *et al.* [76] and by Massoth [77]. However, Massoth *et al.* [78] also report that the initial exotherms and rate of reaction are far in excess of those to be expected on the basis of the results of Weisz and Goodwin [74]. They postulate this initial rate of combustion to be associated with the oxidation of strongly adsorbed hydrocarbons present and the excess isotherm to be the result of the reaction of hydrogen with oxygen. There is evidence [77] that the oxidation of hydrogen and carbon components occurs at different rates and that a parallel gas-solid reaction pathway is required for kinetic description. This will be discussed further in the next section.

B. Intraparticle Processes

There is abundant evidence in the literature that when one passes to the scale of regeneration of coked particles one has to deal with a general class of diffusion-reaction problems. It was shown in early work [79] that coke burning on individual particles was characterized by an apparent dual rate regime in which initial burning occurred at a high rate that subsequently decreased to a lower, pseudosteady rate that persisted until almost all the carbon had been removed. This was interpreted to be the result of initial

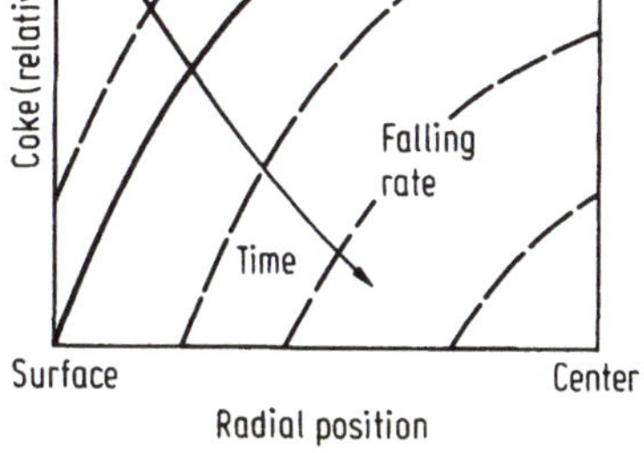

Figure 24. Intraparticle coke profiles as a function of time; constant and falling rate periods

burning on the external surface of the particles, with subsequent reaction limited by the diffusion of oxygen on the combustion zone within the particles. A mathematical analysis of this was given subsequently by Ausman and Watson [80] that envisioned external burning of coke (called the constant rate period) continuing until all surface material was removed, then a second period commenced (falling rate period) in which all reaction was completely within the particle. The evolution of intraparticle coke profiles according to this view is depicted in Figure 24. In this study the carbon burning kinetics were taken to obey equation (68) and an analytical solution to a rather complex moving boundary problem was presented. A similar type of analysis for differences in burning rates has been given by Bondi *et al.* [72] who calculated transients in temperature as a function of burning time. Very large exotherms were indicated, and these were used as a basis for the observed rapid sintering of fresh cracking catalyst upon addition to a fluid bed reactor/regenerator unit.

Dual kinetic regimes for coke burning, however, are not necessarily associated with reaction-diffusion interactions [77, 78]. One recalls that coke is a nominal term used to refer to a variety of materials with empirical formula about (CH). Evidence has been given that the carbon and hydrogen components may not be oxidized at the same rate. Under conditions of chemical control of the oxidation rate, Massoth [77] reported the data shown in

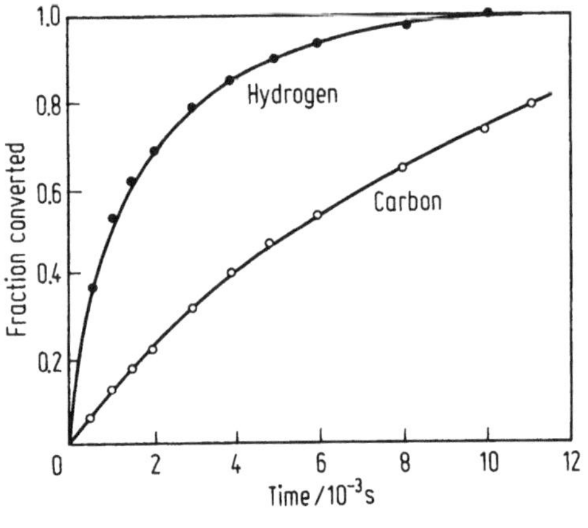

Figure 25. Dual kinetics of coke burning: carbon and hydrogen. (Reproduced with permission from ref. [77])

Figure 25 for the removal of coke from a catalyst of initial coke content 9.6% being regenerated at *ca.* 750 K and 6 mol% oxygen/inert. Clearly there are two parallel reactions here that may interact. A double core model was proposed for the overall oxidation in which two interfaces develop as reaction proceeds. The outer interface is that of oxygen-carbon and the inner that of oxygen-hydrogen. Hydrogen burning rates are viewed as being controlled by oxygen diffusion through an outer layer of carbon to the hydrogen-containing core, while carbon rates are chemically controlled. This would account for the higher initial rates associated with hydrogen as shown in Figure 25.

If one delves into the mathematical details of the approaches discussed above, however, it will be found that they become fairly complex rather rapidly. Fortunately it turns out that a combination of engineering practice and unexpected kindness on the part of mother nature allows considerable simplification in many instances. The unexpected kindness shows up in the fact that at temperature levels where regeneration is normally carried out the kinetics of carbon burning are very rapid. The practical benefit of this is that it allows one to decouple the mass transfer and reaction problem to the extent that the rate of reaction of coke becomes a boundary condition to the intraparticle diffusion of oxygen, *i.e.*, this corresponds to regeneration in the extreme limit of diffusional restriction. Physically then, the process is envisioned as the diffusion of oxygen through a previously regenerated external shell to a core containing coke, with reaction occurring at the interface between the core and the regenerated zone. This is shown in Figure 26.

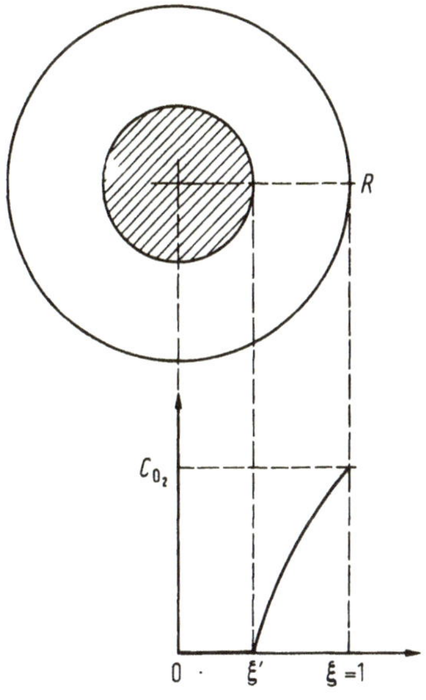

Figure 26. Model of shell-progressive combustion. (Reproduced with permission from ref. [74])

In order to satisfy the requirement for diffusion limitation of the regeneration kinetics, the following criterion must be satisfied [74]

$$\left(\frac{R_p^2}{D_{O_2}}\right)\frac{(-r)_{O_2}}{C_{O_2}} > 1 \qquad (72)$$

In the event that this inequality is not satisfied there will be no appreciable gradients of oxygen and the simplified analysis according to the picture of Figure 26 cannot be used. On the other hand, one may use this expression to determine oxygen consumption rates above which significant diffusion does occur.

One recognizes Figure 26 as being the representation of a shell-progressive process that is entirely analogous to our prior discussion except that now the regenerated zone rather than the poisoned zone is contained in the growing shell. The rate of reaction of oxygen is equal to the diffusion rate at the particle surface, so we may write the following simple balance

$$(-r)_{O_2} = -D_{O_2}\left(\frac{dC_{O_2}}{dr}\right)_{R_p} \times 4\pi R_p^2 = \frac{4\pi R_p D_{O_2} p}{R y_{O_2}}\left(\frac{d\psi}{d\zeta}\right)_{\zeta=1} \tag{73}$$

and the carbon removal is N times this. The fraction of carbon remaining is given by the ratio of the volumes of the core containing coke to that of the total particle, corresponding to radii of ζ' and ζ in Figure 26. The gradient $(d\psi/d\zeta)_{\zeta=1}$ is obtained from solution of the diffusion-without-reaction problem [80], and the final result is the following implicit expression for the fraction of coke remaining, X_F

$$\frac{1}{2}(1 - X_F^{2/3}) - \frac{1}{3}(1 - X_F) = \frac{N D_{O_2} C_{O_2}}{R_p^2 C_c^o}\,\theta \tag{74}$$

In the previous section some experimental data were presented in terms of θ_{85}, the time for 85% coke removal. Substitution of $X_F = 0.15$ in equation (74) gives

$$\theta_{85} = 0.076\,\frac{R_p^2 C_c^o}{N D_{O_2} C_{O_2}} \tag{75}$$

for θ_{85} in minutes. The proportionalities indicated in equation (75) can be directly tested experimentally by varying initial coke content, oxygen concentration, particle radius, and so on. Results were presented for coke combustion at 975 K on 350 m^2 g^{-1} silica-alumina particles differing in initial coke content and diameter under conditions where shell progressive

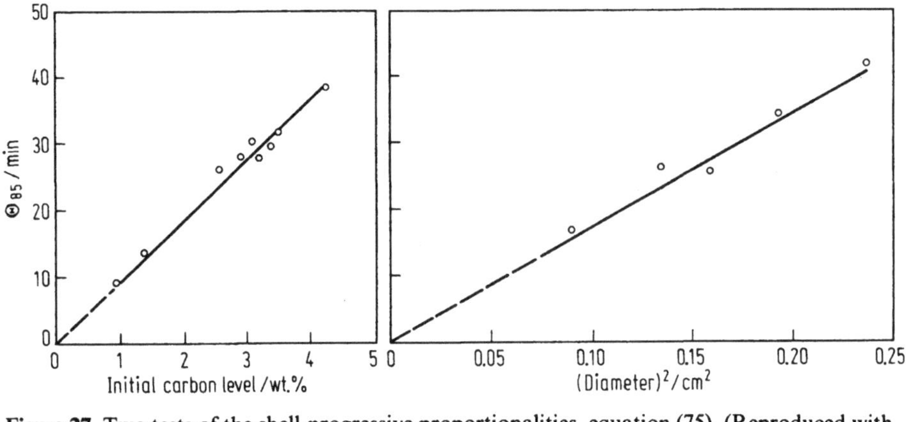

Figure 27. Two tests of the shell-progressive proportionalities, equation (75). (Reproduced with permission from ref. [74])

burning would be expected. Examples of the excellent correlation with the proportionalities predicted by the simple theory are provided in Figure 27 for initial coke content and particle radius.

Thus, the shell-progressive model provides a simple means for the understanding of intraparticle regeneration when in the regime of appropriate diffusion limitation. Pure shell-progressive mechanisms are to be expected for Thiele moduli of the order of 200 for schemes such as (I), (II) or the regeneration considered here [81]. This may be somewhat conservative; close examination of the results of Masamune and Smith [46] would suggest that moduli of the order of tens rather than hundreds give similar overall behavior even though there is an interfacial zone between the regenerated and nonregenerated (or poisoned and nonpoisoned) zones of the particle.

C. Intrareactor Regeneration

The general objectives in analysis of fixed bed regeneration are similar to those in fixed bed deactivation; indeed, the topic can be thought of as a kind of deactivation in reverse. Primary concerns are with the extent of recovery of activity, thermal history of the process, and the time required for a given level of regeneration. As in the case of deactivation, these are all interrelated factors. One is particularly concerned with thermal behavior, since the burning of coke is a very exothermic processs and the thermal waves generated within the fixed bed configuration must be controlled to avoid possible catalyst or reactor damage.

In general the basic kinetics set forth in equation (68) can be used, although the early studies of van Deempter [82] used a constant rate model, while Johnson, *et al.* [83] proposed

$$(-r)_o = k_o p y_{O_2} (C_c/C_c^o) \tag{76}$$

which was claimed to be a good approximation when intraparticle oxygen diffusion is rate limiting. In this case, k_o can be assumed approximately independent of temperature. Three balances are required to describe the regeneration, an overall heat balance and individual mass balances for oxygen and coke. Let us consider the case in which $k_o \neq f(T)$ employing the correlation of equation (76). The oxygen balance in the reactor is

$$\frac{\partial C_{O_2}}{\partial t} + v\frac{\partial C_{O_2}}{\partial z} = a' k_o p y_{O_2} (C_c/C_c^o) \tag{77}$$

where a' is a factor including reaction stoichiometry. The carbon balance is given directly by equation (76), and the heat balance by

$$\overline{\varrho c_p}\frac{\partial T}{\partial t} + c_p G\frac{\partial T}{\partial t} = b k_o p y_{O_2} (C_c/C_c^o) \tag{78}$$

where b is a factor including the heat of reaction. Note that equation (78) presumes adiabatic operation during regeneration. Since the rate constant

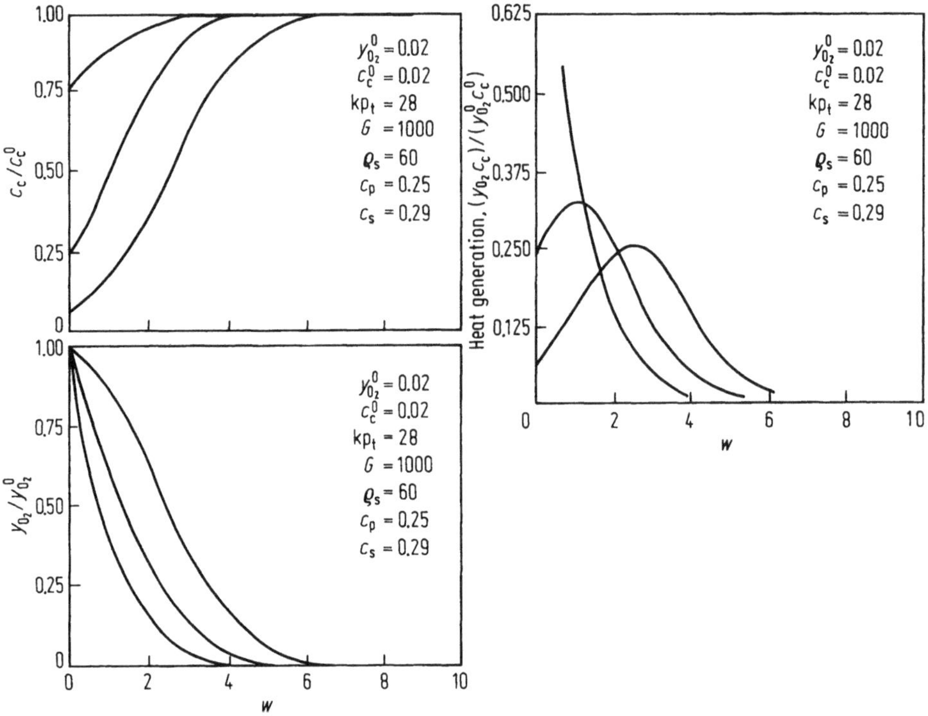

Figure 28. Development of constant pattern profiles. Caron burning with independent kinetics. Notation and units for the calculation:
$y_{O_2}^o$ = initial mol fraction oxygen; C_c^o = initial carbon content, lb per lb catalyst; k = burning rate constant, h^{-1} atm^{-1}; p_t = total pressure, atm; G = mass flow rate, lb hr^{-1} ft^{-2}; ϱ_s = bulk density of bed, lb ft^{-3}; C_p, C_s = heat capacity of fluid and solid, Btu lb^{-1} $°F^{-1}$.
The parameter w is defined by equation (83). (Reproduced with permission from ref. [83])

is independent of temperature, an analytical solution to this set of equations can be obtained [83]. The general form of results is shown in Figure 28 for an example calculation using the parametric values shown. The figure presents a familiar picture that we have seen in the analysis of deactivation — developing waves that assume a constant pattern after an initial transient and then pass through the bed at essentially constant velocity. There is, however, a subtle complexity to the wave behavior here that really doesn't seem to manifest itself so much in the deactivation behavior we have considered. Johnson, *et al.* have shown that the heat generation function of the energy balance equals $(y_{O_2}C_c/y_{O_2}^o C_c^o)$; the behavior of this is shown in Figure 28c. What is not shown is the progression of the thermal wave, that is the progression of T with position and time. The relative motion of these two waves is very important. When the following inequality is satisfied

$$\left(\frac{y_{O_2}}{C_c^o}\right) < \frac{M_g}{M_c}\frac{c_p}{c_s}\,a \qquad (79)$$

it can be shown that the temperature wave travels down the bed faster than the burning zone; conversely, when the inequality is satisfied in the opposite direction the burning zone preceeds the temperature front. In either of these cases there is a type of self-limitation imposed on the regeneration process and there is lessened danger of thermal damage or runaway. Beware, however, if the inequality in either direction is not satisfied. When the temperature wave and burning zone have the same velocity a reinforcement between the two occurs that results in very large temperature rises in the bed. A critical inlet oxygen-carbon ratio can be defined on this basis just from the equality of equation (79)

$$\left(\frac{y_{O_2}}{C_c^o}\right)_{\text{critical}} = a\frac{c_p}{c_s}\left(\frac{M_g}{M_c}\right) \tag{80}$$

where operation either above or below the ratio specified in equation (80) will avoid the problem of reinforcement of the hot spot. Estimation of the regeneration cycle time is provided by the following quasi-steady state solutions for the coke and oxygen profiles

$$\frac{C_c}{C_c^o} = \frac{1}{1 + e^{-aw} + \tau} \tag{81}$$

$$\frac{y_{O_2}}{y_{O_2}^o} = \frac{1}{1 + e^{-aw} - \tau} \tag{82}$$

where a is the stoichiometric coefficient for oxygen consumption ($a = 1$ for combustion to CO_2) and w and τ are dimensionless variables defined as

$$w = z\left(\frac{\varrho_s}{G}\frac{M_g}{M_c}k_o p\right) \tag{83}$$

$$\tau = \left(\frac{y_{O_2}^o}{C_c^o}k_o p\right)\left(t - z\frac{\varepsilon\varrho_g}{G}\right) \tag{84}$$

Subsequent experience, as discussed below, has taught that the general trends identified by this analysis are qualitatively valid, in spite of the assumption of $k_o \neq f(T)$. The interactions between thermal and burning waves are perhaps even underestimated as a result of that assumption. The nature of the temperature profiles and the relative rates of propagation of various gradients are much affected by operating conditions; particularly, for less severe gradients than those illustrated in Figure 28, the zone of combustion is much longer and the corresponding temperature profiles are more representative of what one would consider an adiabatic front. An analysis of this situation has been presented by Shulman [82].

An explicit treatment of fixed bed regeneration with shell-progressive burning has been given by Olsen *et al.* [85], using the Weisz-Goodwin model for kinetics and assuming an initial parabolic coke profile within the individual particles. This approach requires separate balances for the gas and solid phases, in contrast to the pseudohomogeneous approach described above. The results of a parametric numerical investigation are in general accord with those reported by Johnson *et al.* [83], but with one important exception. Very large temperature excursions were found to occur at the beginning of regeneration; these were attributed to the high rates of reaction associated with oxygen-rich feed contacting the unregenerated catalyst (high coke content) in which the diffusion limitation associated with the shell-progressive mechanism had not yet developed. These exotherms are well correlated in terms of the parameter $(pD_e/\sqrt{GR_p})$, which is essentially a measure of internal to boundary layer mass transfer, since the external mass transfer coefficient is proportional to $\sqrt{G}$. Thus, the magnitude of the exotherm increases as this parameter decreases in value. In all other respects, however, the detailed, two phase simulation appears to differ only slightly from the pseudohomogeneous model with $k_0 \neq f(T)$.

These treatments presume high temperature regeneration with diffusional limit. In some cases it may be necessary to conduct regeneration at lower temperatures because of the thermal stability of the catalyst. In such cases the coke burning kinetics are controlled by the intrinsic chemical rate and the reaction is uniform throughout the individual particles. In terms of analysis the major effect is a large change in apparent activation energy — from essentially zero for shell-progressive burning to the *ca.* 160 kJ mol^{-1} associated with the intrinsic rate — hence the regeneration proces is much more temperature sensitive at lower temperature levels. A study of this has been reported by Ozawa [86]. The balance equations are set up in the same general form as equations (77) and (78) with the major difference being the use of intrinsic kinetics in place of equation (76). The major difference in comparison to high temperature regeneration is the possible development of a minimum in the coke profile resulting from a localized maximum burning rate. This, of course, is the product of opposing trends in the reaction kinetics due to increasing temperature and decreasing oxygen concentration with bed length. As one might expect, it is difficult to identify temperature or reaction front waves developing into constant pattern fronts at the lower temperature; here the zone of reaction is the entire bed.

Visualized in terms of "deactivation in reverse", there appear to be some general parallels between these fixed bed regeneration problems and our previous discussion of poisoning or coking. High temperature adiabatic regeneration is very similar to rapid poisoning of exothermic reactions in adiabatic beds. Low temperature regeneration is similar is some respects to coke deposition via scheme (II) when the deposition rate is low and coke is formed throughout the bed.

Notation

a'	reactor/stoichiometry parameter, equation (77); mols O_2 per wt. % coke

a	correlation parameter, equations (1) and (42); stoichiometric coefficient, equation (79); dimensionless
A, B	reactant and product
b	correlation parameter, equation (1) and (41); dimensionless; heat of reaction factor, equation (78); energy per wt. % coke
C_A	concentration of A, mols (volume)$^{-1}$
$C_{(AL)_n}$	surface concentration of growing coke molecule $(A—L)_n$; % wt (area)$^{-1}$
C_c, C_c^o, $C_{c\infty}$	weight percent coke on catalyst, initial and limiting, respectively
$\langle C_c \rangle$	average coke content; % wt.
C_L	concentration at reactor exit; mols (volume)$^{-1}$
C_0, C_1	concentrations at points 0 and 1, Figure 12; mols (volume)$^{-1}$
C_o	reactor inlet concentration; mols (volume)$^{-1}$
C_{O_2}	oxygen concentration; mols (volume)$^{-1}$
C_p, C_p^o, $C_{p\infty}$	poison concentration, inlet and limiting capacity, respectively; mols (volume)$^{-1}$, mols per wt. catalyst (C_∞)
$\langle C_p/C_p^o \rangle$	average defined by equation (44); dimensionless
c_p/c_s	heat capacity ratio, gas/solid; dimensionless
c_p	heat capacity of gas phase; energy mol^{-1} (volume)$^{-1}$ (typically kJ mol^{-1})
D_e	effective diffusivity; (length)^2 (time)$^{-1}$
D_{O_2}	effective diffusivity of oxygen; (length)^2 (time)$^{-1}$
E	activation energy; energy mol^{-1} (typically kJ mol^{-1})
E_A, E_C, E_D	activation energies for reaction of A, coke formation and poisoning; energy mol^{-1} (kJ mol^{-1})
F	ratio of rates, poisoned/fresh catalyst; dimensionless; feed rate, equation (55); weight (time)$^{-1}$
G	mass velocity; weight (time)$^{-1}$ (area)$^{-1}$
h(T)	heat removal term, equation (60); K (time)$^{-1}$
$(-\Delta H)$	heat of reaction; energy mol^{-1} (typically kJ mol^{-1})
i	mixing cell index; dimensionless
k	rate constant on poisoned catalyst, equation (26); (time)$^{-1}$
k_1, k_2, k_3, k_4, k_r	kinetic constants in scheme (IV); mol (time)$^{-1}$ (pressure)$^{-1}$ (k_1, k_4); mol (time)$^{-1}$ (k_2, k_3, k_r)
k_A, $\langle k_A \rangle$	rate constant for A, and average value thereof; (time)$^{-1}$
k_{cB}^o, k_{cD}^o	coking rate constants, butene and butadiene, equation (61); wt. % coke (time)$^{-1}$ (pressure)$^{-q \text{ or } -v}$
k_C, k_D	coking and poisoning rate constants; (time)$^{-1}$
k_o	rate constant on fresh catalyst; (time)$^{-1}$ (first order reaction); coke oxidation rate constant; wt. % coke (time)$^{-1}$ (volume)$^{-1}$ (pressure O_2)$^{-1}$
k_p	poison adsorption rate constant; (time)$^{-1}$
K_A, K_B	adsorption constants for A and B; (pressure)$^{-1}$ (typically atm^{-1})

K_c	constant in coking correlation, equation (24); wt. % coke $(\text{time})^{-1/2}$
K_H	hydrogen adsorption constant, equation (61); $(\text{pressure})^{-1/2}$ (typically $\text{atm}^{-1/2}$)
K_q	adsorption constant for site of heat of adsorption q; $(\text{pressure})^{-1}$
K'	constant in coking rate, equation (22); area $(\text{pressure})^{-1}$ $(\text{time})^{-1}$ $(\text{volume})^{-1}$
L	bed length
L, L_1, L_2	surface site
m	exponent in equation (2); dimensionless
M_c, M_g	molecular weights of carbon and gas phase; weight mol^{-1}
n	exponent in equation (3) and (63); dimensionless
n_q	number of sites with heat of adsorption q; dimensionless
N	number of mixing cells; dimensionless; O_2 stoichiometric coeff, equation (74); wt. % coke $(\text{mol } O_2/\text{volume})^{-1}$
N_t	parameter, equation (47); $N_t = k_p L/v$; dimensionless
p, p_A, p_B	total pressure, partial pressure of A and B; typically atm
P, P_1, P_2	poison
q, q_m	heat of adsorption and maximum heat of adsorption; energy mol^{-1} (typically kJ mol^{-1})
$r_1(C), r_2(T), r_3(s)$	concentration, temperature and activity factors, main reaction; mol $(\text{volume})^{-1}$ $(\text{time})^{-1}$ (overall)
$r_4(C), r_5(T), r_6(s)$	concentration, temperature and activity factors, deactivation reaction; unit activity $(\text{time})^{-1}$ (overall)
$(-r)_c, (-r)_D$	coking and poisoning rates; wt. % coke $(\text{time})^{-1}$ or unit of activity $(\text{time})^{-1}$
$(-r)_o$	oxidation rate of coke; wt. % coke $(\text{time})^{-1}$ $(\text{volume})^{-1}$
$(-r)_{O_2}$	rate of reaction of oxygen; mol $(\text{time})^{-1}$ $(\text{volume})^{-1}$
$(-r)_p$	poisoning rate, equation (59); unit of activity $(\text{time})^{-1}$ $(\text{volume})^{-1}$
$(-r)_s$	deactivation rate; unit of activity $(\text{time})^{-1}$
$(-r)_T$	rate of main reaction; mol $(\text{time})^{-1}$ $(\text{volume})^{-1}$
$(-r)_{TO}$	overall rate on fresh catalyst, equation (25); mol $(\text{time})^{-1}$ $(\text{volume})^{-1}$
$(-r_T)_q$	rate on sites of heat of adsorption q; mol $(\text{time})^{-1}$ $(\text{volume})^{-1}$
R	gas constant
R_p	particle radius; length
$s, \langle s \rangle$	activity variable; average activity; dimensionless
s_q	activity variable for sites of heat of adsorption q; dimensionless
t	time
t_c	time-on-stream, hours (or consistent with α)
t_∞	capacity ratio, equation (47); (mol of poison per weight)/(mol of poison time)

t_R	reactor residence time
T, T_O	temperature, initial temperature
v	velocity; length $(\text{time})^{-1}$
v, q	exponents in equations (19) and (20); dimensionless
w	variable defined via equation (83); dimensionless
W	characteristic dimension of Thiele modulus; weight of catalyst, equation (55)
x, x_A	conversion; conversion of A; mol reacted per mol initial
x_p	fractional uptake of poison; dimensionless
X_F	fraction coke remaining; dimensionless
y_{O_2}	mol fraction of oxygen; dimensionless
z	distance variable; length
α	characteristic decay constant; $(\text{time})^{-1}$; fraction surface poisoned
ε	particle or bed porosity; void volume per total volume
ζ	dimensionless position, equation (73)
θ	holding time per mixing cell, time in coke burning, equation (74); min
θ_A, θ_B	surface coverage of A and B; dimensionless
λ_{eff}	effective bed thermal conductivity; energy $(\text{length})^{-1}$ $(\text{time})^{-1}$ K^{-1}
ϱ, ϱ_c	particle density, weight $(\text{volume})^{-1}$
ϱ_g	density of gas phase, mol $(\text{volume})^{-1}$
$\overline{\varrho c_p}$	average volumetric heat capacity; energy $(\text{volume})^{-1}$ K^{-1}
τ	variable defined via equation (84); dimensionless
φ_o	fresh catalyst Thiele modulus, $W\sqrt{k_o/D_e}$; dimensionless
φ, φ_p	poisoned catalyst Thiele modulus; dimensionless
ψ	dimensionless concentration, equation (73)

References

1. Butt, J. B.: Adv. Chem. **109**, 259 (1972)
2. Billimoria, R. M., Butt, J. B.: Amer. Chem. Soc. Symp. Ser. **72**, 288 (1978)
3. Butt, J. B.: Stud. Surface Sci. and Catalysis **6**, 21 (1980)
4. Froment, G. F.: ibid. **6**, 1 (1980)
5. Hegedus, L. L., McCabe, R. W.: ibid. **6**, 471 (1980)
6. Delmon, B., Grange, P.: ibid. **6**, 507 (1980)
7. Wanke, S. E., Flynn, P. C.: Catal. Revs.-Sci. and Engr., **12**, 93 (1975)
8. Faraday, M.: Phil. Trans. Roy. Soc., **124**, 55 (1834)
9. Maxted, E. B.: Adv. Catal. **3**, 129 (1951)
10. Pease, R. N., Stewart, L.: J. Amer. Chem. Soc. **47**, 1235 (1925)
11. Eley, D. D., Rideal, E. K.: Proc. Roy. Soc. (London) **A178**, 429 (1941)
12. Mills, G. H., Boedeker, E. R., Oblad, A. G.: J. Amer. Chem. Soc. **72**, 1554 (1950)
13. Oudar, J.: Catal. Revs.-Sci. and Engr. **22**, 171 (1980)
14. Berthier, Y., Perdereaux, M., Oudar, J.: Surf. Sci. **36**, 225 (1973)
15. Heegerman, W., Meister, K. H., Berthold, E., Hayek, K.: ibid. **49**, 161 (1975)
16. Herington, E. F. G., Rideal, E. K.: Trans. Faraday Soc. **40**, 505 (1944)

17. Fischer, T. E., Kelemen, S. R.: J. Catal. **53**, 24 (1978)
18. Bonzel, H. P., Ku, R.: J. Chem. Phys. **58**, 4617 (1973)
19. Pines, H., Haag, W. O.: J. Amer. Chem. Soc. **82**, 2471 (1960)
20. Sterba, M. J., Haensel, V.: Ind. Eng. Chem. Prod. Res. Develop. **15**, 2 (1976)
21. Knözinger, H.: Adv. Catal. **25**, 184 (1976)
22. Webb, G., Macnab, J.: J. Catal. **26**, 226 (1972)
23. Levinter, M. E., Panchenko, G. M., Tanatarov, M. A.: Intern. Chem. Eng. **7**, 23 (1967)
24. Muno, W. E., Delgado, S., Butt, J. B.: J. Catal. **37**, 158 (1975)
25. Amelse, J. A., Schwartz, L. H., Butt, J. B.: J. Phys. Chem. **82**, 558 (1978)
26. Voorhies, A., Jr.: Ind. Eng. Chem. **37**, 318 (1945)
27. Ozawa, Y., Bischoff, K. B.: Ind. Eng. Chem. Process Design Develop. **7**, 67 (1968)
28. Eberley, P. E., Jr., Kimberlin, C. N., Miller, W. H., Drushel, H. V.: ibid. **5**, 193 (1966)
29. Appleby, W. G., Gibson, J. W., Good, G. M.: ibid. **1**, 102 (1962)
30. Weekman, V. W., Jr.: ibid. **7**, 90 (1968)
31. Weekman, V. W., Jr.: ibid. **8**, 388 (1969)
32. Weekman, V. W., Jr., Nace, D. M.: AIChE J. **16**, 397 (1970)
33. Nace, D. M., Voltz, S. E., Weekman, V. W., Jr.: Ind. Eng. Chem. Process Design Develop. **10**, 530 (1971); Voltz, S. E., Nace, D. M., Weekman, V. W., Jr.: ibid. **10**, 538 (1971)
34. Gross, B., Nace, D. M., Voltz, S. E.: ibid. **13**, 199 (1974)
35. Voltz, S. E., Nace, D. M., Jacob, S. M., Weekman, V. W., Jr.: ibid. **11**, 261 (1972)
36. Jacob, S. M., Gross, B., Voltz, S. E., Weekman, V. W., Jr.: AIChE J. **22**, 701 (1976)
37. Beuther, H., Larson, O. A., Perrotta, A. J.: Stud. Surface Sci. and Catalysis **6**, 271 (1980)
38. Riley, K. L., Silbernagel, B. G., Butt, J. B.: Heavy Feed Hydroprocessing Catalyst Deactivation; The Chemistry and Impact of Coke and Metals Deposits. 7th North American Meeting, The Catalysis Society, Boston, MA, October 11–15, 1981
39. Szepé, S., Levenspiel, O.: Proc. 4th European Sym. Chem. Reaction Engr., 265, Pergamon Press, Oxford, 1971
40. Butt, J. B., Wachter, C. K., Billimoria, R. M.: Chem. Eng. Sci. **33**, 1321 (1978)
41. Bakshi, K. R., Gavalas, G. R.: AIChE J. **21**, 494 (1975)
42. Önal, I., Butt, J. B.: Stud. Surface Sci. and Catalysis **7(B)**, 1490 (1981)
43. Rudershausen, C. G., Watson, C. C.: Chem. Eng. Sci. **3**, 110 (1954)
44. Wheeler, A.: Adv. Catal. **3**, 250 (1950)
45. Butt, J. B., Weekman, V. W., Jr.: Chem. Eng. Proc. Symp. Ser. (143) **70**, 27 (1971)
46. Masamune, S., Smith, J. M.: AIChE J. **12**, 384 (1966)
47. Sagara, M., Masamune, S., Smith, J. M.: ibid. **13**, 1226 (1967)
48. Murakami, Y., Kobayshi, T., Hattori, T., Masuda, M.: Ind. Eng. Chem. Fundam. **7**, 599 (1968)
49. Lee, J. W., Butt, J. B., Downing, D. M.: AIChE J. **24**, 212 (1978)
50. Corbett, W. E., Jr., Luss, D.: Chem. Eng. Sci. **29**, 1473 (1974)
51. Becker, E. R., Wei, J.: J. Catal. **46**, 365, 372 (1977)
52. Wolf, E. E.: ibid. **47**, 85 (1977)
53. Hegedus, L. L., Petersen, E. E.: Catal. Revs.-Sci. and Engr. **9**, 245 (1974)
54. Anderson, R. B., Whitehouse, A. M.: Ind. Eng. Chem. **53**, 1011 (1961)
55. Wheeler, A., Robell, A. J.: J. Catal. **13**, 299 (1969)
56. Bohart, G., Adams, E.: J. Amer. Chem. Soc. **42**, 523 (1920)
57. Haynes, H. W., Jr.: Chem. Eng. Sci. **25**, 1615 (1970)
58. Froment, G. F., Bischoff, K. B.: Chem. Eng. Sci. **16**, 189 (1961); **17**, 105 (1962)
59. Bischoff, K. B.: Ind. Eng. Chem. Fundam. **8**, 665 (1969)
60. Ozawa, Y.: Chem. Eng. Sci. **25**, 529 (1970)
61. Wentrcek, P. W., McCarty, P. W., Ablow, C. W., Wise, H.: J. Catal. **61**, 232 (1980)
62. Menon, P. G., Sreeramamurthy, R.: J. Catal. **8**, 95 (1967)
63. Pexidr, V., Cerny, J., Pasek, J.: Proc. 4th European Symp. Chem. Reaction Engr., 239, Pergamon Press, Oxford, 1971
64. Blaum, E.: Chem. Eng. Sci. **29**, 2263 (1974)
65. Weng, H.-S., Eigenberger, G., Butt, J. B.: ibid. **30**, 1341 (1975)
66. Price, T. H., Butt, J. B.: ibid. **32**, 393 (1977)
67. Dumez, F. J., Froment, G. F.: Ind. Eng. Chem. Process Design Develop. **15**, 291 (1976)

68. Krishnaswamy, S., Kittrell, J. R.: ibid. **18**, 399 (1979)
69. Butt, J. B., Rohan, D. M.: Chem. Eng. Sci. **23**, 489 (1968)
70. Weekman, V. W., Jr.: Proc. 4th European Symp. Chem. Reaction Engr., 263, Pergamon Press, Oxford, 1971
71. Butt, J. B.: Chem. Eng. J. **2**, 90 (1971)
72. Bondi, A., Miller, R. S., Schlaffer, W. G.: Ind. Eng. Chem. Process Design Develop. **1**, 196 (1962)
73. Haldeman, R. G., Botty, M. C.: J. Phys. Chem. **63**, 489 (1959)
74. Weisz, P. B., Goodwin, R. B.: J. Catal. **2**, 397 (1963); **6**, 227 (1966)
75. Gulbransen, E. A., Andrew, E. A.: Ind. Eng. Chem. **44**, 1034 (1952)
76. Walker, P. L., Jr., Rusinko, F., Jr., Austin, L. G.: Adv. Catal. **11**, 134 (1959)
77. Massoth, F. E.: Ind. Eng. Chem. Process Design Develop. **6**, 200 (1967)
78. Massoth, F. E., Menon, P. G.: ibid. **8**, 383 (1969)
79. Rudershausen, C. G., Watson, C. C.: Chem. Eng. Sci. **3**, 110 (1954)
80. Ausman, J. M., Watson, C. C.: ibid. **17**, 323 (1962)
81. Carberry, J. J., Gorring, R. L.: J. Catal. **5**, 529 (1966)
82. Van Deempter, J. J.: Ind. Eng. Chem. **45**, 1227 (1953); **46** 2300 (1954)
83. Johnson, B. M., Froment, G. F., Watson, C. C.: Chem. Eng. Sci. **17**, 835 (1962)
84. Shulman, B. L.: Ind. Eng. Chem. **55**, 44 (1963)
85. Olson, K. E., Luss, D., Amundson, N. R.: Ind. Eng. Chem. Process Design Develop. **7**, 96 (1968)
86. Ozawa, Y.: ibid. **8**, 378 (1969)

Chapter 2

Catalytic Olefin Polymerization

I. Pasquon[1] *and U. Giannini*[2]

[1] Dipartimento di Chimica Industriale
e Ingegneria chimica "Giulio Natta" del Politecnico
Piazza Leonardo da Vinci 32
I-20133 Milano, Italy

[2] Istituto Guido Donegani
Montedison, V. G. Fauser, 4
I-28100 Novara, Italy

Contents

1. Introduction

Table 1 gives the world production data regarding polyolefins and their copolymers as well as those concerning the other main plastics. As may be inferred from such data, the polyethylenes obtained *via* catalytic processes and polypropylenes (homopolymers and modified products) rank fourth and fifth in terms of quantities produced. Polypropylene fibres and ethylene-propylene based elastomers rank fourth and fifth in their respective classifications.

The commercial scale manufacture of these products began in the mid-fifties and it is interesting to observe that no other group of new products has ever since enjoyed a similar importance in the whole chemicale industry. It should also be noted that, except some polyethylenes prepared with catalysts based on metal oxides, all polyolefins (homopolymers and copolymers) produced commercially today are obtained by the use of catalytic systems known as Ziegler-Natta systems. These catalysts furthermore underlie the discovery of the stereospecific polymerization of olefins higher than ethylene, later extended to various other monomer classes (vinylic, olefinic, aldehydic, etc.) [1, 2].

Table 1. World plastics production (1980) (thousand tons)

Polyethylene	14500
LDPE	9500
HDPE	4500
LLDPE	500
Polyvinylchloride	8500
Polystyrene resins	4500
Polypropylene	4000
Aminic resins	3100
Phenol resins	1200

The most significant catalytic aspects, in terms of catalytic systems employed and reaction mechanisms involved, regarding the polymerization and copolymerization of olefins and cycloolefins are described hereunder. The various types of products manufactured on a commercial scale today and the relevant production processes are also indicated. Many reviews have appeared in the past covering this topic, and will be mentioned when the opportunity offers in the following sections.

2. Catalytic Systems

The catalytic systems used for olefins polymerization and copolymerization may be grouped into:

— Ziegler-Natta systems

— metal oxide-based systems.

The latter find application in ethylene polymerization, whereas the former are employed in the polymerization of ethylene, alpha-olefins and cyclo-olefins.

These systems have meanwhile undergone interesting evolutions and the various cases ought to be examined.

A. Ethylene Polymerization

1. Foreword

Ethylene polymerization by the use of catalysts based on transition metals gives a polymer (HDPE, High Density Polyethylene) exhibiting a greater density and crystallinity than the polymer obtained *via* radical polymerization (LDPE, Low Density Polyethylene). The better properties of the former result from fewer polymer chain ramification. HDPE, however, fails to have thoroughly linear polymer chains, as ethylene oligomers or vinyl bond terminated polymeric chains may copolymerize with ethylene and the sequence of methylene units is thus discontinued by short- or long-chain ramifications.

A number of processes for linear low density polyethylene (LLDPE) manufacture have been developed in recent years, using catalysts based on transition metals. This product is obtained by ethylene copolymerization with small quantities of α-olefins (1-butene, 1-hexene, 1-octene).

2. Catalysts Supported on Oxides

Catalysts prepared by supporting both oxides and organometallic compounds of transition metals on SiO_2 and Al_2O_3 are described hereunder.

a) Standard of Indiana catalyst

In 1951, Zletz [3] filed application for a patent describing the first catalytic system capable of producing HDPE *via* a low pressure process. The catalyst, consisting of Mo_2O_3 deposited by coprecipitation or impregnation over Al_2O_3, requires activation at 773 K and reduction with H_2 at 723–773 K. The Mo valence in the reduced catalyst is lower than VI; metallic Na may be added as activator [4]. The polymerization rate within the temperature range 473–550 K and the ethylene pressure range 4.4 to 7.0 MPa is proportional to monomer and catalyst concentration, but is extremely low (0.5 to 2.5 g PE per g cat. per h.). Better results are obtained by employing partially halogenated Al_2O_3 [5].

These catalysts have later been modified by Tadokoro [6] to produce higher molecular weight polymers.

The process using these catalysts, operating in solution, has failed to be successful and has been employed in three commercial plants only.

b) Phillips catalyst

This catalyst was first described in a patent application filed by Hyan and Banks [7] in January 1953. The processes based on this catalyst have been quite successful commercially and approximately 60% of HDPE world production today employs the Phillips catalyst. A number of publications have appeared on the subject, mostly concerning the oxidation state of Cr active in the catalysis, and some reviews cover either general or specific aspects of ethylene polymerization with these catalysts [8–11].

Catalyst preparation is made by impregnation of silica-alumina (87:13) support with a CrO_3 water solution, followed by drying and air activation in a fluid bed at 773–1073 K for many hours.

Spectroscopic [12] and chemical analysis [13] have proved the formation of a surface chromate and possibly dichromate resulting from CrO_3 reaction with silanol groups according to the following scheme

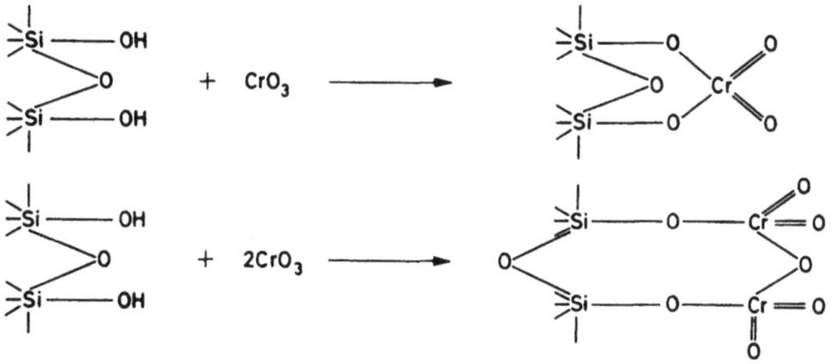

The formation of these compounds ensures the supported Cr (VI) stability at 1073 K, whereas CrO_3 is unstable in air at temperatures above 773 K and is transformed into Cr_2O_3.

A support with a surface area of 600 $m^2\,g^{-1}$ is capable of stabilizing approximatively 5% Cr as Cr(VI). Cr(VI) is reduced to lower valence states by ethylene under polymerization conditions (400 K) and formaldehyde is formed as an oxidation product according to Baker and Carrick [14]. The induction time observed in ethylene polymerization is caused by this reaction and the displacement of the oxidation products. Clark [15] has found that ethylene absorption conversely occurs immediately with a catalyst prereduced with CO at 573 K.

The efficiency of chromium in polymerization decreases by increasing the Cr loading in the catalyst. At 373 K, for ethylene concentrations of 1.5 mol dm^{-3}, productivity decreases from 7×10^6 to 6×10^5 g PE $(g\,Cr)^{-1}\,h^{-1}$ with a Cr content increasing from 0.01 to 1% by weight.

According to Clark [16, 17], the polymer molecular weight lowers as the catalyst activation temperature rises in the range 673–1123 K and increases by 45% approximatively as the Cr content in the catalyst decreases from 0.75 to 0.001% by weight. Lower polymer molecular weights have been obtained with supports [18] having smaller pore diameters.

For LLDPE synthesis, chromium oxide-based catalysts have been developed, supported on SiO_2—Al_2O_3 and modified with small quantities of TiO_2. Preparation is generally made by impregnation of the support with Ti alcoholates, followed by a heat treatment at high temperature [19, 20]. In commercial processes utilizing Phillips type catalysts, polymerization is conducted either in cyclohexane solution (T = 400–450 K; p = 2.8–3.5 MPa) or in slurry in low boiling isoparaffins (particle form process) at a temperature of 350–373 K.

c) Organometallic compounds of transition metals supported on oxides

The catalytic activity of relatively stable organometallic (σ,η^3-allyl, cyclopentadienyl-Cp- and arene types) compounds of transition metals (Ti, Zr, Hf, Nb, Cr) which is very low if operating in solution (*c.f.* section 2.A.3.b.) increases by 2–3 orders of magnitude if the same compounds are supported on SiO_2 or Al_2O_3 (*c.f.* Table 2 [21, 23]). The course of reaction between organometallic compounds and support depends on the ratio between the reagents and on the thermal activation of the support. The oxide is actually

Table 2. Ethylene polymerization with transition metals organometallic compounds

Catalyst	Temperature/ K	Polymerization rate/ kg PE (g. at. M[a])$^{-1}$ × atm^{-1} h^{-1}	Reference
Cr(allyl)$_3$	353	0.3	[21]
Cr(allyl)$_3$ on Al_2O_3	353	58	[22]
Zr(allyl)$_4$	353	2	[21]
Zr(allyl)$_4$ on Al_2O_3	353	280	[22]
Hf(allyl)$_4$	433	1	[21]
Hf(allyl)$_4$ on SiO_2	353	25	[22]
Zr(benzyl)$_4$	353	0.8	[21]
Zr(benzyl)$_4$ on Al_2O_3	353	390	[22]
Ti(benzyl)$_4$	353	0.2	[21]
Ti(benzyl)$_4$ on Al_2O_3	353	170	[22]
Cr(cyclopentadienyl)$_2$	333	0	[23]
Cr(cyclopentadienyl)$_2$ on SiO_2	333[b]	238	[23]

[a] Transition metal
[b] Temperature corresponding to maximum in activity

dehydrated prior to reaction with the organometallic compound (MR_n). Concentration and type of surface reactive groups [24] depend on dehydration temperature. Such groups may react with MRn as follows [25]

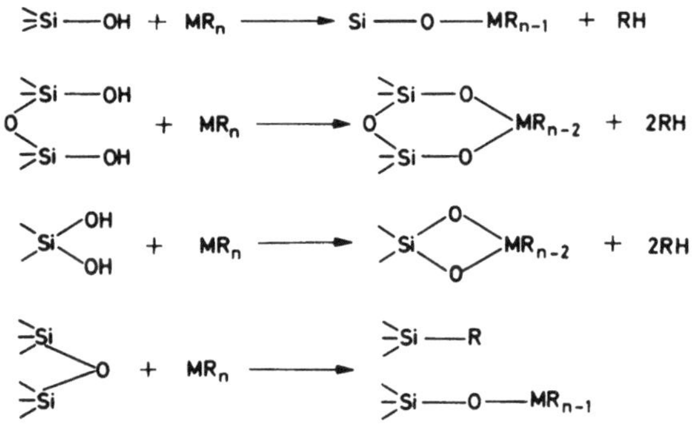

Support type and activation temperature therefore affect the nature of surface complexes and, as a consequence, catalyst activity [21, 26, 27].

These catalysts are active also in the absence of aluminum alkyls and may undergo activation by reaction with hydrogen [28], UV or gamma irradiation [22]. In the case of supported catalysts based on Zr compounds, hydrogen treatment causes the formation of Zr—H bonds and the polymerization rate is proportional to the content of zirconium hydrides in the catalyst [28].

It is interesting noticing that the catalysts obtained by supporting Ti and Zr benzyl or allyl derivatives on SiO_2 or Al_2O_3 maintain a high activity even at temperatures above 473 K. The reason for such behaviour has been attributed to the stability at high temperature of the metal-hydrogen bond formed through cleavage of the bonds between the transition metal and the growing polymer chains by β-hydrogen elimination. The insertion of an ethylene molecule in the metal-hydrogen bond regenerates a M—C bond and the site continues to be active in the polymerization.

Union Carbide has developed a gas phase process for ethylene polymerization [23, 29, 30] by employing chromocene and bis(triphenylsilyl) chromate supported over silica-alumina, as catalysts. The chromocene-based catalyst is preferably activated by alkyl aluminum alcoholates. The polymerization rate is a function of the monomer pressure and the molecular weight may be easily controlled by means of hydrogen. At a 2.1 MPa monomer pressure and at temperatures of 361–364 K, in the presence of Et_2AlOEt, yields of 7000 g polymer (g catalyst)$^{-1}$ h^{-1} are reported with reference to a typical catalyst containing 2.5% chromium.

3. *Soluble Catalysts*

After discovering that, by reaction between $(\eta^5C_5H_5)_2TiCl_2$ and aluminum alkyls, soluble catalysts are obtained, active in ethylene polymerization [31, 32], several soluble catalytic systems have been studied. These may be grouped into two categories: 1) Ziegler-Natta catalysts and 2) organometallic compounds of transition metals.

The interest for these catalysts generally originates from the aim of developing models for heterogeneous systems, which may be easily investigated and characterized by spectroscopic techniques, rather than from the possible use of these catalysts in polyethylene manufacture.

a) Ziegler-Natta soluble catalysts

The catalytic system Cp_2TiCl_2—$AlEt_2Cl$ is thoroughly soluble and, after a short induction time, exhibits an elevated efficiency in the initial 2–3 minutes (1.5×10^{-3} mol dm^{-3} s^{-1}; [Ti] = 3×10^{-3} mol dm^{-3}; [Al] = 6×10^{-3} mol dm^{-3}; [C_2H_4] = 10.4×10^{-2} mol dm^{-3}) followed by quick deactivation probably due to side reactions [33].

Table 3. Ethylene polymerization with soluble Ziegler-Natta calatysts

Catalyst	Co-catalyst	Polymerization rate/ kg PE (g. at. M[a])$^{-1}$ h^{-1} atm^{-1}	Temperature/ K	Reference
Cp_2TiCl_2	$AlMe_2Cl$	12	303	[33]
Cp_2TiCl_2	$AlEt_2Cl$	35[b]	288	[32]
Cp_2TiCl_2	$AlMe_2Cl—H_2O$	104	303	[35]
Cp_2TiMe_2	$AlMe_3—H_2O$	3×10^3	285	[37]
Cp_2ZrMe_2	$(AlMeO)_n$[c]	1×10^5	293	[39]
$VO(Acac)_2$	$AlEt_2Cl$	190	293	[45]
$Cr(Acac)_3$	$AlEt_2Cl$	154	303	[45]

[a] Transition metal
[b] In the presence of O_2
[c] Cyclic methyl aluminoxane

The activity of these catalytic systems is considerably enhanced by the addition of small quantities of water to the Al alkyl compound (Table 3).

Reichert [34] and Breslov [35] have studied the $Cp_2Ti(Et)Cl—EtAlCl_2$-(Al/H_2O = 20–100) and $Cp_2TiCl_2—AlMe_2Cl$(Al/H_2O = 3) systems, respectively. The greater activity is ascribed to the formation of a stabilized catalytic complex

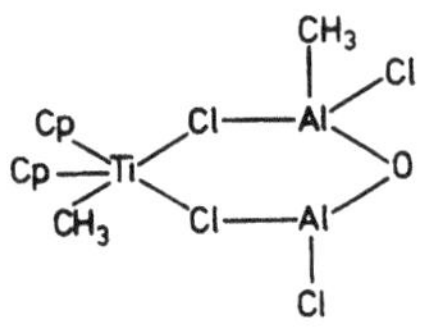

However, the most surprising results have been obtained with catalytic systems containing no halogens and based on the product of reaction between $Cp_2M(CH_3)_2$ (M = Ti or Zr) and cyclic methyl aluminoxanes $[(CH_3)AlO]_{10-14}$, [36, 39]. Such catalysts maintain an elevated activity even after 5 days. Yields as high as 250×10^3 kg PE (g Zr)$^{-1}$ and activity of 12.8×10^3 kg PE (g Zr)$^{-1}$ h^{-1} are reported. The molecular weight may be controlled by polymerization temperature:

$$\bar{M}_n = 1.5 \times 10^6 \text{ at } 283 \text{ K}$$
$$1.8 \times 10^5 \text{ at } 323 \text{ K}$$
$$9.0 \times 10^4 \text{ at } 363 \text{ K}$$

At 373 K, α-olefins with an even number of C atoms (10 to 30) are obtained

Many other catalytic systems described as soluble are also known; however, whether the solutions involved are true or colloidal may sometimes be hard to tell, especially after the onset of the polymerization process which takes place in heterogeneous phase. Apparently soluble catalysts are obtained by reaction of $Ti(OR)_4$ with AlR_3 [40, 41]. 1-Butene or high molecular weight polyethylene is obtained in variable amounts with Al/Ti ratios ranging between 10 and 50. Similar results are attained by replacing Ti alkoxide with $Ti(NR_2)_4$ [42].

Soluble catalytic systems are also obtained by reaction of $Ti(acac)_3$ [43] and $Cr(acac)_3$ [44] with $AlEt_3$ as well as by reaction of $Cr(acac)_3$ and $VO(acac)_2$ with $AlEt_2Cl$ in the presence of triethyl phosphite [45]. In the last case, the catalytic activity reaches its maximum at Al/V ratio = 50. Under these conditions, up to 67% vanadium is in the bivalent oxidation state, the balance being trivalent; both would be active.

Carrick [46, 48] describes soluble catalysts, active in ethylene polymerization, prepared by reaction of AlR_3 with a complex between VCl_2 and $AlBr_3$ soluble in cyclohexane thus confirming that vanadium may be active also in the II valence state.

Hexane soluble complexes from by reaction of polybutadienyllithium and $TiCl_4$ when the molecular weight of the organolithium compound is above 5,000 and the Li/Ti ratio is between 1 and 2. Ethylene polymerization brings about the formation of partially crosslinked PB-PE block copolymers [49].

b) Organometallic compounds of transition metals

Relatively stable organometallic compounds of Ti, Zr, Hg, Th, V, Nb and Cr, have been employed, in the absence of Al alkyls, in ethylene polymerization.

For benzyl derivatives, the greatest activity is exhibited by Zr complexes [50, 51]. The presence of halogen atoms bonded to the metal enhances the catalytic activity which varies in the series $TiBz_4 < TiBz_3I < TiBz_3Br < TiBz_3Cl$ and increases with the number of halogen atoms [51]. The replacement of benzyl groups also with other electronegative ligands as in the following cases

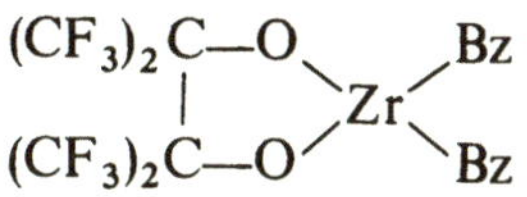

and

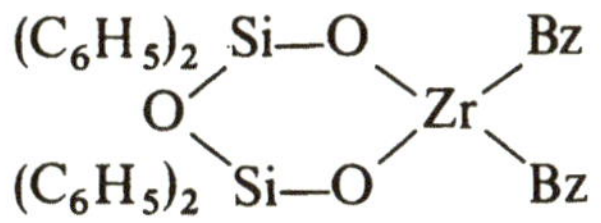

enhances the positive charge on the transition metal thus favouring the coordination of olefin with Zr and increasing the Zr—C bond polarization; consequently the bond becomes more active for insertion of the coordinated olefin [52]. The activity of Zr benzyl derivatives is enhanced by UV irradiation likely owing to the formation of hydride species [51].

Other σ organometallic derivatives of Ti and Zr, containing such ligands as $-CH_2-Si(CH_3)_3$, $-CH(C_6H_5)\,Si(CH_3)_3$, $-CH_2C(CH_3)_3$ and $-CH_2-O-CH_3$ are active in ethylene polymerization. Wilke *et al.* [53] first mention ethylene polymerization by tris η^3-allyl chromium. Allyl derivatives of Ti, Zr, Hf, V and Nb have also later been found [54] to be active in polymerization. The catalytic activity increases for the Zr compounds and decreases for the Cr compounds when the allyl ligands are partly replaced with halogens.

Watson [55] recently described also soluble catalysts based on cyclopentadienyl derivatives of Lu and Yb. $(\eta^5\text{-}C_5Me_5)_2\,Lu(Yb)CH_3$-ether polymerize ethylene, in the absence of Al alkyls, with high activity (66 g PE $(\text{mmol Lu})^{-1}\,h^{-1}\,atm^{-1}$ at 313 K). Lower activities are reported by Ballard *et al.* [56] for dimeric lanthanide alkyl complexes.

The ethylene polymerization was observed [57] also in the presence of π complexes of the type COT MC_4H_6 (M = Ti, Zr, Hf; COT = cyclooctatetraene).

4. *Ziegler-Natta Heterogeneous Catalysts*

In 1953, Ziegler *et al.* [58, 59] found that high density polyethylene was easily made at low pressure and within a wide temperature range (195 to 473 K) with binary mixtures of metal alkyls of groups I to III and compounds of transition metals of groups IV to VIII. The combinations of the two components which are active in ethylene polymerization are numberless [60, 62]. As organometallic compounds, they have been used, in mixture with $TiCl_4$ or $TiCl_3$, alkyl or hydride derivatives of Al, Li [63], Na [64], K [65], Zn [66, 67], Cd [68], Ga [69], and B [70].

Almost all transition metal halides form active catalysts in the presence of aluminum alkyls. Most studies however concern Ti, Zr, V, Nb and Cr compounds.

Other derivatives of transition metals such as alcoholates, dialkylamides, acetylacetonates, carboxylates, phosphates, oxides, etc. have also been used besides halogenides. Despite the great variety of catalytic mixtures developed, the catalysts which have found industrial application were mostly based on AlR_3 or AlR_2Cl mixtures with $TiCl_4$ or $TiCl_3$. These catalytic systems too have almost thoroughly been abandoned at present and replaced with high yield supported catalysts (*c.f.* section 2.A.5.).

Heterogeneous catalysts are formed in the reaction of metal alkyls with the halogenides of transition metals in their highest valence stage (Ti^{IV}, V^{V}, Cr^{VI}). Through an alkyl-halogen exchange, the initial reaction brings about the formation of an unstable alkyl derivative of the transition metal which decomposes to give a lower valence halogenide and an alkane-alkene mixture.

The degree of reduction of the transition metal depends on the nature of reagents, on their molar ratio, on their concentration, as well as on reaction time and temperature. All these parameters affect the catalytic activity [71, 74]; the resulting catalysts are thus unstable and hardly reproducible and conflicting experimental results have therefore been often reported by different authors.

The activity of catalyst *n*-BuLi—$TiCl_4$ is maximum with Li/Ti ratio 1,5:1 when ethylene is absent during catalyst formation. When catalyst preparation takes place in the presence of the monomer, the ratio 4:1 has proved as the most effective [75].

As the ratio between the two components varies, also the catalytic systems based on $AlEt_3$ or $AlEt_2Cl$ and $TiCl_4$ exhibit a quick enhancement of activity up to a maximum, followed by a fall down to an almost stationary value. If catalytic systems based on Al alkyls and $TiCl_3$ are employed, the polymerization rate is much less affected by the Al/Ti ratio; it actually increases up to a given Al alkyl concentration and then it remains constant over a wide range [76]. Table 4 shows the maximum polymerization rate referring to various types of $TiCl_3$ and Al alkyls.

LLDPE synthesis *via* ethylene copolymerization with alpha-olefins at high pressures (150 MPa) and temperatures (473–523 K) employing a mixture

Table 4. Maximum rate of catalyst systems based on $TiCl_3$ in ethylene polymerization

Catalyst	Co-catalyst	Maximum polymerization rate/ g PE (mmol Ti)$^{-1}$ h^{-1} × atm^{-1}	Temperature/ K	Reference
α-$TiCl_3$	$AlEt_3$	35	348	[77]
δ-$TiCl_3 \cdot 0.3\ AlCl_3$	$AlEt_3$	72	348	[77]
δ-$TiCl_3 \cdot 0.3\ AlCl_3$	Al(*i*-Bu)$_3$	96	348	[77]
$TiCl_3$	$AlEt_2Cl$	32.5	323	[78]

of $TiCl_3$. 0.3 $AlCl_3$ with $Al(n\text{-}C_8H_{17})_3$ or $Me_2EtSiOAlEt_2$ has been described recently [79, 80]. Activity results are very high and polymer purification from catalytic residues may be avoided.

5. High Yield Supported Catalysts

The name is given to supported catalytic systems of the Ziegler-Natta type exhibiting an activity sufficiently high to avoid the removal of catalytic residues in manufacturing processes. As a consequence, yield must be extremely elevated both in respect of the transition metal and of the support if it contains easily hydrolizable metal-halogen bonds [81].

Other characteristics are also required for commercial utilization [82]. The control of the polymer molecular weight and, possibly, of the molecular weight distribution is actually necessary. Ethylene must furthermore be capable of copolymerizing with alpha-olefins so as to allow the control of polymer density over a wide range. Finally, the polymer resulting from slurry and gas-phase processes is required to have a sufficiently high bulk density.

A number of catalytic systems [83, 87] meeting most of the mentioned requirements were developed towards the end of the sixties and such systems are now being employed practically by all polyethylene manufacturers in lieu of the former traditional Ziegler-Natta catalysts.

a) Catalysts obtained by reaction of transition metal compounds with supports containing surface hydroxyl groups

This class includes the catalysts obtained by reaction of transitional metal compounds (mostly $TiCl_4$) with supports containing surface hydroxyl groups capable of reacting according to the following scheme:

$$(MX_n) - OH + TiCl_4 \rightarrow (MX_n) - O - TiCl_3 + HCl$$

Catalysts obtained by reaction of $TiCl_4$ with SiO_2 and Al_2O_3 [88, 91] are scarcely interesting on account both of insufficient activity and of the difficulties involved in controlling the molecular weight of the polymer with hydrogen. During catalyst reaction with aluminium alkyl, the destruction of the surface complexes and the formation of $TiCl_3$ in the dispersed phase, constituting the true catalyst, have been ascertained [92, 93].

Catalysts exhibiting a much higher activity [94–97] are obtained by reaction of $TiCl_4$ with Mg hydroxylated derivatives such as partially hydrated MgO, $Mg(OH)_2$, MgOHCl, $MgCO_3$ (hydroxylated), $MgSO_4$. 3 $Mg(OH)_2$, etc.

By means of IR spectroscopy and thermogravimetric techniques, Eley *et al.* [98] studied the reaction between MgO partially hydroxylated surface and $TiCl_4$ followed by the reaction with $AlEt_3$, and proposed the following reaction scheme:

$$-Mg - OH + TiCl_4 \rightarrow -Mg - O - TiCl_3 + HCl$$

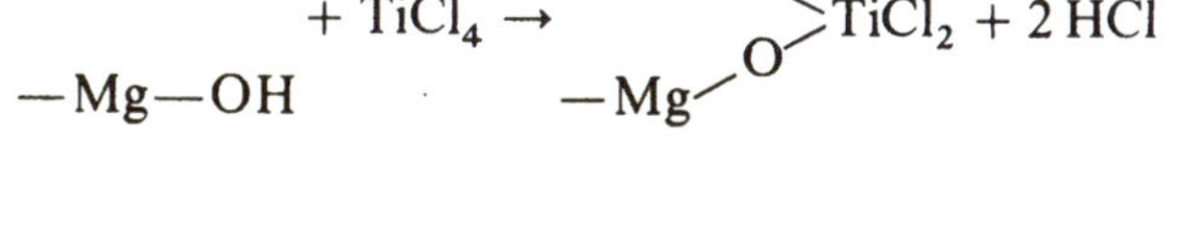

$$-\text{Mg}-\text{O}-\text{TiCl}_3 + \text{AlEt}_3 \rightarrow -\text{Mg}-\text{O}-\text{TiEtCl}_2 + \text{AlEt}_2\text{Cl}$$
$$-\text{Mg}-\text{O}-\text{TiEtCl}_2 \rightarrow -\text{Mg}-\text{O}-\text{TiCl}_2 + (\text{Et.})$$
$$-\text{Mg}-\text{O}-\text{TiCl}_2 + \text{AlEt}_3 \rightarrow -\text{Mg}-\text{O}-\text{TiEtCl} + \text{AlEt}_2\text{Cl}$$

The alkylated derivative of Ti^{III} is assumed [98] to be the active species, whereas Ti^{IV} is deemed to be present in the active centres by Soga *et al* [99]. Ethylene polymerization with these catalytic systems has been studied by many authors [90, 100, 102] who report the attainment of very high yields (Table 5) and confirm a greater efficiency of AlR_3 as cocatalyst in respect of halogenated Al alkyls as is the case of all catalytic systems supported on Mg compounds.

Table 5. Ethylene polymerization with catalysts based on $TiCl_4$ supported on oxides[b,c]

Catalyst	Ti/ %	S^a/ $m^2\,g^{-1}$	Yield/ kg PE $(g\ Ti)^{-1}$	Maximum polymerization rate/ $dm^3\ C_2H_4\ (mol\ Ti)^{-1}\ s^{-1}$
$TiCl_4-MgO$	0.2	22.5	484.2	205.5
$TiCl_4-SiO_2-Al_2O_3$	3.6	8	19.5	5.4

[a] Specific surface area
[b] Reference [90]
[c] Polymerization conditions: $[C_2H_4] = 0.216\ mol \cdot dm^{-3}$; cocatalyst $= Al(C_2H_5)_3$; 343 K; time = 2 h

b) Catalysts obtained by reaction of transition metal compounds with bivalent metal halogenides in active form

A number of Montedison patents [103, 104] describe the synthesis of various catalysts by comilling of transition metal compounds (generally halogenides, halogen alcoholates and halogen amides of Ti) with bivalent metal halogenides ($MgCl_2$, $MgBr_2$, MgI_2, CaI_2, $MnCl_2$), all characterized by layer structure.

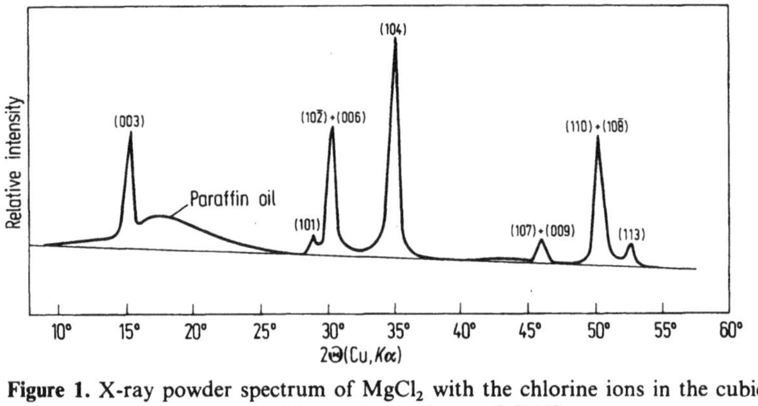

Figure 1. X-ray powder spectrum of $MgCl_2$ with the chlorine ions in the cubic close packed arrangement. (Reproduced with permission from ref. [105])

The structure of the usual crystalline form of $MgCl_2$ is quite similar to that of γ-$TiCl_3$ and corresponds to a cubic packed stacking (ABC—ABC) of double chlorine layers with hexacoordinated interstitial Mg^{+2} ions. The layer structure, of the $CdCl_2$ type, leads to a characteristic diffraction spectrum exhibiting a strong reflection (104) at $d = 2.56$ Å resulting from the cubic close packing of the chlorine atoms (Figure 1). Especially in the presence of Ti salts, during the milling, the planes of chlorine atoms undergo translation and rotation producing the destruction of the crystal order in the stacking direction (*c* axis) [105, 108]. A disordered structure forms as a result, as in the case of δ-$TiCl_3$. Owing to the increasing disorder, a gradual disappearance of the (104) reflection is noticed in the X-ray diffraction spectrum, along with the presence of a halo, centred at $d = 2.65$ Å, which occurs between the spacing typical of the cubic and of the hexagonal successions (Figure 2). Concomitantly, an increase of the specific surface (50–200 $m^2\ g^{-1}$) is also observed as well as a reduction of crystallite

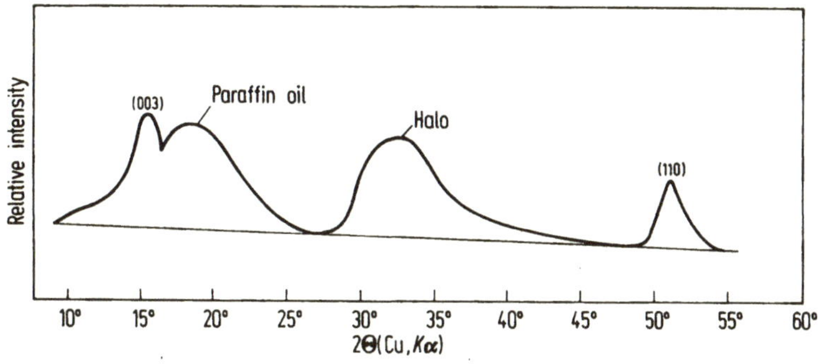

Figure 2. X-ray powder spectrum of activated $MgCl_2$. (Reproduced with permission from ref. [105])

size which may attain very low values (50–80 Å) in the two directions (110) and (001). As a consequence of reduced crystallite dimensions and of the disordered structure many Mg ions, coordinatively unsaturated and located on the edges and at the most probable cleavage surfaces (110) and (101) of the crystallites, become available. Such Mg ions might interact with Ti halides forming double halogen bridge bonds [105] or ionic complexes similar to K_2TiCl_6 [108].

According to some authors [109], ESR analysis of catalysts reduced with $AlEt_3$ reveals the presence of isolated Ti^{III} ions in chloride environment and of alkylated Ti^{III} ions; a considerable quantity of Ti^{III} is however not revealed by ESR signals owing to the formation of aggregates. The authors assume active centres may result from both isolated Ti^{III} ions and Ti ions belonging to aggregates.

These catalysts deserve special attention and are being described in detail since they alone, among all high yield catalytic systems available for ethylene polymerization, have been employed successfully in propylene polymeri-

Table 6. Ethylene polymerization with $MgCl_2$—$TiCl_4$($TiCl_3$) supported catalysts

Catalyst	Ti %	Co-catalyst	Temperature K	Polymerization rate p PE $(mmol\ Ti)^{-1}\ h^{-1}\ atm^{-1}$	Reference
$MgCl_2$—$TiCl_4$	3.6	$Al(C_2H_5)_3$	353	1650	[87]
$MgCl_2$—$TiCl_4$	3.6	$Al(C_2H_5)_2Cl$	353	330	[87]
$MgCl_2$—$TiCl_3$	3.7	$Al_2(C_2H_5)_3Cl_3$	353	20	[87]
$MgCl_2$—$TiCl_3$	3.7	$Al(C_2H_5)_3$	353	430	[355]

zation after proper modification with Lewis bases. Table 6 compares the activities of $MgCl_2$—$TiCl_4$ and $MgCl_2$—$TiCl_3$ comilled catalysts, in the presence of several Al alkyls, for ethylene polymerization.

Yermakov *et al.* [110] state that polyethylene prepared with catalysts supported on $MgCl_2$ has a molecular weight lower than polyethylene prepared under the same conditions with conventional $TiCl_3$-based catalysts; moreover, a lower H_2 concentration is needed for molecular weight control. As a matter of fact, with supported catalysts, the chain transfer reaction with H_2 is first order in respect of H_2 concentration, whereas a reaction order of 0.5 was found for polymerization with $TiCl_3$.

c) Catalysts obtained by reaction of $TiCl_4$ with adducts between $MgCl_2$ and a Lewis base

Very active catalysts [87, 111] are prepared by reaction of a $TiCl_4$ excess with adducts between $MgCl_2$ and a wide range of Lewis bases (alcohols, acids, esters, amines). In the presence of a Lewis base excess, stoichiometric complexes may be prepared [112] which may sometimes be isolated also in the crystalline state. In such complexes, Mg and Ti are bound through a double chlorine bridge [113] (Figure 3), or belong to ion lattices [114]. In the

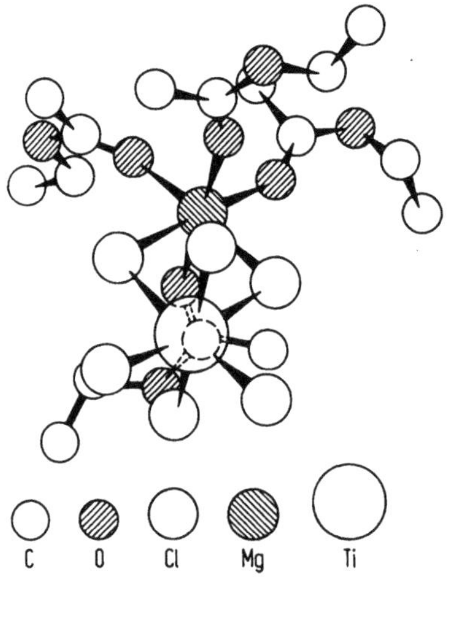

Figure 3. Crystal structure of the $TiMgCl_6 \cdot 4\ CH_3COOC_2H_5$ adduct. (Reproduced with permission from ref. [113])

presence of AlR_3, these compounds are very active in ethylene polymerization.

d) Catalysts obtained by reaction of Mg alkoxides with transition metal compounds

Catalysts obtained by reacting Ti halogenides or halogen alcoholates with Mg alkoxides [82, 115–117], mainly developed by Hoechst, exhibit an elevated activity in the presence of AlR_3. Böhm [117] describes the polymerization of ethylene with a catalyst obtained by reaction of $TiCl_4$ with $Mg(OEt)_2$. This product contains 8.5% by weight Ti and has a density of $1 \cdot 335$ kg m^{-3}, a specific surface (BET) of 60 m^2 g^{-1}, and a pore volume of 1.16 cm^3 g^{-1}. High porosity would favour catalyst disintegration during polymerization. No induction time is observed, and the polymerization rate keeps constant for approximately 1 hour and then declines gradually. The initial rate ($T = 358$ K, ethylene pressure 100 kPa) is a function of the Al/Ti ratio; it reaches a peak value of 115 g mol^{-1} s^{-1}, with a ratio between 10 and 20, and decreases down to 30 g mol^{-1} s^{-1} at an Al/Ti ratio of 1000.

As compared with the traditional $TiCl_3$—$AlEt_3$ catalytic system, a lower H_2 amount is required in this case for molecular weight control. The presence of H_2 however causes a considerable reduction of the polymerization rate which falls by 80% at a 0.18 ratio between H_2 and C_2H_4 partial pressures.

e) Catalysts obtained by reaction of transition metal compounds with Mg alkyls

Ethylene polymerization by means of catalysts prepared by reaction between $TiCl_4$ and Mg organometallic compounds is described by several patents and scientific papers [78, 118–126]. Grignard reagents as well as Mg dialkyls and diaryls are employed. The reaction is conducted at varying temperatures (195–373 K) and Mg/Ti ratios (0.15 to 2) in the presence or in the absence of solvating compounds. Ti is reduced to the trivalent and bivalent state; the ratio between the two forms is dependent on the Mg/Ti ratio as well as on the reaction temperature. The catalytic activity keeps high also when 40% of Ti is reduced to Ti^{II} species [121].

Both AlR_3 and AlR_2Cl are used as cocatalysts. By operating in the absence of H_2, yields of 1612 kg PE (g Ti)$^{-1}$ in 3 hours at 323 K and at a pressure of 1 MPa have been obtained [118]. The high activity of these catalysts is ascribed to the high number of active centres (60% of total Ti).

If the catalytic system $TiCl_4$ — $AlEt_2Cl$ — $Mg(C_6H_5)_2$ is employed, components may be joined just prior to polymerization without isolating the reaction product between $TiCl_4$ and MgR_2 [124, 125]. With this catalyst, hydrogen is able to control the molecular weight at high temperatures only (403 K), and, under such conditions, it also enhances the catalytic activity. Best results are attained with the following ratios: Mg/Ti = 7.6, Al/Ti = 7.40, C_2H_4/H_2 = 7–10. Productivity is 19 kg PE (g Ti)$^{-1}$ in 15 minutes at a total pressure of 1 MPa in isooctane solution. A variant in the preparation of this type of catalyst consists in allowing Mg alkyl [119] or

$TiCl_4$ [120] to react with SiO_2, Al_2O_3 or MgO surface hydroxyl groups prior to reaction with the other component of the catalytic system. The presence of oxides as supports, however, does not seem to bring about a significant improvement of the catalytic activity.

f) Catalysts obtained by immobilizing transition metal compounds in functional polymeric carriers

A class of catalysts [127] obtained by fixing $TiCl_4$ or VCl_4 on a polymeric support consisting of ethylene-propylene diene rubber grafted with 4-vinylpyridine, allyl alcohol or methacrylic acid, has recently been developed. The catalytic systems containing supported $TiCl_4$ polymerize ethylene in the presence of AlR_3 with constant reaction rate (15 g PE (g Ti)$^{-1}$ h^{-1} atm^{-1}) for several hours also operating at 413 K.

If the polymerization is carried out in solution the polymer diffuses continuously from the swollen catalyst particles, goes into solution and does not contain impurities resulting from the transition metal. Higher activities (70–100 g PE (g Ti)$^{-1}$ h^{-1} atm^{-1}) are obtained by reaction of the support with RMgX and $TiCl_4$.

Another method for the preparation of supported catalysts consists in performing $TiCl_4$ reduction with Al alkyls in the presence of finely pulverized polyolefins [128].

B. Propylene and Other α-Olefins Polymerization

1. Foreword

Unlike ethylene, α-olefins may give rise to various types of stereoisomeric modes of enchainment forming polymers with quite different characteristics (structure, melting point, mechanical resistance, etc.) [1, 2].

When the successive head to tail monomeric units, having different configurations, are randomly distributed along the chain, a *atactic* polymer is formed, whereas a stereoregular head-to-tail enchainment of monomeric units causes the growing of *tactic* polymers. Among the latter, the *isotatic* and *syndiotactic polymers* consist of a succession respectively of all *d* or *l* units and of *d* and *l* alternating units (*c.f.* Figure 4).

The macromolecules of isotatic polymers have a helicoidal structure (*c.f.* Figure 5) and generally crystallize into larger structures. Among these polymers, isotactic polypropylene aroused the greatest interest in industry on account both of propylene availability and product characteristics (*c.f.* Figure 6).

Some practical interest is also offered by isotactic poly-1-butene and poly-4-methyl-1-pentene (other isotactic polymers of various other α-olefins [1, 2] have been prepared, but are not being considered hereunder). These polymers are crystalline and their polymeric chains are helicoidal. The structure of syndiotactic polypropylene [132] chains is shown by Figure 7. A planar structure [133] is also known. Syndiotactic polypropylene, however, has not found practical application.

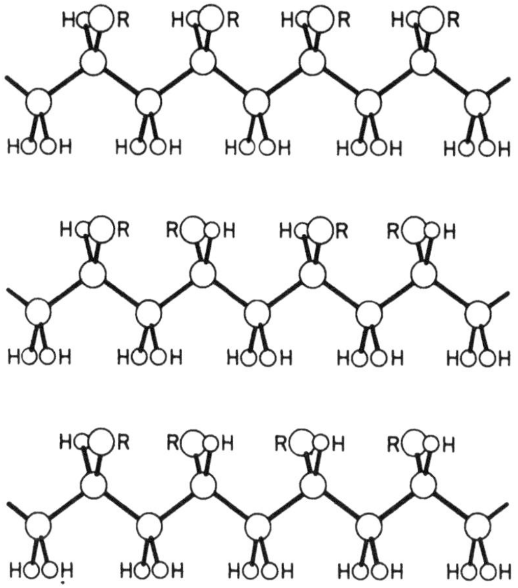

Figure 4. Chains of different stereoisomers of polymeric α-olefins (main chain arbitrarily stretched on a plane)

Concerning the industrial interest of the catalytic systems, besides activity and stereospecificity, the amount of catalytic residues (Ti and Cl in particular) present in the raw polymer at the end of polymerization is a very important feature. Therefore, the following are to be regarded as most significant factors to determine the reliability of any catalytic system for the polymerization of alpha-olefins:

- *yield*, to be understood as polymer weight obtainable in respect of catalysts weight or, better, of transition metal (Ti) weight consumed;
- *stereospecificity*, to be understood as the percentage of stereoregular polymer obtained; in the case of isotactic polypropylene, it is represented by the percentage of polymer insoluble in boiling *n*-heptane [134].
- *percentages* of residual metals and halogen in the raw polymer.

Propylene polymerization is mostly considered hereunder, no substantial differences being involved with regard to the catalytic aspects of isospecific polymerization of the various α-olefins (with the exception of the polymerization rate which lowers with higher α-olefins).

2. *Isospecific Polymerization with Conventional Catalysts*

a) Foreword

The first system utilized by Natta, Pino and Mazzanti in 1954 [134, 135] for propylene polymerization to partially isotactic polymer was based on $TiCl_4$ and $Al(C_2H_5)_3$, previously developed by Ziegler for ethylene polymerization [58, 59]. The system, however, exhibited a modest catalytic activity and low

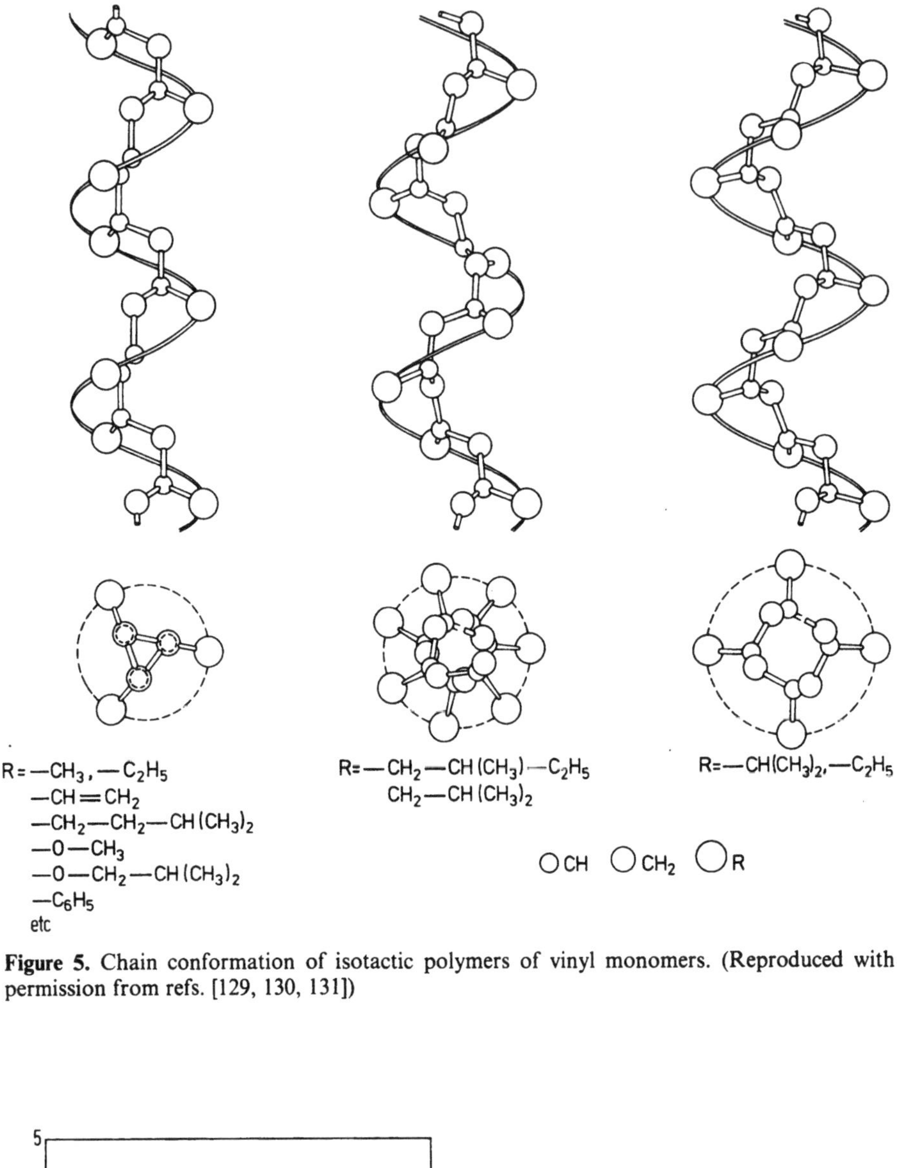

Figure 5. Chain conformation of isotactic polymers of vinyl monomers. (Reproduced with permission from refs. [129, 130, 131])

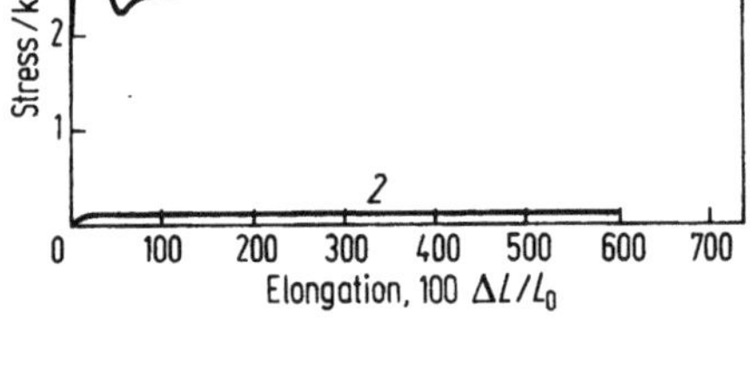

Figure 6. Stress-elongation curves of:
1-isotactic polypropylene
2-atactic polypropylene

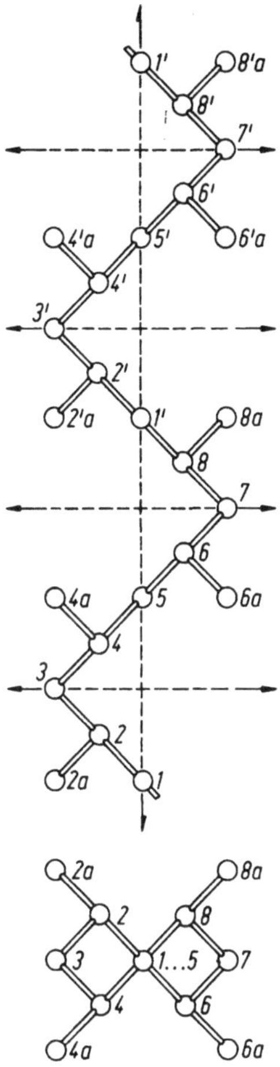

Figure 7. Chain conformation of syndiotactic polypropylene. (Reproduced with permission from ref. [132])

stereospecificity (I.I.[1] = 30–40%) [134]. More stereospecific catalytic systems (I.I. > 85%) were soon introduced by Natta who, assuming stereospecificity to be connected to the regularity of the heterogeneous catalyst surface, employed crystalline violet $TiCl_3$ instead of $TiCl_4$ (liquid) for the preparation of the catalyst [136, 137].

Since then and for many years, violet $TiCl_3$ has been the basic component of the catalysts employed on a commercial scale for isotactic polypropylene manufacture. Still today, most commercial plants operate with systems of this type, whereas new catalysts, also based on titanium chlorides, but substantially differing from the former, have been developed in quite recent times (*c.f.* sect. 2.B.3.). As there are various types of violet $TiCl_3$, synthesis methods and the most significant features of these compounds will be briefly illustrated.

---

[1] I.I. ≡ isotacticity index

b) $TiCl_3$ crystalline modifications and preparation methods

As demonstrated by Natta *et al.*, four crystalline modifications of $TiCl_3$ are known, named α, γ, δ [138–143] and β [142]. The first three correspond to violet $TiCl_3$ and the fourth to brown $TiCl_3$. The three violet $TiCl_3$ forms have

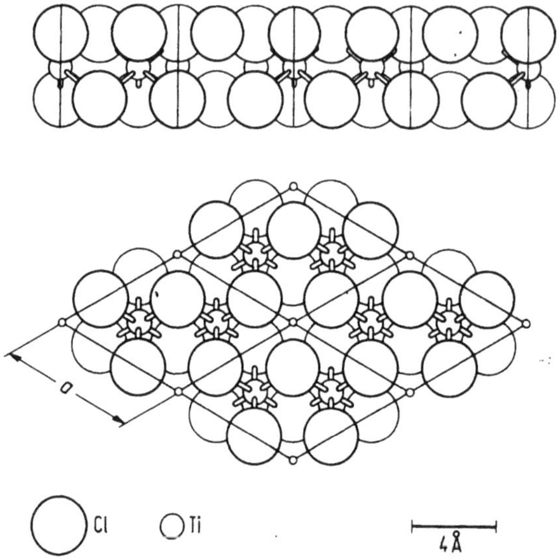

Figure 8. Stereochemical model of the structural layer which characterizes the layer modifications (α, γ and δ) of $TiCl_3$. (Reproduced with permission from ref. [141])

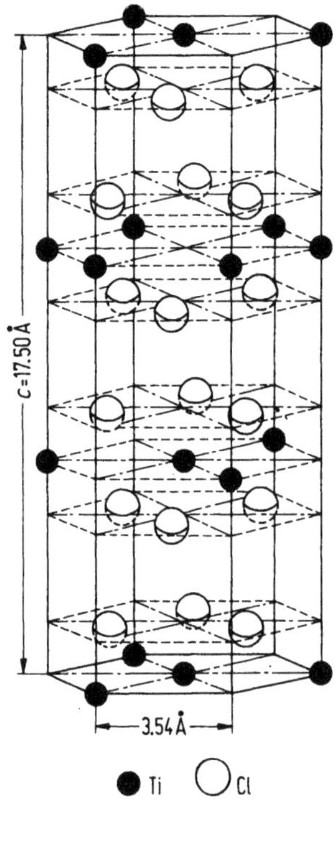

Figure 9. α-$TiCl_3$ crystalline lattice. (Reproduced with per mission from ref. [308])

a layer structure consisting in a stacking of the same type of sandwich layer of Ti between two layers of Cl (Figure 8). The structure of α-$TiCl_3$ corresponds to a hexagonal close packing of chlorine anions (Figure 9). In the γ form the stacking leads to a cubic close packing. In the δ form the stacking is irregular. The β modification has a linear structure (Figure 10). α-$TiCl_3$ was synthesized by reduction of $TiCl_4$ with H_2 at 1070 K approximately [144]. Low temperature reduction in the presence of silent electric discharges produces modification β [145], while the γ modification is obtainable by

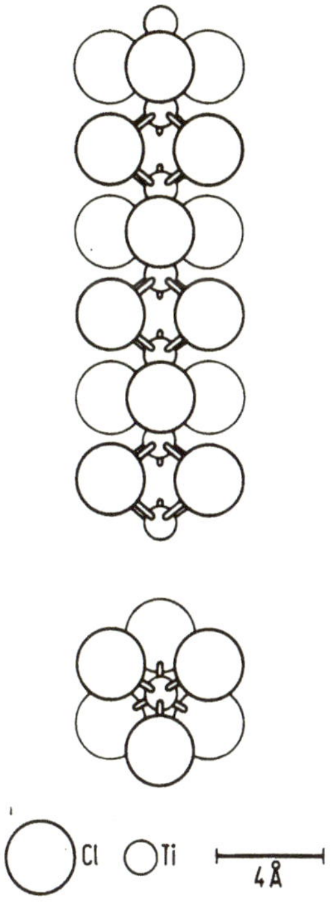

Figure 10. Stereochemical model of the structural unit in β-$TiCl_3$. (Reproduced with permission from ref. [142])

heating β-form above 423 K [143]. Modification δ is obtained by prolonged dry milling of either the α and γ forms [146]. Instead of using H_2, $TiCl_4$ can be reduced with metals or metal alkyls. Most commercial catalysts [147] are usually synthesized by reduction of $TiCl_4$ with aluminum alkyls or aluminum metal [148]; in this case, the reaction product is a solid solution of $AlCl_3$ and $TiCl_3$ the composition of which is mostly close to $TiCl_3$. 0.3 $AlCl_3$ [149]. The catalyst systems obtained by this route are more active than those prepared by reduction with H_2. As described by Tornquist *et al.* [146], considerable further increments of activity (up to 10 times approximately) may be attained by dry milling of these products. Increments are due both to an increase of the surface area (from less than 10 m^2 g^{-1} to 40 m^2 g^{-1}) and to the modification of the crystalline structure (from α or γ to δ) [146].

c) First generation catalytic systems with two components — monometallic systems

The first catalytic systems used in polypropylene industrial production were based on violet $TiCl_3$ and on a organometallic compound of aluminum.

Systems based on δ-$TiCl_3$ (with or without $AlCl_3$ in solid solution) and on $Al(C_2H_5)_2Cl$ are still largely employed today. A decaying-type polymerization rate curve was observed with all these systems, more significant decrements occurring in the presence of $Al(C_2H_5)_3$. In addition to the nature of $TiCl_3$ and its preparation method, the catalyst activity depends on the kind of organometallic compound as well as on polymerization temperature. Concerning γ-$TiCl_3$, activity lowers in the sequence: $Be(C_2H_5)_2 \cong Al(C_2H_5)_3 > Al(C_2H_5)_2Cl > Al(C_2H_5)_2Br > Al(C_2H_5)_2I > Zn(C_2H_5)_2$ [150]. Unlike δ-$TiCl_3$, α-$TiCl_3$-based catalytic systems generally exhibit an almost constant activity for many hours after an initial induction period [151].

As described by Natta *et al* [152], the stereospecificity of violet $TiCl_3$-based systems is generally very high irrespective of the modification type (α, γ, or δ), provided that the same organometallic compound is employed (see Table 7), and decreases in the sequence: $Al(C_2H_5)_2I > Al(C_2H_5)_2Br \cong Be(C_2H_5)_2 > Al(C_2H_5)_2Cl > Al(C_2H_5)_3$. Such results, however, are attainable only when using properly purified $TiCl_3$.

Table 7. Stereospecificity of catalytic systems prepared starting from $TiCl_3$[a] in propylene isospecific polymerization[c]

Organometallic compound	Stereospecificity Index[b] at polymerization temperatures of	
	288 K	343 K
$Al(C_2H_5)_2I$	99–100%	96–98%
$Al(C_2H_5)_2Br$	97– 98%	94–96%
$Al(C_2H_5)_2Cl$	96– 98%	91–94%
$Al(C_2H_5)_3$	80– 85%	80–85%
$Be(C_2H_5)_2$	94– 96%	93–95%

[a] α-, γ-, or δ-$TiCl_3$ with or without $AlCl_3$ in a solid solution
[b] Expressed as percentage of highly isotactic polypropylene insoluble in boiling *n*-heptane
[c] Reference [152]

Catalytic systems based on the β-$TiCl_3$ (brown) modification are less stereospecific (*ca.* 55%) [150]. Higher stereospecificities (*ca.* 85%) are obtained in the presence of $Al(C_2H_5)_2I$, but the catalytic activity is considerably reduced [150].

Table 8 collects selected examples of two-component catalytic systems based on $TiCl_4$, $TiCl_3$, $TiCl_2$, $VOCl_3$, VCl_4 and of metal alkyl-free catalysts.

The stereospecificity of catalysts based on Ti, V and Cr halides and some organometallic compounds is indicated under Table 9.

Table 8. Unsupported two-components catalytic systems and monometallic systems active in propylene isospecific polymerization

Catalytic systems	Remarks	Reference
α-$TiCl_3$(violet) + $Al(C_2H_5)_3$	medium activity, high stereospecificity; first commercial system	[137], [152], [154]
α- or δ-$TiCl_3$(violet) + $Al(C_2H_5)_2Cl$	high activity and stereospecificity; commercial systems	[137], [152], [154]
α-, γ-, or δ-$TiCl_3$ (violet) with various Al or Be organometal compounds	medium to high activity, high stereospecificity	[137], [152], [154]
$TiCl_2$ + AlR_3, AlR_2X or BeR_2	low activity, high stereospecificity	[137], [150]
$TiCl_4$ or $VOCl_3$ or VCl_4 + AlR_3, AlR_2X, BeR_2, LiR, ZnR_2	medium to high activity, low stereospecificity; activity depends considerably on the ratio between the catalytic system components	[137], [153], [155], [156]
VCl_3 or β-$TiCl_3$ + AlR_3, AlR_2X, BeR_2	high activity, medium-low stereospecificity	[150], [153], [156], [157]
$Cp_2Ti(C_6H_5)_2$ + $TiCl_4$	low activity and stereospecificity	[158]
$TiCl_3$ + $TiCl_3CH_3$		[48], [159]
Ti + C_2H_5Br		[160]
Ti + $TiCl_3$(violet) + $TiBr_4$		[160]
Ti + I_2		[160]
$TiCl_2$ milled		[161]
β-$TiCl_3$ milled		[161]
$TiCl_3$(violet) containing or not containing $AlCl_3$ + NR_3, in the presence or in the absence of H_2	low activity and stereospecificity	[162], [163]
$TiCl_4$ or $VOCl_3$ + chromium allyl		[164]

Table 9. Stereospecificity of catalytic systems in propylene polymerization at 343 K [a, b]

Transition metal halogenide	Organometallic compound					
	$Al(C_2H_5)_2I$	$Be(C_2H_5)_2$	$Al(C_2H_5)_2Br$	$Al(C_2H_5)_2Cl$	$Ga(C_2H_5)_3$	$Al(C_2H_5)_3$
$TiCl_3$(violet)	99	96	95	92	86	83
β-$TiCl_3$	86	62	—	70	62	56
$TiCl_2$	84	91	62	71	81	75
$TiBr_3$	—	42	72	77	61	49
VCl_3	50	88	48	37.5	86	67
$CrCl_3$	84	15	—	60	—	17

[a] Reference [165]
[b] Stereospecificity in percent

The possibility of obtaining isotactic polypropylene (though with a very low activity and stereospecificity) with monometallic catalytic systems (*c.f.* Table 8) is of significant scientific importance because it has allowed interesting conclusions to be drawn regarding the nature of the catalytic complexes (*c.f.* sect. 3.A.6).

d) Modified Ziegler-Natta catalysts

The addition of modifiers (both during the preparation of the transition metal compound and during polymerization) to the described catalytic systems has been essentially suggested by practical reasons in order to:
— control the molecular weight;
— increase the stereospecificity;
— enhance the catalytic activity.

The first modifiers to be studied have been molecules able to control (reduce) the molecular weight of the polymer obtained. Among these, the most important are H_2 proposed by Vandenberg [166] and Bua and Luciani [167, 168] and Zn alkyls proposed by Natta, Pasquon and Giachetti [169, 170]. There are some doubts that such substances should only be considered as true chain transfers in that they cause also à reduction of the propylene polymerization rate [168–170].

The addition of various types of additives to Ziegler-Natta catalysts for propylene isospecific polymerization is extensively described in the literature [37, 60, 153, 163, 171–177]. In particular, Lewis bases (esters of carboxylic acids [174], amines [60], cycloheptatriene [178] and fluoride ions [179, 180] added to $TiCl_3$ or to the catalytic system in proper quantities increase the stereoregulating ability and sometimes also the activity of the catalyst. It should be noted in any case that the stereospecificity attainable in the aforesaid way seldom exceeds that achieved with the same type of $TiCl_3$ properly purified [152].

In another class of modified catalysts, the $TiCl_3$—$Al(C_2H_5)Cl_2$ system is activated by addition of inorganic halides such as KCl and NaF [152] or Lewis bases such as $[(CH_3)_2N]_3PO$ [176] and $HN(C_2H_5)_2$ [181]. Zambelli *et al.* [182–184] and Watt [185] suggest that upon mixing with the third component $Al(C_2H_5)Cl_2$ is converted according to the following reaction scheme to $Al(C_2H_5)_2Cl$ which is the true Al alkyl interacting with $TiCl_3$ to form the active catalytic complexes:

$$2\,AlC_2H_5Cl_2 + [(CH_3)_2N]_3PO \rightarrow Al(C_2H_5)_2Cl + AlCl_3[(CH_3)_2N]_3PO$$

It follows that these systems do not differ substantially from those described above. In fact it may be observed that they exhibit the same stereospecificity as the systems prepared with $Al(C_2H_5)_2Cl$ and with the same type of $TiCl_3$ previously purified, *e.g.* by means of washing with boiling aromatic hydrocarbons [152, 181, 186].

For practical purpose more interesting results are obtained by adding the Lewis base during the synthesis of $TiCl_3$. The first result in this way was obtained by Hermans and Henrioulle [187]. According to these authors, $TiCl_4$ is first reduced at approximately 273 K, in a hydrocarbon diluent, by $Al(C_2H_5)_2Cl$, to give a solid having a composition corresponding to about β-$TiCl_3 \cdot 0.3\,AlCl_3$ and a surface area of about 1 $m^2\,g^{-1}$. The product thus obtained is then treated, for instance, with diisoamyl ether at 308 K. By this operation most $AlCl_3$ is removed from the solid solution. The residue is then treated with $TiCl_4$ at 338 K to produce violet δ-$TiCl_3$ containing small amounts of aluminum halide and of ether (less than 0 · 1 mol per mol $TiCl_3$).

This product has a very high surface area (up to 150 $m^2 g^{-1}$) and polymerizes propylene in mixture with $Al(C_2H_5)_2Cl$ with higher stereospecificity and higher activity (4–5 times) than the catalyst based on conventional δ-$TiCl_3$.

3. *Isospecific Polymerization with Supported High Yield Catalysts*

The numerous improvements brought to the traditional Ziegler-Natta catalysts for propylene polymerization have not been however sufficient to enhance their activity as much as to allow the elimination of one of the burdensome stages of the production process, namely the removal of catalyst residues from the polymer. In order to achieve this goal, mainly industrial research laboratories have been actively engaged to develop supported high activity catalysts.

A suitable catalyst must meet the following requirements:

a) the transition metal content in the resulting polymer must be lower than 10 ppm to avoid colour problems;
b) the chloride residuals, in the form of metal halides, tolerable in the resulting polymer, may vary as a function of the halogenide stability to hydrolysis, but must in any case be lower than 50–100 ppm; the use of AlR_3 instead of AlR_2Cl is therefore recommendable as cocatalyst to avoid the presence of easily hydrolizable Al—Cl bonds in the polymer;
c) the high activity must be joined to a high stereospecificity (I.I. $\geqq$ 95%), an aim that had already been attained by the most developed catalytic systems based on $AlEt_2Cl$ and on particular types of $TiCl_3$ [187].

Taking into account that only a part of the titanium present on the side surfaces of $TiCl_3$ crystallites is deemed to be active in propylene polymerization [188], research has endeavoured to obtain a better utilization of the Ti halogenide by trying to fix it on the surface of proper supports. The great industrial interest is evidenced by the many patent applications filed, whereas a comparatively low number of studies has appeared in scientific journals. Early attempts to utilize for propylene polymerization the supported catalysts developed previously with success for ethylene polymerization, did not lead to interesting results since the high activity sometimes attained was always accompanied by a very low stereospecificity.

Soga *et al.* [99] studied propylene polymerization with catalytic systems based on AlR_3 and the reaction products of $TiCl_4$ with Mg, Ba, Cd, Sr, Cr, Mn, Co, Ni, Fe, Al, Si and Cu hydroxides and hydroxychlorides. Although the activity values reported for these systems are quite low, the most active and stereospecific catalysts (4.1 g PP (g cat)$^{-1}$ h^{-1} atm^{-1}) are obtained by employing $Mg(OH)_2$ and Mg(OH)Cl as support. Unlike the $TiCl_4$—$AlEt_3$ system, these catalysts exhibit an almost constant overall rate of polymerization for at least 2 hours, consistent with a stabilization of the active centres by the support. A relationship has been found between tacticity of the produced polymer and the ionic radii of the metals used as supports with a peak value (I.I. = 55%) for an ionic radius of near 0 · 70 Å. Baulin *et al.* [90] studied the polymerization of propylene and of higher α-olefins at 343 K using $AlEt_3$ and catalysts derived from the interaction of $TiCl_4$ with MgO and SiO_2/Al_2O_3. Polymer yields were respectively 6.7 and 0.9 kg PP $(g\,Ti)^{-1}$ and isotacticity indexes of 7.3 and 9.9, respectively. Catalysts obtained by reaction of Ti benzyl and cyclopentadienyl derivatives with Mg(OH)Cl have been investigated by Chien [189] in propylene polymerization in the presence of $AlEt_2Cl$. The catalytic activity is low (about 2 g PP $(mmol\,Ti)^{-1}\,h^{-1}\,atm^{-1}$), but the isotacticity index of the obtained polymers may attain also values of 70 to 90. $TiCl_4$ supported on SiO_2 [91] fails to give satisfactory results.

A family of catalysts [119, 190], obtained by reduction of $TiCl_4$ with Grignard reagents at varying ratios between the two components and very active in ethylene polymerization, exhibits a low activity (22.5 — 100 g PP $(mmol\,Ti)^{-1}\,h^{-1}\,atm^{-1}$) and stereospecificity (I.I. = 35–55%) in propylene polymerization in the presence of AlR_3. Worse results are obtained in the presence of $AlEt_2Cl$.

A class of very interesting catalysts is obtained by supporting Ti halides on activated $MgCl_2$. Catalysts resulting from co-milling of $MgCl_2$ and $TiCl_4$ [103] exhibit a high activity in propylene polymerization in the presence of AlR_3 [87, 191, 192] accompanied however, by a low stereospecificity (Table 10).

Table 10. Propylene polymerization with $MgCl_2—TiCl_4$ supported catalysts[b,c]

Catalyst	Ti/ %	Co-catalyst	Activity/ g polypropylene $(mmol\,Ti)^{-1} \times h^{-1}$	I.I.[a]/ %
$MgCl_2—TiCl_4$	3.6	$Al(C_2H_5)_3$	870	49.2
$MgCl_2—TiCl_4$	3.6	$Al(C_2H_5)_2Cl$	45	33
$TiCl_3 \cdot 0.3\,AlCl_3$	24	$Al(C_2H_5)_3$	23	75.9

[a] Isotacticity index
[b] Reference [87]
[c] Polimerization conditions: $[C_3H_6] = 0.3\,mol\,l^{-1}$; 343 K

The discovery of catalysts [193, 194] supported on activated $MgCl_2$ and modified by Lewis bases has overcome the problems connected with the low stereospecificity exhibited by so many supported catalysts. An intense research activity has led to the development of catalysts endowed with high activity and stereospecificity which are to-day being employed in industrial processes. These catalytic systems [105, 195–202] generally consist of Ti halides supported with a Lewis base on activated $MgCl_2$. Typical catalysts exhibit high specific surface area (50 to 300 $m^2\,g^{-1}$) especially after reaction with aluminum alkyl and contain from 0.5 to 5% Ti by weight. In a typical preparation [199, 203, 204] anhydrous $MgCl_2$ and ethyl benzoate are co-milled for 20 h and the resulting solid is then treated with excess $TiCl_4$ at

353–408 K. After filtration the precipitate is washed with hydrocarbons and dried. AlR_3 or its mixtures with AlR_2Cl are employed as cocatalyst. The addition of a Lewis base during polymerization is recommended, particularly in the presence of a high aluminum alkyl concentration to enhance the catalyst stereospecificity. Owing to the small quantities of catalyst employed and to the need for maintaining a minimum aluminum alkyl concentration to protect the catalyst from the impurities present in the system (solvent, monomer and reactor), high Al/Ti ratios are generally used; varying this ratio over a wide range, however, does not appear to affect the catalyst activity considerably.

The best catalytic systems allows, in about three hours in bulk polymerization, yields of 1000 to 2000 kg (g Ti)$^{-1}$ of polypropylene containing 95–98% isotactic polymer. The initial polymerization rate is very high (2–2.5 kg PP (mmol Ti)$^{-1}$ h^{-1} atm^{-1}) and decreases more or less gradually according to the catalyst type. The polymer molecular weight may be controlled by means of hydrogen as in the case of traditional catalysts [193].

On the basis of the ^{13}C-NMR analysis of the polymers prepared with these catalysts, Doi *et al.* [199] conclude that both the number and type of steric irregularities present in the fraction insoluble in boiling *n*-heptane are similar to those of the polymers obtained with the $TiCl_3$—$AlEt_2Cl$ catalytic system.

4. *Isospecific Polymerization with Soluble Catalysts*

Whereas there are a number of soluble catalytic systems active in ethylene polymerization (*c.f.* section 2.A.3), a few only are capable of polymerizing propylene. The most studied soluble catalyst, obtained by reaction of AlR_2Cl and $Al_2R_3Cl_3$ with vanadium compounds, will be dealt with under section 2.B.5 and is capable of producing syndiotactic polypropylene.

Propylene and 4-methyl-1-pentene isotactic polymers have been obtained, with a low yield, by means of Ti and Zr benzyl [51] and allyl [21] derivatives. As it is interesting to observe, these systems have for the first time allowed the stereospecific polymerization of α-olefins in homogeneous phase.

Some authors [206, 207] though acknowledging that no colloidal particles may be identified in the solution of these catalysts, suggest that the polymerization proceeds on the surface of small particles formed by $ZrBz_4$ decomposition during the course of polymerization. Such hypothesis seems to be supported by the fact that both stereospecificity and the $\bar{M}_w/\bar{M}_n$ ratio of the polymer have been noticed to increase with polymerization time.

Homogeneous catalytic systems based on presumable cyclic oligomers of methylaluminoxane and on $(\eta^5—C_5H_5)_2\,M(CH_3)_2$ (M = Ti, Zr) are capable of polymerizing propylene too with high yields [39, 208]. The catalytic activity keeps almost constant also for more than 24 h and productivities up to 550,000 g PP (g Zr)$^{-1}$ may also be attained. The polymer is thoroughly atactic but, a regiospecificity above 90% [209] may be observed on the basis of ^{13}C-NMR spectra.

5. Syndiospecific Polymerization of Propylene

Among α-olefins, only propylene has thus far been polymerized to syndiotactic polymer [210, 211]. This polymer was first separated by Natta, Pasquon, Zambelli *et al.* from mostly isotactic polypropylenes obtained at 343 K with heterogenous catalysts based on α- or γ-$TiCl_3$ and $Al(C_2H_5)_2F$ or LiC_4H_9 and on $TiCl_4$ and LiC_4H_9 [210, 212]. The fraction soluble in boiling *n*-hexane (in these cases generally only 1–10% of the raw polymer) is the richest in syndiotactic polypropylene. By using the catalytic systems γ-$TiCl_3$—$Al(C_2H_5)_2F$ at room temperature, the hexane fraction exhibits crystallinity only for the crystalline form of syndiotactic polypropylene [210]. Higher though moderate percentages of syndiotactic polymer are obtained with $TiCl_4$—Na alkyls [212].

The best catalytic systems for this type of polymerization are the homogeneous ones, prepared and utilized at low temperature (*e.g.* 195 K), and based on a vanadium compound (VCl_4, $VOCl_3$, VA_3, A = acetylacetone residue, $VO(OC_2H_5)_2Cl$) and aluminium dialkyl monochloride ($Al(C_2H_5)_2Cl$, Al $(i\text{-}C_4H_9)_2Cl$, etc.). Most effective and syndiospecific catalysts are obtained when the Al/V ratio is between 2 and 10 (generally 5) and when a weak Lewis base such as anisole is added in a molar ratio 1:1 with vanadium compound [213–217].

Table 11. Propylene syndiospecific polymerization[j]

Vanadium compound	Organometallic compound	Al:V:An[a] (mol.)	Solvent	Temperature/ K	IS[b] of the polymer obtained[c]
VCl_4[d]	$Al(C_2H_5)_2Cl$	5:1:1	toluene	195	1
VCl_4[d]	$Al(C_2H_5)_2Cl$	10:1:1	toluene	195	0.90
VCl_4[d]	$Al(C_2H_5)_2Cl$	2:1:1	toluene	195	>0[e]
VCl_4[d]	$Al(i\text{-}C_4H_9)_2Cl$	5:1:1	toluene	195	1.30
VCl_4[d]	$Al(neo\text{-}C_5H_{11})_2Cl$	5:1:1	toluene	195	1.85
VCl_4[f]	$Al(i\text{-}C_4H_9)_2Cl$	5:1:1	*n*-heptane	195	2.4
VCl_4[d]	$Al(C_2H_5)_2Cl$	5:1:1	*n*-heptane	195	1.9
VCl_4[g]	$Al(C_2H_5)_2Cl$	5:1:1	*n*-heptane	236	0.5[h]
VCl_4[d]	$Al(C_2H_5)_2Cl$	2:1:0	*n*-heptane	195	1.28
VCl_4[d]	$Al(C_2H_5)_2Cl$	5:1:0	*n*-heptane	195	1.15
VA_3[i]	$Al(C_2H_5)_2Cl$	5:1:0	toluene	195	0.65
VA_3[i]	$Al(C_2H_5)_2F$	5:1:0	toluene	195	0.80

[a] An = anisole; Al = aluminum; V = vanadium
[b] IS = crystallinity index due to syndiotactic polymer; determined by IR methods [214]
[c] Polymerization time 18–20 h
[d] 1 mmol in 100 cm^3 solvent
[e] Very low polymer yield and IS
[f] 2.66 mmol in 100 cm^3 solvent
[g] 0.3 mmol in 100 cm^3 solvent
[h] Polymerization time 1 h
[i] 1.4 mmol in 100 cm^3 solvent
[j] References [211], [213]

The syndiospecificity increases as the polymerization temperature lowers and as the steric hindrance of the alkyl group of AlR_2Cl increases [213] (*c.f.* Table 11). Higher syndiospecificity indexes are attainable when operating in *n*-heptane instead of in toluene (*c.f.* Table 11). Stereospecificity is also affected by the nature of the halogen contained in these organometallic compounds. Under the same reaction conditions (T = 195 K, Al/V ratio = 5, no anisole), among the VCl_4—$AlEt_2X$ systems, only when X is Cl syndiospecific catalysts are formed [214]. Syndiotactic polymer is however obtainable with the system VA_3—$Al(C_2H_5)_2F$ employed under the same conditions [211]. The viscosimetric molecular weight of the polymers obtained in the presence of anisole increases almost linearly with the polymerization time [214] and is in the order of some tens of thousands of units (intrinsic viscosity determined in tetralin at 408 K = 0.5–1). The polymerization rate is rather slow; for the conditions shown under Table 11, the resulting polymer amounts only to some percent of the monomer charged.

C. Olefins Copolymerization

1. Random Copolymers

Amorphous random copolymers based on ethylene and propylene, prepared by Natta *et al.* soon after the discovery of stereospecific polymerization, are today largely used for the preparation of elastomers. Two types of products are prepared: EPM and EPDM, resulting from ethylene and propylene copolymerization respectively in the absence and in the presence of a non-conjugated diolefin introduced to allow sulphur vulcanization of the copolymer obtained.

The catalysts utilized for the preparation of these products are the Ziegler-Natta ones active in propylene polymerization [218–223]. The best results are attained by operating at low temperature with catalysts soluble or colloidally dispersed in hydrocarbon solvents and based on a vanadium compound in mixture with an aluminum alkyl, at least one of the two being halogenated [223, 224] (*c.f.* Table 12).

The more recent high yield catalytic systems based on supported titanium compounds [231] also appear very interesting. The polymerization can be carried out either in solution, in aliphatic or aromatic hydrocarbons or in slurry using excess propylene as suspending medium.

Among the heterogeneous systems (*e.f.* Table 12), best results with reference to the copolymer composition, are obtained by utilizing aluminium alkyls compounds containing long-chain alkyl groups (*e.g.* — C_6H_{13}) capable of performing a greater dispersion of the catalyst in the hydrocarbon phase.

For some catalytic systems (*e.g.* VA_3—$Al(C_2H_5)_2Cl$) [218], the molecular weight of the resulting copolymer is higher with aged catalysts. The molecular weight may be controlled by transfer agents such as H_2 [168] or ZnR_2 [169, 170]. The copolymers used for the synthesis of elastomers contain 25 to 55% by weight propylene and 2 to 5% by weight termonomer [232]. Terpolymers containing up to 10% termonomer are also prepared. The ter-

Table 12. Catalytic systems based on a vanadium compound and high yield systems based on a titanium compound for ethylene-propylene copolymerization in the presence or in the absence of a non-conjugated diolefin

Catalytic system	Molar ratio Al/V	Polymerization temperature/K	Reference
$VOCl_3$—$Al(C_6H_{13})_3$ — heterogeneous	3	298	[225]
VCl_4—$Al(C_6H_{13})_3$ — heterogeneous	3	298	[226]
VCl_3—$Al(C_6H_{13})_3$ — heterogeneous	3	298	[227]
VA_3[a]—$Al(C_2H_5)_2Cl$ — colloidal — homogeneous	5	253	[218]
VCl_4—$HCl_2AlO(C_2H_5)_2$ — heterogeneous	2.5	298	[228]
$VO(OR)_3$, $VO(OR)_2Cl$, $VO(OR)Cl_2$ + AlR_2Cl or $Al_2R_3Cl_3$[b] — colloidal	—	303	[224], [229], [230]
$VOCl_3$, VCl_4—AlR_2Cl — colloidal — homogeneous	—	—	[229]
High yield catalyst — heterogeneous	—	—	[231]

[a] A = acetylacetone ligand
[b] Any combination between vanadium compound and aluminum compound

monomers mostly utilized in practice are [220] 5-ethylidene-2-norbornene (ENB) [233], 1,4-hexadiene (1,4-HD) [234] and dicyclopentadiene (DCPD) [223, 235, 236]. Other termonomers are reported in literature [220, 224, 237].

The activity of all the catalysts previously described quickly decreases during the polymerization and the kinetics of the decay depends on the nature of the catalyst components, on their concentrations, on temperature, etc. This behaviour has been often ascribed to the formation of inactive V^{II} species.

The addition of substances [238, 239] deemed to be able to oxidize the inactive V^{II} species produces a considerable enhancement of the catalytic activity. For this purpose O_2 [240] has been employed as well as nitrobenzene [241], alcohols [242] and mainly chlorinated organic substances containing activated Cl atoms, such as benzotrichloride and esters of chlorinated carboxylic acids [243, 244]. The catalytic activity of homogeneous systems based on vanadium compounds (VA_3—$AlEt_2Cl$) may also be improved through the addition of aluminoxanes [245, 246].

Catalysts based on vanadium compounds such as for instance VCl_4—$Al(C_6H_{13})_3$ and VCl_3—$Al(C_6H_{13})_3$ have been used also in ethylene-1-butene [247], propylene-1-butene [248] and ethylene-propylene-1-butene [249] copolymerization. Random ethylene-butadiene and propylene-butadiene copolymers have been prepared with catalytic systems based respectively on $TiCl_4 \cdot 2\,P(C_2H_5)_3$—$Al(C_2H_5)_2Cl$ (used at 313 K) [250], and on VCl_4 (or $VOCl_3$)—$Al(C_2H_5)_3$—$Al(C_2H_5)_2Cl$ (used at 243 K [251, 252].

2. *Alternating Copolymers*

Internal olefins such as for instance 2-butene or 2-pentene and cycloolefins with more than four C atoms in the ring fail to homopolymerize with the conventional Ziegler-Natta catalysts. These olefins may however copolymerize with ethylene owing to the limited removal of steric hindrances at the transition metal-carbon site exhibited by this monomer.

The best results (obtained by Natta *et al.*) with regard to the formation of copolymers with sufficiently high internal olefin contents, are attained by operating for instance at 243 K with catalytic systems containing a vanadium compound such as the VCl_4—$Al(C_6H_{13})_3$ system colloidally dispersed in hydrocarbon medium (*c.f.* Table 13) [253, 257]. Owing to the high ethylene reactivity as compared with that of the internal olefin, a considerable excess of the latter monomer is required to avoid the presence of long sequences of ethylene units in the polymer chains.

From the resulting copolymer, always containing a percentage of ethylene units above 50% by moles, fractions containing 50% by moles of each monomer may be separated by extraction with solvents. These results suggest that the monomer units deriving from the internal olefin add only after an ethylene unit and that the products containing 50% by moles of each monomer are alternating copolymers. The alternating copolymers formed by ethylene and *cis*-2-butene (*c.f.* Figure 11) or by ethylene and cycloolefins with an odd number of carbon atoms in the ring are crystalline with *erythro*-diisotactic structure [258, 259]. *Trans*-2-butene [253] and cycloolefins with an even number of carbon atoms [257] conversely supply amorphous copolymers only.

Alternating crystalline copolymers have been separated also from products obtained by ethylene copolymerization with butadiene conducted for instance

Table 13. Alternating crystalline copolymers from olefins

Monomers	Catalytic system	Polymerization temperature/K	Types of polymer and butadiene units	Reference
ethylene-cis-2-butene	VA_3[a]—$Al(C_2H_5)_2Cl$-toluene		erythro-diisotactic	[253]
	VCl_4—$Al(C_6H_{13})_2Cl$-*n*-heptane Al/V = 2.5 to 5 (mol)	243	erythro-diisotactic	[253]
ethylene-cyclopentene	idem	243	erythro-diisotactic	[255]
ethylene-butadiene	VCl_4—$Al(i\text{-}C_4H_9)_3$-$Al(i\text{-}C_4H_9)_2Cl$ anisole (1:2:2:2 mol) in toluene[b]	248	1,4-trans	[260]
propylene-butadiene	VCl_4 or $VOCl_3$-$Al(C_2H_5)_3$	243 to 233	1,4-trans	[261]
	VCl_4 or $VOCl_3$—$Al(C_2H_5)_3$—$Al(C_2H_5)_2Cl$		1,4-trans	[262]
	$VOCl_3$—$Al(C_2H_5)_3$ $Al(C_2H_5)_2Cl$ (1:3:3 mol)		1,4-trans	[263]
	$VOCl_3$—$VO(OC_2H_5)_3$—$Al(i\text{-}C_4H_9)_3$		1,4-trans	[264]
	$VO(OR^c)_2Cl$—$Al(i\text{-}C_4H_9)_3$ (0.8:5 mol)	223	1,4-trans	[266]

[a] A = acetylacetone ligand. [b] Catalyst prepared at 195 K. [c] R = branched chain group

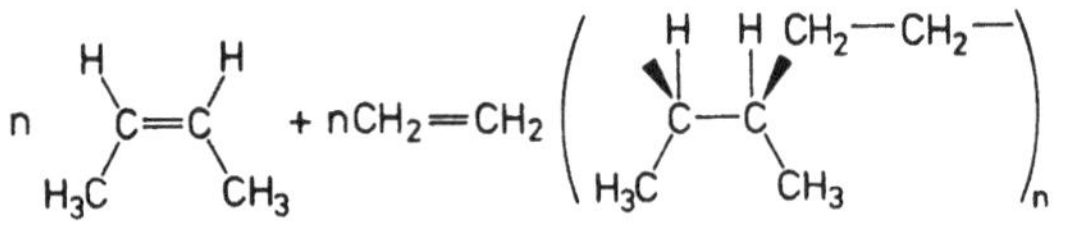

Figure 11. Alternating ethylene-*cis*-butene-2-copolymer with *erythro*diisotactic structure. (Reproduced with permission from ref. [258])

at 248 K in the presence of homogeneous catalytic systems based on VCl_4, AlR_3, AlR_2Cl and anisole [260] (*c.f.* Table 13).

These polymers contain 1,4-*trans* butadiene units and the yields in alternating copolymer are very low. Alternating copolymers of butadiene with propylene and other higher α-olefins are obtained at low temperature with catalysts based on a vanadium compound and AlR_3 alone or in mixture with $Al(C_2H_5)_2Cl$ [261–166] (*c.f.* Table 13).

3. Heteroblock Copolymers — Polyallomers

The growing polymeric chains in the Ziegler-Natta polymerization remain bound to the catalytic site, in the absence of monomer in particular, for a time sufficiently long to allow the preparation of polymeric chains formed at least by two segments (heteroblock copoymers), each containing a monomer differing from any of the adjacent segments. Each monomer must be added sequentially and the last added olefin must be consumed or removed before the introduction of the next one.

The possibility of obtaining heteroblock copolymers consisting of more than two blocks depends on the life time of the growing polymeric chains. The question is quite controversial (*c.f.* section 3.A.2). Various trials have been made to prepare ethylene-propylene heteroblock copolymers. Very likely, the products mentioned in the literature [219] contain heteroblocks constituted by two types of segments only. These copolymers have been obtained with the conventional catalytic systems used for propylene isospecific polymerization [151, 268–270] (*c.f.* section 2.B.2). The same systems have been employed also for the synthesis of propylene-1-butene/heteroblock copolymers [271]. Ethylene-propylene copolymers consisting of two blocks have also been obtained with catalytic systems promoting propylene syndiospecific polymerization [214].

To the same class of products belong polyallomers [272]. These products are manufactured commercially by the use of the catalytic systems promoting propylene isospecific polymerization. Propylene, ethylene-propylene mixtures, and sometimes ethylene alone are fed in consecutive steps. Most likely, these products largely consist of polypropylene and ethylene-propylene copolymer mixtures since the first polymerization stage takes place in the presence of a chain transfer such as hydrogen.

D. Stereoregular Polymerization of Cycloolefins

Cycloolefins may polimerize through double bond opening, thus maintaining the ring in the structural unit, or through ring opening, thus maintaining the double bond in the structural unit. By double bond opening polymerization, Wiberg and Bartley [273] obtained cyclopropene homopolymers with radical initiators, whereas Natta *et al.* prepared cyclobutene [274–277] and 3-methylcyclobutene homopolymers [277, 278] with Ziegler-Natta catalytic systems. Only in some cases and with very low yields, they have also obtained cyclopentene and 3-methylcyclohexene homopolymers containing cyclic units in the polymer chains [277, 278].

The second type of polymerization can be obtained catalytically for several cycloolefins [279].

1. Double Bond Opening Polymerization of Cyclobutene and 3-Methylcyclobutene

When polymerization of cyclobutene occurs by double bond opening two types of crystalline polymers melting at 483 K and 423 K respectively can be prepared. These polymers are believed to be polycyclobutylenamer-2-*erythro*-diisotactic (I) and polycyclobutylenamer-2-*erythro*-disyndiotactic (II).

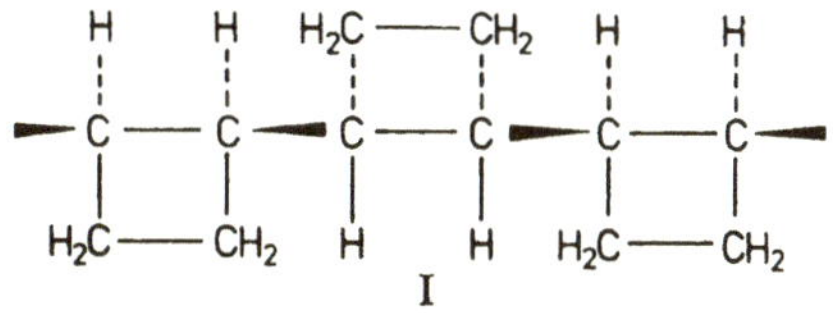

I

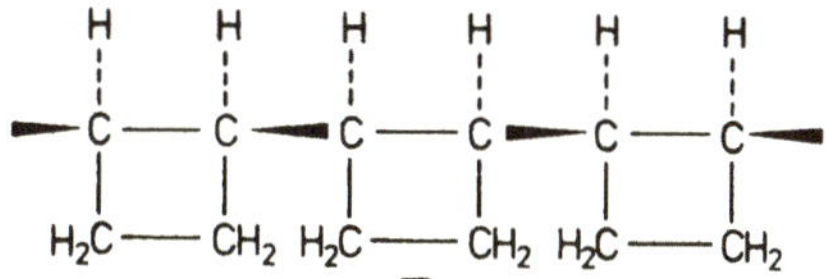

II

The more convenient catalytic systems for obtaining type (I) polymer are based on vanadium or chromium compounds and an organometallic compound of Al or Be employed in a hydrocarbon diluent [277, 280] (*c.f.* Table 14). The activity and stereospecificity of these systems are strongly conditioned by the nature of the organometallic compound [277]. Besides these systems, other catalysts based on $RhCl_3$ [280] or η^3-allyl-Ni-bromide [278] (*c.f.* Table 14) are active in cyclobutene polymerization to type (I) polymers. With these systems, 3-methylcyclobutene too polymerizes through double bond opening, supplying non-stereoregular polymers.

Type (II) polymer is obtained by polymerization of cyclobutene with homogeneous systems based on a vanadium compound and AlR_2Cl [276] (*c.f.* Table 14).

2. Ring Opening Polymerization of Cycloolefins

Ring opening of a cycloolefin leads to an unsaturated structural unit, the double bond of which may have *cis-* or *trans* configuration. The most significant results obtained in the ring opening polymerization of some cycloolefins (cyclobutene, cyclopentene, cycloheptene, *cis*-cyclooctene and *cis*-cyclodecene) are shown in Table 15. Also some substituted cycloolefins such

Table 14. Stereospecific polymerization of cyclobutene to crystalline polymers through Double Bond Opening

Catalytic System		Polymerization temperature/K	Polymer type	Reference
VCl_4, VCl_3 or $VOCl_3$—$Al(C_2H_5)_3$-*n*-heptane	heterogeneous	...	(I)	[276]
$VOACl_2$[a], VA_2Cl_2[a], $VOCl_3$—$Al(C_2H_5)_2Cl$-toluene	colloidal Al/V = 5	253	(I)	[276]
CrO_2Cl_2, $CrCl_3$ or CrA_3[a]—$Al(C_2H_5)_3$ or $Al(C_2H_5)_2Cl$[b]-hydrocarbon	heterogeneous	253	(I)	[276]
Ni-π-allylbromide-ethanol			(I)	[278]
$RhCl_3$—H_2O			(I)	[280]
VA_3[a], $VO(OC_4H_3)_3$, VOA_2[a] or VOA_2Cl[a]-—$Al(C_2H_5)_2Cl$-toluene	homogeneous Al/V $\geqq$ 5	$\leqq$253	(II)	[276]

[a] A = acetylacetone residue
[b] One component at least must be halogenated

as for instance 3-methylcyclobutene, 1-methylcyclobutene and 3-methyl- or 3-phenylcyclooctene may polymerize through ring opening, whereas cyclohexene fails to polymerize under the same conditions. As may be inferred from Table 15, in some cases the stereospecificity is also affected by the solvent nature. Furthermore, a given catalytic system does not produce the same type of polymer with the various cycloolefins, as may be observed for the systems based on WCl_6 and on $MoCl_5$.

In the ring opening polymerization of higher membered cycloolefins, molybdenum based systems exhibit only a poor activity. Moreover, catalysts such as $MoCl_5$—Al(*n*-hexyl)$_3$ polymerize higher membered cycloolefins to high *trans* (and not to high *cis* as for cyclopentene) polyalkenamers.

An interesting case is supplied by the system WF_6—$Al_2(C_2H_5)_3Cl_3$. Depending on the Al/W ratio, *cis*-polypentenamer (Al/W = *ca.* 1), *trans*-polypentenamer (Al/W = *ca.* 7) or intermediate *cis*/*trans* ratios are obtainable [283]. The dependence of the stereospecificity on the nature of the organometallic compound for WCl_6 based catalytic systems has been studied by Gunther, Haas *et al.* [283, 284].

Calderon *et al.* [285] have found that the catalytic activity at room temperature increases in the order $AlEtCl_2$ > $AlEt_2Cl$ > $AlEt_3$ for WCl_6 based systems.

$MoO_2(CH_3COCHCOCH_3)_2$—$AlEt_2Cl$, $MoCl_5$—AlH(*i*-Bu)$_2$, WCl_6—$BeEt_2$ systems are only poorly or not at all stereospecific and yield polymers with large amounts of both *cis* and *trans* pentenamer units [279].

Catalytic systems based on WCl_6-aluminium alkyls are activated and stabilized by a preliminary addition to the W compound of oxygenated

Table 15. Polymerization of cycloolefins through ring opening

Cycloolefin and catalytic system	Polymerization temperature/K	$-CH=CH-(CH_2)_n-$ trans %	$-CH=CH-(CH_2)_n-$ cis %	$-CH-CH-$ $[CH_2]_n$ %	Polymer structure	Reference
cyclobutene						
$TiCl_4-Al(C_2H_5)_3$ [a]	253				amorphous	[275]
$TiCl_4-Al(C_2H_5)_3$ [b]	253	65	30	5	crystalline	[275]
$WCl_6-Al(C_2H_5)_3$ [b]	253	40	30	30	amorphous	[275]
$MoCl_5-Al(C_2H_5)_3$ [b]	253	30	30	40	amorphous	[275]
$RuCl_3-H_2O$	323	50	50	0	amorphous	[275]
$RuCl_3-C_2H_5OH$	293	100	0	0	crystalline	[275]
cyclopentene						
$TiCl_4-Al(C_2H_5)_3$ [c]	243	97	3	0	crystalline	[275]
$WCl_6-Al(C_2H_5)_3$ [c]	243	90	10	0	crystalline	[275]
$MoCl_5-Al(C_2H_5)_3$ [c]	243	1	99	0	crystalline	[275]
$Cr(\eta^3\text{-allyl})_3-WCl_6$		27.4				[283]
$BCl_3-Li_3[W(C_2H_5)_6]$		93.8				[283]
$TiCl_4-Al[W(C_2H_5)_5]$		60				[283]
$WF_6-Al_2(C_2H_5)_3Cl_3$ [e]		according to the Al/W ratio				[284]
cycloheptene						
$WCl_6-Al(C_2H_5)_2Cl$ [d]	293	94	6	0	crystalline	[275]
$MoCl_5-Al(C_2H_5)_3$ [d]	293	93	7	0	crystalline	[275]
cis-cyclooctene						
$WCl_6-Al(C_2H_5)_3$ [d]	293	85	15	0	crystalline	[275]
$WCl_6-AlC_2H_5Cl_2$	298				crystalline	[281], [285]
cis-cyclodecene						
$WOCl_4-Al(C_2H_5)_2Cl$ [d]	293	75–85	15–25	0	crystalline	[282]
$WCl_6-Al(C_2H_5)_2Cl$ [d]	293	75–85	15–25	0	crystalline	[282]

[a] In *n*-heptane
[b] In toluene
[c] Without solvent
[d] Catalyst prepared at 243 K
[e] If Al/W ratio $\cong 1$, *cis*-polypentenamer is obtained; with Al/W ratio $\cong 7$, *trans*-polypentenamer is obtained

compounds [286] such as peroxide, molecular oxygen, phenols, *t*-butylhypochlorite, 1,3-dinitro-2,5-dichloro-benzene [279]. More active appear to be some halogenated alcohols and phenols [279].

Heterogeneous systems based on molybdenum oxides and alumina polymerize cyclopentene through ring opening however at rates lower than the systems shown under Table 15 [287]. Ring opening polymerization of cyclopentene at room temperature to atactic polymers can be induced by tungsten halides without cocatalysts or promoters [279].

High polymer yields from cyclopentene, cycloheptene, cyclooctene, and cyclododecene are obtained with WCl_6—$AlBr_3$, WCl_6—$AlCl_3$ anisol, $MoCl_5$—$AlBr_3$ catalytic systems [279, 288, 289].[3]

Other systems proposed for ring opening cycloolefin polymerization, based on Ta, Nb, Zr, Rh, Re, W and Zr-allyl compounds, etc. are described in the literature [279].

3. Mechanism, Kinetic Aspects, Catalytic Complexes

A. Isospecific Polymerization of α-Olefins

1. Behaviour of Catalytic Systems — Overall Kinetics

The behaviour of the Ziegler-Natta catalytic systems for propylene and other α-olefins isospecific polymerization depends on a variety of factors. Typical of the systems prepared starting from a soluble transition metal compound in a high oxidation state (*e.g.* $TiCl_4$, VCl_4) is a significant dependence of the activity on the Al/Ti or Al/V ratio, exhibiting a pronounced peak (*c.f.* Figure 12).

For α-$TiCl_3$ based systems, Natta, Pasquon *et al.* [151] have noticed practically no dependence of the propylene polymerization rate on the $Al(C_2H_5)_3/TiCl_3$ ratio over a wide range of such ratio. The non-dependence

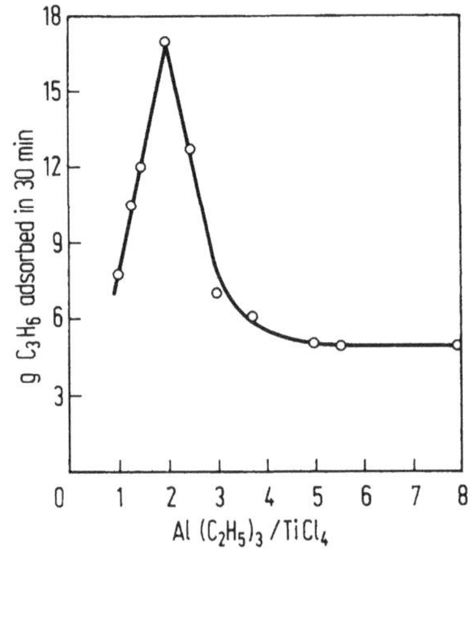

Figure 12. Propylene consumption during the first 30 min. in dependence of $Al(C_2H_5)_3/TiCl_4$ molar ratio. (Reproduced with permission from ref. [137])

of the catalytic activity on the Al/Ti ratio, at least for organometallic compound concentrations in the range 10 to 100 mmol dm^{-3}, has also been observed by Schnecko *et al.* as well as by Keii *et al.* for catalytic systems $TiCl_3$—$AlEt_3$ [290, 291], $TiCl_3$—$AlEt_2Cl$ [292, 293] and $TiCl_3$—$ZnEt_2$ [294]. However, a dependence of the reaction rate on the metal component ratio has been observed by Tait [295] in 4-methyl-1-pentene polymerization in the presence of the VCl_3—AlR_3 system and by Zakharov *et al.* [296] in propylene polymerization with a δ-$TiCl_3$, 0.3 $AlCl_3$—$AlEt_3$ catalyst. In the latter case, the effect seems to be reversible and is more pronounced at low temperature. It must however be remarked that extremely higher $AlEt_3$ concentrations (0.3 mol dm^{-3}) have been employed than are usual in Ziegler-Natta catalysis (*ca.* 0.005 mol dm^{-3}). By way of explanation, these authors introduced kinetic models of the Langmuir-Hinshelwood type with reversible adsorptions of aluminium alkyl on the transition metal halogenide surface. Similar models and phenomena have been proposed by Saltman [297], Vesely *et al.* [298], and Reich and Stivala [299]. These models are not applicable to all cases [300], and fail always to take into account that organometallic compounds react also irreversibly, especially when employed at high concentrations, both with $TiCl_3$ [301–305] and with VCl_3. Zambelli, Pasquon *et al.* [306, 307] have studied the reducing power of various organometallic compounds towards VCl_3 and have observed it to increase as follows:

$$AlEt_2I \leqq AlEt_2Cl < GaEt_3 < AlEt_3 < BeEt_2 < ZnEt_2 < CdEt_2 < LiBu$$

The extent of the reduction is also affected by the type of transition metal halogenide on which such action is exerted and is probably greater for δ-$TiCl_3$ than for α-$TiCl_3$.

Other differences in behaviour between the investigated catalytic systems concern the dependence of the polymerization rate on time. With the α-$TiCl_3$—$Al(C_2H_5)_3$ or $Al(C_2H_5)_2Cl$ catalytic systems, the activity increases until it reaches a value which keeps constant for several hours [156, 308] (Figure 13).

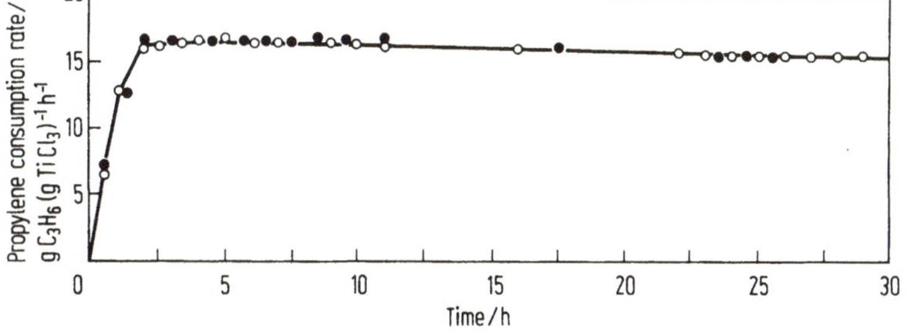

Figure 13. Polymerization rate-time curves for the polymerization of propylene with α-$TiCl_3$ and $Al(C_2H_5)_3$ at 343 K and at propylene pressure of 193 Pa. $TiCl_3$: ○, 0.80 g l^{-1}; ●, 1 g l^{-1}. $Al(C_2H_5)_3$: ○, 44.5 mmol l^{-1}; ●, 29.4 mmol l^{-1}. (Reproduced with permission from ref. [308])

If the polymerization is interrupted by removing the monomer when the steady-state rate was reached, and then after several hours the initial monomer concentration is re-established, the reaction starts again at once and at the same steady-state rate.

Decaying-type kinetic curves have been observed for the polymerization of propylene with catalysts based on Al alkyls and wet-milled α- and γ-$TiCl_3$ or δ-$TiCl_3$. After an induction period a maximum rate is reached and then the rate declines with time [151, 308, 309] (*c.f.* Figure 14). The decrement is greater when operating with $AlEt_3$ instead of $AlEt_2Cl$.

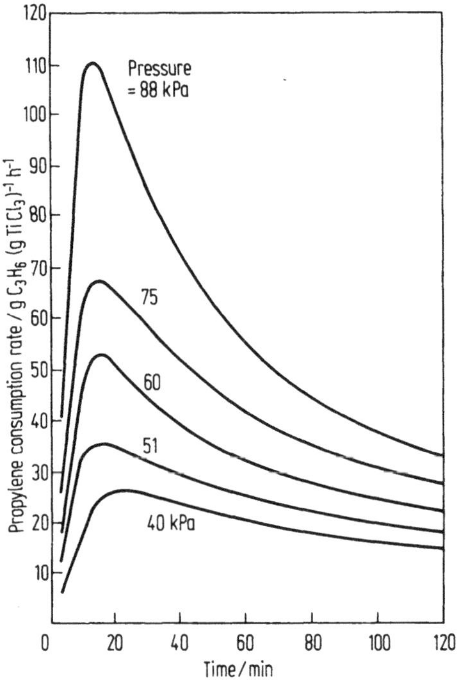

Figure 14. Polymerization rate-time curves for the polymerization of propylene with δ-$TiCl_3 \cdot 0.3\ AlCl_3$ and $Al(C_2H_5)_3$ at 313 K and various pressures. (Reproduced with permission from ref. [290])

Behaviour discrepancies may be noticed for the different systems. It is interesting to observe, that in some high yield-low stereospecificity supported catalysts (*e.g.* based on $TiCl_4/Mg(OH)Cl/AlEt_3$), the polymerization rate keeps practically constant at least for some hours [99] (*c.f.* Figure 15). Conversely, in high yield catalytic systems based on $MgCl_2/TiCl_4/C_6H_5COOEt/AlEt_3$, a gradual decrease in activity may be observed from the first minute of polymerization [105, 197] (*c.f.* Figure 16).

In several cases, at least when not-premilled catalysts are involved, a break up of the aggregates of the catalyst due to mechanical action of the growing polymer chains seems to occur during the initial polymerization phase, thus allowing new active centers to form [99, 151, 309–311]. Increments of surface area from 11 to 70–80 $m^2\ g^{-1}$ caused by catalyst dis-

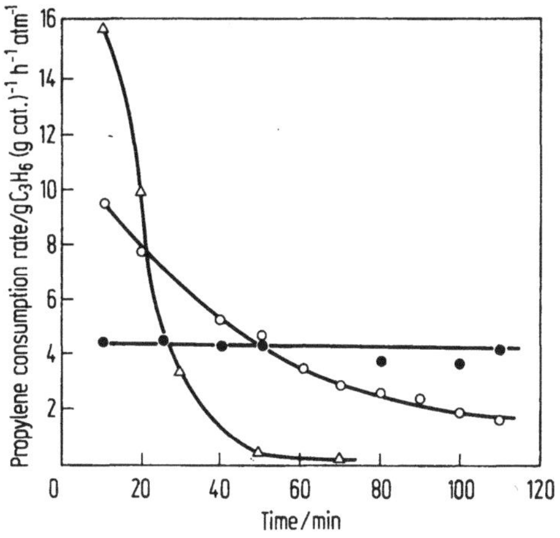

Figure 15. Comparison of the polymerization rate-time curves for the polymerization of propylene, obtained with conventional and supported catalysts. △, $TiCl_4$—$Al(C_2H_5)_2Cl$; ○, $TiCl_4$—$Al(C_2H_5)_3$; ●, $TiCl_4/Mg(OH)Cl$—$Al(C_2H_5)_3$. (Reproduced with permission from ref. [99])

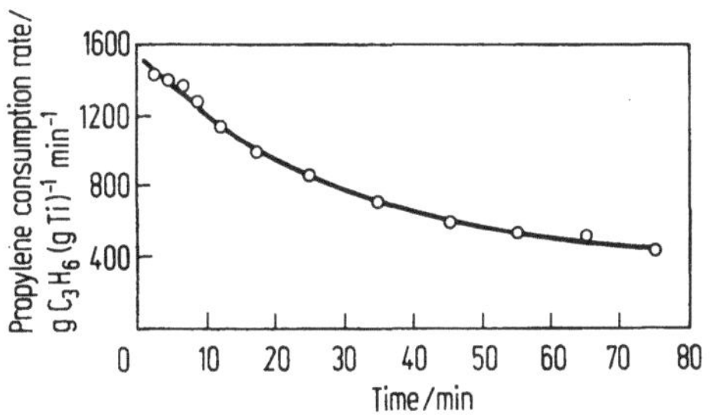

Figure 16. Polymerization rate-time curve for the polymerization of propylene initiated by a stereospecific $MgCl_2$-supported catalyst. 333 K, monomer concentration 0.82 mol l^{-1}. (Reproduced with permission from ref. [105])

gregation have been observed for a system based on δ-$TiCl_3 \cdot 0.3\ AlCl_3$ [313]. However, the extent of this effect depends on the nature of the formed polymer, and particuarly on its crystallinity and solubility. In other cases (*e.g.* VCl_3—AlR_3 in 4-methyl-1-pentene polymerization) the initial increment of activity has been ascribed to the desorption of AlR_2Cl from the catalyst surface [314].

Some authors have tried to correlate the catalytic activity to the surface area of the transition metal halogenide. Natta and Pasquon [151], by utilizing samples of moderately milled and unmilled α-$TiCl_3$, have observed the same steady-state rate after an initial adjustment time, almost irrespective of the initial size of the $TiCl_3$ crystals. Vecchi *et al.* [315] have instead observed the catalytic activity to increase concomitantly with surface area measured according to the BET method. The same conclusion has been reached also by Keii *et al.* [309] and Tornquist [146]. The failure of BET methods to give, in some cases, a good correlation between surface area and activity, may be ascribed to a disagregation of the catalyst which causes an increase of the catalyst specific surface during the polymerization. A better correlation between catalytic activity and surface area has been found by Yermakov and Zakharov [188] who estimated surface area on the basis of the average dimensions of crystallites, determined *via* a X-ray technique.

The decline in the rate of polymerization has been attributed by some authors to a shortage of monomer in the vicinity of the active centers, caused by a diffusional barrier of polymer encapsulating the catalyst [202, 316, 320]. This phenomenon, however, does not appear to be determinant in most cases and many authors [60, 290, 309] have attributed the decline in polymerization rate to an irreversible deactivation of the active centers. However, the various authors agree substantially insofar as concerns the influence of other variables on the behaviour of different types of heterogeneous catalytic systems; in particular:

(i) linear dependence of the reaction rate on the monomer concentration [151] when concentration values are not too low [322];
(ii) linear dependence of the reaction rate on the concentration of the transition metal salt (*e.g.* $TiCl_3$), at least for moderate variations of the Al/transition metal ratio;
(iii) energies of activation of the overall process ranging between 42 and 59 kJ mol^{-1} for the systems based on a Ti compound.

Polymerization does not seem to involve kinetic chain termination phenomena proper, but polymer chain transfer processes making the active site available to initiate a new polymer chain. Such processes may be associated with [151]:

(i) a chain transfer process with the monomer:

$$[Cat]-PH + M \rightarrow [Cat]-MH + P;$$

(ii) a transfer process (alkyl exchange), with the organometallic compound present in solution:

$$[Cat]-PH + AlR_3 \rightarrow [Cat]-R + AlR_2PH$$

Natta, Pasquon and Giachetti have observed a decrease in the chain transfer activity in the sequence [169]

$$ZnR_2 > AlR_3 > AlR_2Cl$$

There are some doubts, however, as to these substances having to be considered as mere chain transfers since, for instance, at ZnR_2 concentrations above a given value, significant reductions in polymerization rate take place. A parallelism seems in particular to exist between the reducing power of the organometallic compounds towards the transition metal halogenide and their capacity of lowering the molecular weight of the polymer [306, 307]. There is evidence at any rate that many polymer chains are bound to aluminium at the end of polymerization [170].

(iii) a spontaneous termination process by β-hydride extraction

$$[Cat]-CH_2-CH(CH_3)-P \rightarrow [Cat]-H + CH_2=C(CH_3)-P$$

(iv) a chain transfer process by H_2

$$[Cat]-P \xrightarrow{H_2} [Cat]-H + PH$$

when operating in the presence of H_2 as molecular weight regulator.

When employing highly purified catalytic systems, solvents and monomers, the molecular weight of the resulting polymer generally attains high constant values just at the beginning of the polymerization, unless operating at sufficiently low temperatures with low reactive monomers (*e.g.* 1-butene) [321].

Reaction rates and molecolar weights are deeply affected by a variety of impurities (water, oxygen, polar compounds, $TiCl_4$ in $TiCl_3$, Al hydride in aluminium alkyl, etc.).

The foregoing considerations clearly evidence the difficulties that are encountered not only when endeavouring to draw conclusions conveying a sufficiently general validity, but also when endeavouring to obtain sufficiently reliable experimental data. Specialized literature ought to be consulted for a better knowledge of the matter [22, 188, 295, 309, 323, 325].

2. *Concentration and Location of Catalytic Centres in Heterogeneous Systems — Propagation Rate and Average Lifetime of Poly-α-olefins and Polyethylene Chains*

Various studies have been conducted in order to evaluate the concentration of active sites and their location, in particular, with reference to the catalytic systems based on crystalline components such as violet $TiCl_3$ and VCl_3. Such studies are facilitated by the fact that the first polymer chains forming on active sites preserve a „trace" of their origin as well as by the fact that the variation of their molecular weight with polymerization time depends on the number of active sites. On the other hand, a correct determination of the active sites is made complex by the nature of the phenomena occurring during the polymerization.

Two parameters that are strictly bound to the concentration of active sites are the polymerization rate and the growing time (average lifetime) of the polymer chains. The various methods used for enquiries on the subject have been outlined by Tait [326], Schnecko *et al.* [327] and by Yermakov [328], and are summarized hereunder with some observations.

(i) Variation of the molecular weight as a function of polymerization time, used by Natta, Pasquon *et al.* as well as by other authors [75, 151, 329, 331]. The method supplies average values affected by possible change in the number of active centres during the initial phase of the polymerization process. Only by operating at sufficiently low temperatures and at low monomer conversions, significant variations of the molecular weight with the polymerization time are obtained.

(ii) Determination of the number of labelled alkyl groups bound to the polymer chains of polymers obtained with catalysts prepared in the presence of ^{14}C-labelled Al alkyls, proposed by Natta, Pasquon *et al.* [151]. If the catalyst is prepared by contacting the transition metal halide (*e.g.* $TiCl_3$) with the radioactive organometallic compound (*e.g.* $Al(C_2H_5)_3$) and then the excess of the Al alkyl is removed and substituted by a non-radioactive organometallic compound, there may be uncertainties regarding the validity of the experimental data thus obtained. Conversely, if the polymers are obtained in the presence of excess radioactive organometallic compound during the polymerization process, chain transfer with the metal alkyl must be taken into account. Further uncertainties may derive from the formation, by decomposition of the Cat-$^{14}C_2H_5$ bonds, of labelled ethylene which is then polymerized [332].

A similar technique has been used by Ayrey and Mazza [333].

(iii) Method proposed by Chirkov [334], based on the assumption of a particular kinetic model and therefore affected by the validity of the model.

(iv) Method proposed by Kissin *et al.* [335], based on the inhibiting effect of methanol. Since it is most unlikely for methanol to react with active sites only, the method predictably supplies excess values.

(v) Method proposed by Burns, Caunt and Tait [326, 336] based on the inhibition of the active centers by allene; despite the complexity, the method seems to supply interesting results.

(vi) A method based on allene adsorption isotherm on $TiCl_3$ preactivated with Al alkyls [337].

(vii) Various methods based on the reaction between the metal-carbon bonds present in the polymerization system and a labelled quenching agent; such methods may be subdivided into two groups:
(a) non-selective methods
(b) selective methods.

The former, (a), utilize quenching agents ($^{131}I_2$: Chien *et al.* [338–340]; tritiated alcohols: Chien, Feldman, Perry *et al.* [269, 294, 338, 341, 348]; tritiated water: Ballard *et al.* [349] and deuterated methanol and water: Rodriguez and Yan Looy [350]) reacting with all the metal-carbon bonds (*e.g.* Al—C, Ti—C) present in the polymerization system. In this case, it is therefore necessary to take into account the presence of "dead" polymer chains containing metal-carbon bonds, resulting from the transfer processes with the organometallic compounds present in the catalytic system. Especially when employing tritiated compounds, utmost attention must be paid to polymer contamination due to isotopic exchange reactions catalyzed by the components of the catalytic system [351].

The latter methods, (b), utilize quenching agents (^{14}CO and $^{14}CO_2$: Zakharov *et al.*, and others [105, 188, 189, 326, 352, 353]) reacting selectively with the transition metal-carbon bonds (*e.g.* Ti—C) present only in the polymer chains still bound to active sites. These methods too, however, provide results susceptible to criticism [326, 353, 354].

In addition to the estimation of the concentration of active sites and of the average lifetime of the growing polymer chains, the methods described above have found application also in the calculation of propagation and chain transfer constants, activation energies, and frequency factors, thus allowing interesting comparisons between different types of catalytic systems. The most significant results obtained from these studies and mostly reported by Tait [326], Zakharov *et al.* [355] and by Pasquon *et al.* [306, 307] are

Table 16. Concentrations of active sites C* and propagation constants k_p relating to catalytic systems based on Ti halogenides used in propylene polymerization

Catalytic system	Temperature/K	C*/mmol (mol Ti)$^{-1}$	k_p/dm^3 mol^{-1} s^{-1}	Method[e] — Authors	Reference
α-$TiCl_3$—$AlEt_3$[a]	343	7–17		MW and ^{14}C — Natta, Pasquon	[151]
α-$TiCl_3$—$AlEt_2Cl$[a]	343	3–6		MW and ^{14}C — Natta, Pasquon	[151]
α-$TiCl_3$—Al(*i*-Bu)$_3$	353	5.4	2.9	K — Chirkov	[334]
α-$TiCl_3$—$AlEt_3$	343	5.2	—	I — Kissin and coll.	[335]
α-$TiCl_3$—$AlEt_2Cl$	333	4	—	I_2 — Caunt	[339]
α-$TiCl_3$—$AlEt_2Cl$	343	1	40	TQ — Cooper *et al.*	[347]
α-$TiCl_3$—$AlEt_2Cl$	343	36	1.1	^{14}C — Chien	[356]
α-$TiCl_3$—$AlEt_2Cl$	343	0.03	71	^{14}CO — Zakharov *et al.*	[355]
δ-$TiCl_3$—$AlEt_2Cl$	343	28	124	TQ — Kohn *et al.*	[346]
δ-$TiCl_3$·0.3$AlCl_3$—$AlEt_3$	343	5.8	100	^{14}CO — Zakharov *et al.*	[355]
δ-$TiCl_3$·0.3$AlCl_3$—$AlEt_2Cl$	343	1.7	90	^{14}CO — Zakharov *et al.*	[355]
δ-$TiCl_3$·0.3$AlCl_3$—$AlEt_2Cl$	333	12	13	^{14}CO — Varzelhan *et al.*	[354]
δ-$TiCl_3$ · 0.3 $AlCl_3$—$AlEt_2Cl$		13.7–20	5.9–8.2	^{14}CO and allene — Tait *et al.*	[357]
δ-$TiCl_3$ · 0.3 $AlCl_3$—Al(*i*-Bu)$_3$	343	8	90	^{14}CO — Zakharov *et al.*	[355]
$TiCl_3$ Solvay—$AlEt_2Cl$	333	7.9–27.3	7.6–17.2	^{14}CO and allene — Tait *et al.*	[357]
$TiCl_4$/$MgCl_2$(P)—$AlEt_3$[b,c]	343	5.2	740	^{14}CO — Zakharov *et al.*	[355]
	343	42	800	^{14}CO — Zakharov *et al.*	[355]
$TiCl_4$EB/$MgCl_2$(P)—$AlEt_3$[b,c]	343	2.7	870	^{14}CO — Zakharov *et al.*	[355]
		6.8	1250	^{14}CO — Zakharov *et al.*	[355]
$MgCl_2$/$TiCl_4$/EB/$AlEt_3$[c]	338	80	500	^{14}CO — Giannini	[105]
$MgCl_2$/$TiCl_4$/EB/$AlEt_3$[c]	318	25	970	MW — Keii *et al.*	[197]

[a] Milled $TiCl_3$
[b] $MgCl_2$ obtained from magnesium alkyl halogenides; (P) organic polymer residue present in $MgCl_2$
[c] Data relating to polymer fractions insoluble in *n*-heptane
[d] EB = ethylbenzoate
[e] MW = molecular weight variation method; K = kinetic method; I = inhibitor method; I_2 = non-radioactive iodine method; ^{14}C = radioactive alkyl method; TQ = tritiated quenching agent method; ^{14}CO = radioactive carbon monoxide method

Table 17. Concentrations of active sites C* and propagation constants k_p relating to catalytic systems based on VCl_3 used in propylene polymerization

Organometallic compound	Temperature/K	C*/mmol $(mol\ Ti)^{-1}$	$k_p/dm^3\ mol^{-1}\ s^{-1}$	Method[a]-Authors	Reference
$Al(i\text{-}C_4H_9)_3$	333	7.9	23.3	K – Novokshonova *et al.*	[358]
$Al(i\text{-}C_4H_9)_3$	343	4.7	46.5	K – Chirkov	[334]
$Al(i\text{-}C_4H_9)_3$	343	0.2	370	^{14}CO-Zakharov *et al.*	[355]

[a] K = kinetic method; ^{14}CO = radioactive carbon monoxide method

shown in Tables 16, 17 and 18. As may be inferred from the tables, there are considerable discrepancies between the results obtained by the various authors. Even the data obtained by the same author with different catalytic systems and monomers do not always appear to be exempt from possible criticism. It is therefore impossible to draw quantitative deductions.

Table 18. Concentrations of active sites C* and propagation constants k_p relating to catalytic systems based on $TiCl_3$ used in ethylene polymerization

Catalytic system	Temperature/K	C*/mmol $(mol\ Ti)^{-1}$	$k_p/dm^3\ mol^{-1}\ s^{-1}$	Method[a]-Authors	Reference
$\gamma\text{-}TiCl_3\text{—}AlEt_2Cl$	313	15	76.5	MW – Grieveson	[359]
$\alpha\text{-}TiCl_3\text{—}Al(i\text{-}Bu)_3$	343	5.4	226	K – Chirkov	[334]
$\delta\text{-}TiCl_3 \cdot 0.3\ AlCl_3\text{—}AlEt_2Cl$	333	6.9	78.1	TQ – Schnecko *et al.*	[327]
$\delta\text{-}TiCl_3 \cdot 0.3\ AlCl_3\text{—}AlEt_3$	353	0.52	12000	^{14}CO – Zakharov *et al.*	[355]
$TiCl_4/MgCl_2(P)\text{—}AlEt_3$	353	12–44	10000	^{14}CO – Zakharov *et al.*	[355]
			17000	^{14}CO – Zakharov *et al.*	[355]
$TiCl_4/MgO\text{—}AlEt_3$	343	700	2380	TQ – Chirkov *et al.*	[101], [102]

[a] MW = molecular weight variation method; K = kinetic method; TQ = tritiated quenching agent method; ^{14}CO = radioactive carbon monoxide method

Nevertheless it appears evident that an increase in the number of active sites takes place with catalytic systems based on Ti halogenides, in the sequence:

$$\alpha\text{-}TiCl_3 < \delta\text{-}TiCl_3 < MgCl_2\text{—}TiCl_4 \quad \text{supported catalysts.}$$

From the data shown by Figures 17–20 concerning l-butene polymerization and from other data, it seems it may be concluded that the polymer chain growing rate does not depend on the nature of the organometallic component. The same propagation rate is noticed in systems based on violet $TiCl_3$ (α, γ, and δ containing or not containing $AlCl_3$). Higher rates are reported for systems based on VCl_3 and lower for those based on β-$TiCl_3$. Also Burfield and Tait [343] have noticed that in 4-methyl-1-pentene polymerization the polymer chain propagation rate is independent from the type of aluminium alkyl employed. The same conclusions are reached by Youngman and Boor [342], Bier *et al.* [269], Chumaevski *et al.* [360], and Chien and Hsieh [189].

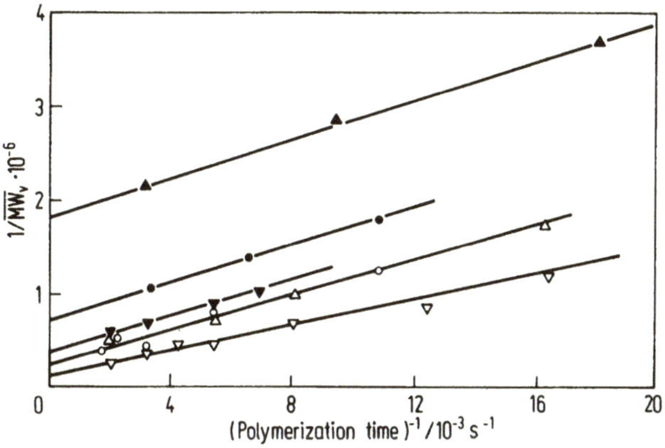

Figure 17. Reciprocal of the viscosity average molecular weight *vs.* the reciprocal of polymerization time for isotactic polybutenes (pentanic residues) obtained at 273 K by different catalyst systems based on violet $TiCl_3$. C_4H_8:333 g l ', toluene. ▲, δ-$TiCl_3 \cdot 0.3\ AlCl_3$—$BeEt_2$; ●, δ-$TiCl_3$—$BeEt_2$; ▼, δ-$TiCl_3 \cdot 0.3\ AlCl_3$—$AlEt_2Cl$; △, δ-$TiCl_3 \cdot 0.3\ AlCl_3$—$AlEt_2I$; ○, δ-$TiCl_3$—$AlEt_2Cl$; ▽, δ-$TiCl_3 \cdot 0.3\ AlCl_3$—Ali-Bu_2Cl. (Reproduced with permission from ref. [165])

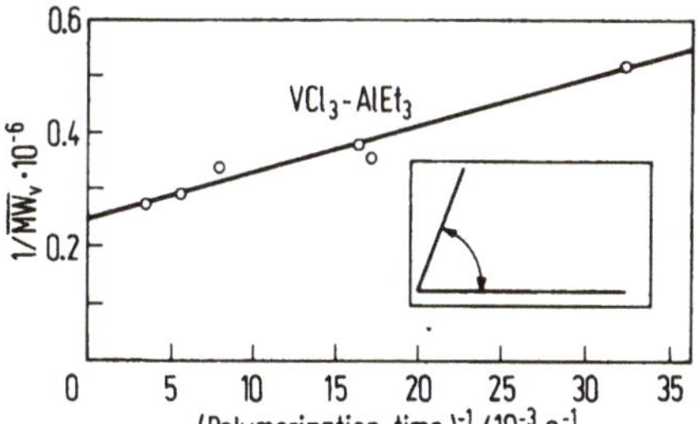

Figure 18. Reciprocal of the viscosity average molecular weight *vs.* the reciprocal of the polymerization time for isotactic polybutenes (pentanic residues) obtained at 273 K by VCl_3—$Al(C_2H_5)_3$. Comparison with the systems based on violet $TiCl_3$. C_4H_8:333 g l^{-1}, toluene. (Reproduced with permission from ref. [165]). Insert: average slope for systems based on violet $TiCl_3$

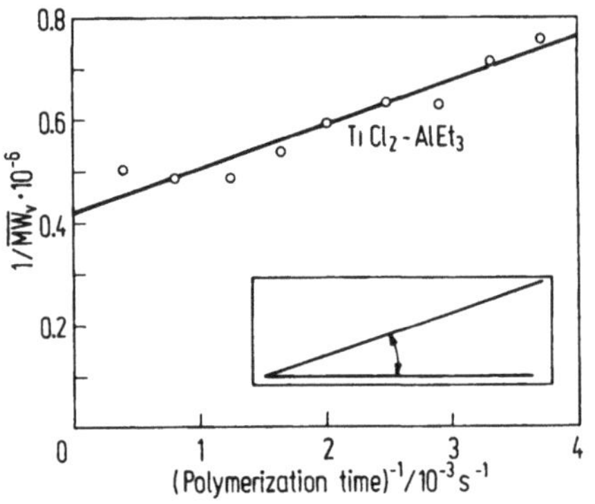

Figure 19. Reciprocal of the viscosity average molecular weight *vs.* the reciprocal of the polymerization time of isotactic polybutenes obtained at 273 K by $TiCl_2—Al(C_2H_5)_3$. Comparison with the systems based on violet $TiCl_3$. C_4H_8: 333 g l^{-1}, toluene. Insert: average slope for systems based on violet $TiCl_3$. (Reproduced with permission from ref. [165])

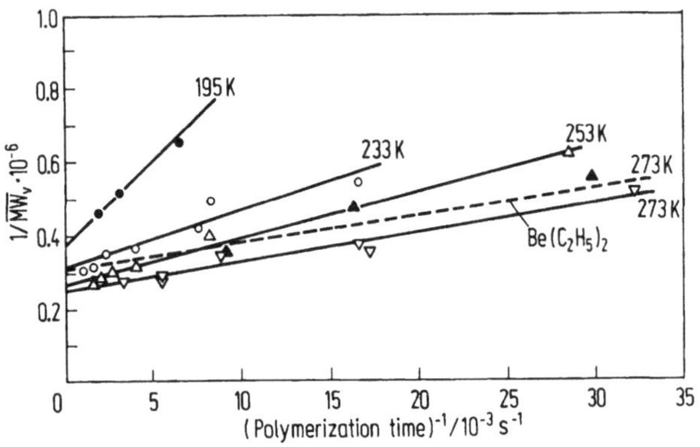

Figure 20. Behaviour of the catalyst system $VCl_3—Al(C_2H_5)_3$ in 1-butene polymerization at various temperatures and comparison with $VCl_3—Be(C_2H_5)_2$ at 273 K. (Reproduced with permission from ref. [165]). ▲, $VCl_3—Be(C_2H_5)_2$ at 273 K; △, $VCl_3—Al(C_2H_5)_3$ at 253 K; ▽, $VCl_3—Al(C_2H_5)_3$ at 273 K; ○, $VCl_3—Al(C_2H_5)_3$ at 233 K; ●, $VCl_3—Al(C_2H_5)_3$ at 195 K

On the basis of the variation of the molecular weight of isotactic polybutene with the polymerization time, Natta *et al.* [165] estimated the activation energies of the propagation reaction, obtaining the following results:

Catalyst based on violet $TiCl_3$, E_p = 18.9 kJ mol^{-1}
Catalyst based on VCl_3, E_p = 9.6 kJ mol^{-1}

The above data agree with those shown in Table 19 from which the propagation rate relating to VCl_3-based systems turns out to be higher than that relating to $TiCl_3$-based systems. Furthermore, the results obtained by Zakharov *et al.* [355] give the following values for the activation energies of the propagation reaction:

Ethylene polymerization with:

Catalyst based on $TiCl_3$, E_p = 12.6 kJ mol^{-1}
Catalyst based on VCl_3, E_p = 2.5 kJ mol^{-1}

Propylene polymerization with:

Catalyst based on $TiCl_3$, $E_p = 23.0\ kJ\ mol^{-1}$

Table 19. Average propagation rate of isotactic polybutene-1 chains (catalytic systems based on different organometallic compounds)[a, b]

Transition metal halogenide	Propagation rate: monomer units s^{-1}
VCl_3	3100
δ-$TiCl_3$	275
$TiCl_2$	280
β-$TiCl_3$	54

[a] 273 K; $[C_4]$ = 33 g per 100 cm^3 toluene; data by viscometry

[b] Reference [165]

Table 20. Comparison between reducing power and catalytic activity of some organometallic compounds in VCl_3 based catalytic systems

Scale of the reducing power towards VCl_3 at 363 K
$Al(C_2H_5)_2I$ ($\cong 0$) < $Al(C_2H_5)_2Cl$ < $Ga(C_2H_5)_3$ < $Al(C_2H_5)_3$ < $Be(C_2H_5)_2$ < $Zn(C_2H_5)_2$ < $Cd(C_2H_5)_2$ < LiC_4H_9

Scale of catalytic activity in propylene polymerization with VCl_3 at 363 K
$Zn(C_2H_5)_2$ < LiC_4H_9 < $Cd(C_2H_5)_2$ < $Al(C_2H_5)_2Cl$ < $Al(C_2H_5)_2I$ < $Ga(C_2H_5)_3$ < $Be(C_2H_5)_2$ < $Al(C_2H_5)_3$

Considering that the activation energies relating to the overall polymerization process have higher values, it must be concluded that the concentration of active sites increases with temperature.

There are, however, discrepancies between the conclusions reached by the different authors as to the following points:

(i) the dependence of the number of active sites on the nature of the monomer;

(ii) the constancy or not of the propagation rate constant (k_p) for a given monomer by using a conventional or a supported catalyst;

(iii) the presence or not on the surface of the same catalyst of stereospecific active centers of different types and with different k_p values;

(iv) the higher growing rate of isotactic polypropylene chains as compared with atactic chains.

As regards the average lifetime of growing polypropylene chains at 343 K with violet $TiCl_3$-based systems, Natta and Pasquon [151] have calculated values ranging between 360 and 600s, whereas Zakharov *et al.* [355] have found 160s under comparable conditions, despite the fact that the two groups of authors have found quite different concentrations of active centres with the same types of catalytic systems. Values of *ca.* 5 s have been found by Keii *et al.* [197] for catalytic systems supported on $MgCl_2$ and operating at 318 K. For ethylene polymerization with violet $TiCl_3$-based systems, Zakharov *et al.* have calculated an average chain lifetime of 25 s. Values of the same order of magnitude have been found by Natta *et al.* in measuring the variation of the molecular weight of poly-1-butene with the polymerization time, at 273 K, in the presence of different catalytic systems.

Other indications that may be inferred from the data reported in this section concern the location of the active sites on the surface of the $TiCl_3$ crystals, and the kinetic mechanism involved in the polymer chain propagation step. Many authors [151, 302, 350, 355, 361–364] believe the active sites to be located on the side faces of violet $TiCl_3$ crystallites, on the edges between side faces and basal planes (001), on the corners and on the defects of basal planes (001), even though some [335, 365] believe basal planes too may take part in the catalytic process.

The experimental evidences of Rodriguez *et al.* [350], agree with the first assumption; these authors showed by electron microscopy the growth of the polymer along the spiral dislocations and on surface defects of the $TiCl_3$ crystal faces. Moreover, the same authors studied by I.R. spectroscopy the reaction between $TiCl_3$ and AlR_3 and concluded that AlR_3 interacts with crystal faces (001) without however originating any active site [301–304]. According to Zakharov *et al.* [355], 10 to 15% of the Ti atoms present on the surface of $TiCl_3$ crystals takes part in the formation of active complexes which are believed to be located on side faces.

3. Addition of Monomeric Units and Chemical Properties of Catalytic Centres

The obtainment of radioactive polymers by using catalytic systems prepared in the presence of ^{14}C-labelled organometallic compounds demonstrates that monomer units insert into a metal-carbon bond of the catalyst. The presence of radioactivity also in polymers obtained after quenching with ^{14}CO or $^{14}CO_2$ suggests that the metal-carbon reactive bond of the catalytic complex is more correctly a transition metal-carbon bond since the mentioned reagents do not react with the Al—C bonds.

These conclusions agree with the results of IR studies on the nature of the end groups of the polymer chains [366] as well as with the results obtained by Zambelli *et al.* [367] by ^{13}C-NMR spectroscopic investigations on the first insertion step in propylene polymerization in the presence of $TiCl_3$ and ^{13}C enriched Al alkyls. Such results confirm also that the addition of monomer units takes place to a metal-carbon σ-bond. The conclusion regarding the growth of the polymer chain on a transition metal-carbon bond is supported by the existence of monometallic isospecific catalytic systems (*c.f.* section 3.A.3) containing the transition metal only (*e.g.* Ti).

This experimental evidence has led some authors also to suggest that the catalytic centers present in bicomponent systems containing an organometallic compound (*e.g.* AlR_3, BeR_2) are monometallic. The assumption is also supported by the fact that practically identical values have been found for the chain propagation constant when using, in the presence of the same metal salt, different Al organometallic compounds and also in the absence of Al alkyls. The hypothesis of a monometallic active centre, on the other hand, is not shared by many authors [350, 368, 369] who justify the activity and/or the stereospecificity of the catalyst only by admitting the presence of the aluminum alkyl as ligand of the transition metal bearing the growing chain. In favour of this assumption, the following experimental data may be mentioned:

(i) some organometallic compounds of transition metals such as CH_3TiCl_3 and $Cp_2Ti(CH_3)_2$ are thoroughly inactive, whereas they exhibit a high catalytic activity in the presence of Al alkyls;

(ii) catalytic activity and stereospecificity are considerably affected by the nature of the organometallic compound. Furthermore the scale of the alkylating and reducing power of different organometallic compounds towards the same transition metal halogenide does not correspond to the scale of activity of catalytic systems prepared with such compounds (see table 20)

(iii) the addition of $AlEt_3$ to a scarcely active system prepared from α-$TiCl_3$ and a metal alkyl having a high alkylating and reducing power such as $ZnEt_2$ immediately makes the system much more active in propylene isospecific polymerization (*c.f.* Figure 21) [305]. Similar results have been obtained by Keii *et al.* [368] by replacing the organometallic compound during polymerization.

In catalytic systems prepared starting from a transition metal halogenide and an Al organometallic compound, isospecific catalytic centres are therefore likely to be of the bimetallic type.

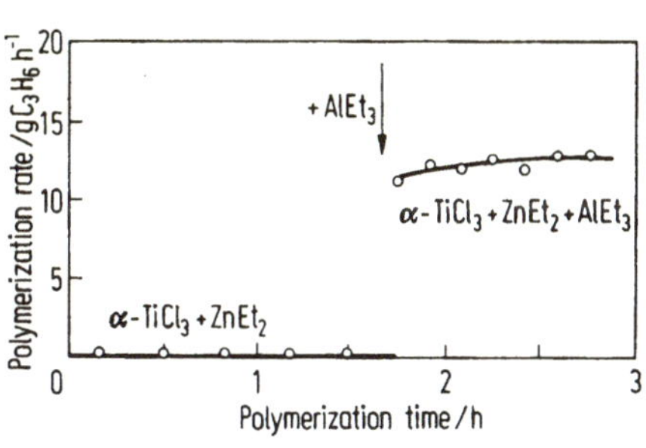

Figure 21. Rate of propylene polymerization in the presence of a catalyst system α-$TiCl_3$—$Zn(C_2H_5)_2$ before and after addition of $Al(C_2H_5)_3$. Reagents used: *n*-heptane: 250 cm^3; α-$TiCl_3$: 0.30 g; $Zn(C_2H_5)_2$: 0.70 cm^3; $Al(C_2H_5)_3$; 0.45 cm^3; $p_{C_3H_6}$: 127 kPa, T = 293 K. (Reproduced with permission from ref. [307])

Insofar as it concerns the kinetic mechanism of monomeric unit addition, most authors believe it to take place in two steps:

(i) monomer coordination on the transition metal with formation of a π complex;

(ii) insertion of the coordinated monomer into a transition metal-carbon bond present at the catalytic centre.

With reference to the two elementary steps described above, many authors believe the kinetically determinant one is monomer complexation to the transition metal. In such a case, both the overall polymerization rate and the growing rate of the single polymer chain must linearly depend on the monomer concentration. Such dependence of the overall polymerization rate is well established for a variety of catalytic systems, as well as for the growing rate of the single polymer chains observed by Natta *et al.* in, for instance, 1-butene isospecific polymerization (*c.f.* Figure 22) [165].

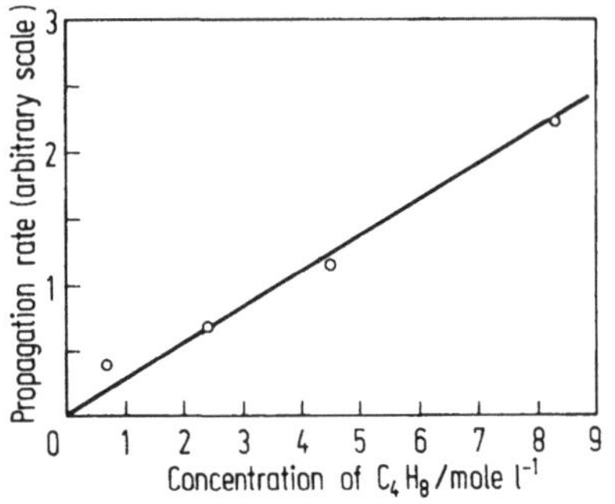

Figure 22. Rate of propagation of the isotactic polymer chains of polybutene *vs.* concentration of the monomer. $TiCl_3ARA—Al(C_2H_5)_2Cl$. T = 273 K. (Reproduced with permission from ref. [165])

The hypothesis that the coordination of the olefin to the transition metal is the rate determining step agrees with the values of the frequency factor calculated by Zakharov *et al.* [355] for the various steps on the basis of the experimental data for k_p and E_p.

A further result, confirming the same hypothesis, has been obtained by Danusso and Sianesi [371] who have observed that the polymerization rate of differently *para*-substituted vinyl aromatic monomers, in the presence of isospecific catalytic systems for propylene polymerization, increases concomitantly with the electronic density on the double bond. Taking into account the polarity of the catalytic centre metal-carbon bond ($Ti^{\delta+}—C^{\delta-}$) and the type of monomer insertion into this bond (*c.f.* section 3.A.5), the result described above proves that, at least in this case, the rate determining step of the addition process is the interaction of the olefin at a positive site. The conclusion that the rate determining step is the monomer complexation to the transition metal agrees with the following experimental data:

(i) the polymer chain growth rate is independent of the nature of the organometallic compound and is the same for the various systems prepared for instance from violet $TiCl_3$ (*c.f.* section 3.A.2)

(ii) the reactivity ratios in copolymerization reactions depend on the nature of the transition metal and not on that of the organometallic compound [370].

As pointed out by Pino and Mülhaupt [209], it must be recognized that experimental evidences for the existence of olefin-transition metal π complexes in the catalytic systems are very poor. This may, however, be justified, taking into account the extremely low number of active centres and assuming the formation of such complexes to be the kinetically determinant step, in which case, the concentration of the complexes on the catalyst surface will necessarily be very low.

4. *Stereoselectivity and Stereoelectivity — Polymerization with Optically Active Monomers and Catalytic Systems*

The polymerization of asymmetric α-olefins by means of Ziegler-Natta catalysts has been thoroughly studied by Pino *et al.* [205, 372–377]. The most significant results obtained in the field may be briefly summarized as follows.

The isospecific polymerization of racemic olefins with violet $TiCl_3$-based catalysts is stereoselective; namely, each macromolecule contains a prevalence of one of the two optical antipodes. This conclusion was reached first on the basis of the observed structural identity of the isotactic polymers obtained respectively from (R) (S) and (S)-4-methyl-1-hexene. This fact suggests that the polymer resulting from a racemic monomer is not a random copolymer of the two enantiomers, but it is a mixture of macromolecules each of which consists of units mostly deriving from the same enantiomer. The polymer obtained from the racemic mixture may be fractionated by chromatography into two optically active fractions having opposite signs of rotation. Stereoselectivity has been observed to be strongly affected by the position of the asymmetric carbon atom with respect to the double bond. Very high stereoselectivities have been noticed with 4-methyl-1-hexene and 3,7-dimethyl-1-octene racemic mixtures. In these monomers, the asymmetric carbon atom is in β and α positions respectively with regard to the double bond. With 5-methyl-1-heptene, where the chiral centre is in the γ position, no stereoselectivity is observed [378].

Polymerization may also be stereoselective even though scarcely stereospecific as it has been demonstrated through examination of the acetone soluble fraction (which is amorphous, being scarcely stereoregular) of poly-(R)(S)-3,7-dimethyl-1-octene. On the basis of these results, Pino *et al.* have concluded that stereoselectivity is to be ascribed to chirality of the catalytic centres rather than to asymmetric induction exerted by the growing chain. The conclusion is also confirmed by the results obtained in the copolymerization of racemic α-olefins with ethylene [379], demonstrating copolymerization still to be stereoselective although the last asymmetric carbon atom in the growing polymer chain is at least in a δ position with respect to the metal atom bearing the growing chain.

Experiments conducted with model compounds have shown that an optically active olefin coordinates with a metal preferably in one of the two possible ways, as a result of the asymmetric induction exerted by the asymmetric carbon atom present in the ligand of the metal. A conclusion may thus be drawn: if it is stereospecific, a polymerization process must also be stereoselective [380].

The isospecific polymerization of α-olefins is stereoelective in the sense that asymmetric catalysts (prepared from optically active organometallic compounds or containing optically active Lewis bases) give optically active polymers starting from racemic monomers, thus suggesting that one enantiomer is preferentially polymerized in respect of the other. These results have been obtained in polymerizations carried out with (R)(S)-3,7-dimethyl-

1-octene, (R)(S)-4-methyl-1-hexene and (R)(S)-3-methyl-1-octene in the presence of catalysts prepared starting from $TiCl_4$ or violet $TiCl_3$ and an organometallic compound of Al, Be, Zn, etc., containing asymmetric alkyl groups (S)$-CH_2CH(CH_3)C_2H_5$; (S)$(CH_2)_2CH(CH_3)C_2H_5$; (S)$-CH_2CH(CH_3)C_6H_5$. Only when the asymmetric carbon atom of the metal alkyl is in a β position with respect to the metal does stereoelection occur [380, 382].

The stereoelectivity of systems prepared with Li or Be alkyls is negligible [382]. Only racemic olefins with the asymmetric carbon atom in a α or β position with respect to the double bond have been stereoelectively polymerized, and the preferentially polymerized enantiomer has the same chirality of the optically active group present in the metal alkyl.

The stereoelectivity of asymmetric catalytic systems is rather modest: the optical purity of isotactic polymers obtained by the use of most stereospecific catalytic systems does not exceed 10.5%, and the value of the relative polymerization rate of one enantiomer with respect to the other does not seem to exceed 1.2. On the basis of these results, stereoelectivity is to be ascribed to a modification of the enantiomeric catalytic centre (one or more asymmetric alkyl groups are bound to the metal of the catalytic centre) rather than to a prevalence of centres having the same type of chirality. The presence of asymmetric alkyl groups brings about a preferential coordination of the enantiomer having the same chirality as the asymmetric alkyl.

Interesting results have been obtained by Pino *et al.* [376, 377] by polymerizing racemic 4-methyl-1-hexene and racemic 3,7-dimethyl-1-octene in the presence of supported catalysts based on $MgCl_2-TiCl_4-Al(C_4H_9)_3$ and (−)menthyl-anisate as an asymmetric Lewis base. These authors have obtained two polymeric fractions, one amorphous and the other crystalline, having optical activity of opposite sign, and have come to the conclusion that two classes of active centres are present in the catalytic system, one of which is constituted by centres having high Lewis acidity, low stereospecificity, and low stereoelectivity. The centres belonging to the other class are weaker Lewis acids having high stereospecificity and stereoelectivity. Stereoelectivity may derive from the complexation of a molecule of the optically active base to centres exhibiting two vacant coordination sites and/or from a selective poisoning of (1) or (d) chiral centres by the same base. The assumption is confirmed by Pino *et al.* [376, 383] who have observed that the quantity of amorphous polypropylene obtained with the same catalytic system decreases considerably as the Lewis base/AlR_3 ratio increases, whereas the quantity of isotactic polypropylene undergoes a slight decrement only. These results seem to suggest that, for the systems concerned, stereospecific centres are less acidic than aspecific ones, and are less affected by the presence of a Lewis base. The stereospecificity and the activity of such systems appear therefore to be influenced also by equilibrium conditions between Lewis acids and bases present in the systems both in a homogeneous phase and on the catalyst surface.

5. Regiospecificity, Stereochemistry and Steric Control of Monomeric Unit Insertion

The polymerization of *cis* and *trans*-1-deutero-propylenes has demonstrated that the opening of the double bond at the moment of its insertion into the metal—carbon bond of the catalytic complex is of the *cis* type Figure 23) [384, 386]. As a matter of fact, *cis*-1-deutero-propylene gives an

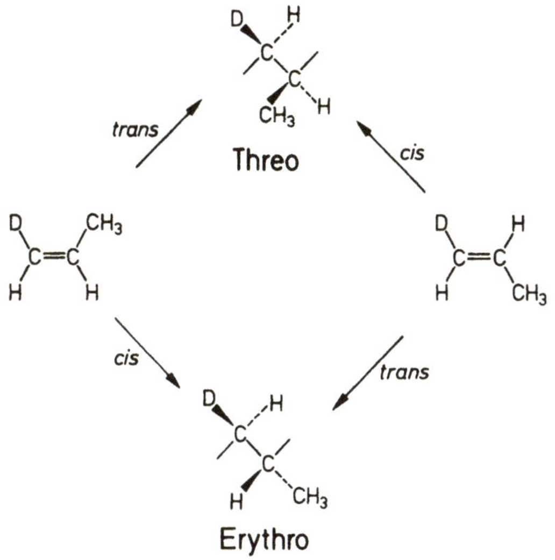

Figure 23. Stereochemical structures from *cis* and *trans*-addition to the double bond of *cis*-(1d_1) and *trans*-(1d_1)-propylene. (Reproduced with permission from ref. [384])

erythro-diisotactic polymer, whereas *trans*-1-deutero-propylene supplies a *threo*-diisotactic polymer. Further evidence of this type of addition have come from the analysis of the structure of alternating *cis*-2-butene/ethylene copolymers [253, 387] and of NMR spectra of deutero-polypropylenes [214].

As pointed out by Pino and Mülhaupt [209], the *cis*-addition of an α-olefin to a catalytic centre whereon a monomer unit has already become inserted may originate eight different situations according to the linkage type, namely to the regioselectivity of the insertion reaction as well as to the selectivity of the α-olefin prochiral face. Thorough isotacticity is obtained with a similar type of insertion, *e.g.* primary:

$$\text{(a)}\ \text{(Cat)—P} + \underset{\displaystyle \text{R}}{\text{CH}_2\text{=}\overset{}{\text{CH}}} \rightarrow \text{(Cat)—CH}_2\text{—}\underset{\displaystyle \text{R}}{\text{CH}}\text{—P}$$

or secondary:

$$\text{(b)}\ \text{(Cat)—P} + \underset{\displaystyle \text{R}}{\text{CH}}\text{=CH}_2 \rightarrow \text{(Cat)—}\underset{\displaystyle \text{R}}{\text{CH}}\text{—CH}_2\text{—P}$$

and with a constant attack of the same prochiral face, *e.g.*:

(c) 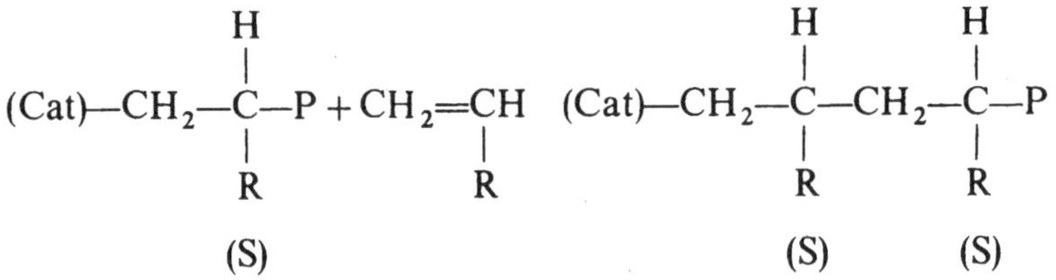

The presence of terminal groups $CH_2{=}C(CH_3){-}P$ in the chains of both isotactic and amorphous polypropylenes and the observed crystallinity of isotactic polymers obtained with catalytic systems based on titanium halogenides and aluminium alkyls, have demonstrated since the initial researches of Natta *et al.* [388] that, in these polymers the enchaiment of the monomer units was regular (of the head-to-tail type) with primary (or 1—2) insertion. The matter was further investigated by Takegami and Suzuki [389–391] through chemical tests as well as by Zambelli *et al.* through ^{13}C-NMR analysis on isotactic polypropylenes and propylene copolymers obtained with isospecific catalysts [392–395]. The results obtained by these authors show that, in the presence of isospecific catalysts, the addition of monomer units is essentially of the primary type and that some polypropylene macromolecules at least are thoroughly regioregular [392]. ^{13}C-NMR analysis of ethylene-propylene copolymers obtained with these catalytic systems shows the absence of sequences of 4 methylene groups [393] which demonstrates that regiospecificity is not conditioned by the last monomeric unit (ethylene or propylene) of the growing chain. A further evidence of this type of insertion derives from the presence of isobutyl terminal groups with a ^{13}C enriched methyl group in polypropylenes obtained with catalytic systems containing active [Cat]—$^{13}CH_3$ bonds [367].

A number of experimental data finally demonstrate that the constancy in attack of a same prochiral face of α-olefin is due to the chirality of the catalytic centres present in isospecific systems and not to that of the last monomeric unit of the growing chain. In other words, in isospecific polymerization, steric control is due to the catalytic centre chirality. In fact ^{13}C-NMR investigations conducted by Zambelli *et al.* on ethylene-propylene copolymers obtained with isospecific systems prove that the addition of a propylene unit after an ethylene unit continues to be isospecific although the latter unit is achiral [395, 396]. The examination of fractions of isotactic polypropylene (insoluble in *n*-heptane) [395] suggests the same conclusion. As a matter of fact, in these highly isotactic fractions the main structural defects are syndiotactic dyads isolated between blocks of isotactic dyads. This indicates that the addition of monomeric units after an inversion continues to be isospecific and fails therefore to be conditioned by the chirality of the last monomeric unit of the growing chain whereas it is conditioned by the chirality of the catalytic centre. The stereoselectivity verified by Pino *et al.* is also in agreement with such assumption.

From a ^{13}C-NMR spectroscopic study performed on propylene and on 1-butene polymers prepared with the following catalytic systems:

violet $TiCl_3-Al(^{13}CH_3)_3$
violet $TiCl_3-Al(^{13}CH_2CH_3)_3$
violet $TiCl_3-Al(^{13}CH_2CH_3)_2I$

Zambelli *et al.* [394] have concluded that the steric control of the first monomeric unit insertion in the presence of isospecific systems, besides catalytic center chirality, requires also the presence of an alkyl group, with a steric hindrance higher than of the CH_3 group, as ligand of the transition metal. These experimental data agree with the conclusions come to by Corradini *et al.* [397] who, on the basis of detailed studies concerning non-bonded interactions on the catalytic sites located on the corners and at side faces (110) of α-$TiCl_3$ crystallites, have found that the stereospecifity of sites may derive from a fixed chiral orientation of the first C—C bond of the growing chain, due to interaction with the surface. It ought to be finally noted that the different types of enchaiment of the monomeric unit which may take place in Ziegler-Natta catalytic systems obviously depend on the relative values of the corresponding activation energies. Quantitative aspects of these phenomena have been gone into by Pino *et al.* [205, 209] as well as by Zambelli *et al.* [398].

6. *Models of Catalytic Centres*

Various models of catalytic centres and of monomeric unit addition mechanisms have been proposed to interpret the isospecific polymerization of α-olefins with Ziegler-Natta catalytic systems [60, 399]. One of the most complete presentations, made by Pino and Mülhaupt [209], is shown by Figure 24.

One of the first model types, proposed by Cossee and Arlmann [363, 400] and widely accepted, is based on monometallic catalytic centres with hexacoordinated transition metal (Figure 25).

Allegra [364] has proposed a monometallic model which however avoids the back-jump of the Ti—R bond required by the Cossee—Arlmann mechanism. According to this author, the Ti atoms close to the $TiCl_3$ surface have a vacant octahedral site, one chlorine ligand singly bonded and four chlorine ligands bridge bonded with neighboring Ti atoms. By reaction with the organometallic compound, the single bonded chlorine atom is substituted by an alkyl group with the consequent formation of a Ti—C bond. The olefin is complexed on the vacant site with the double bond parallel to an octahedral axis. Two orientations are therefore possible (*c.f.* Figure 26). Structure (a), which corresponds to that provided for by Cossee and Arlmann, is deemed to be improbable by Allegra for steric reasons. Of the four possible positions for the C—CH_3 bond with regard to structure (b), the outwards *trans* only is probable in not involving unacceptable interatomic distances. After monomeric unit insertion, the Ti—R bond enters the vacant

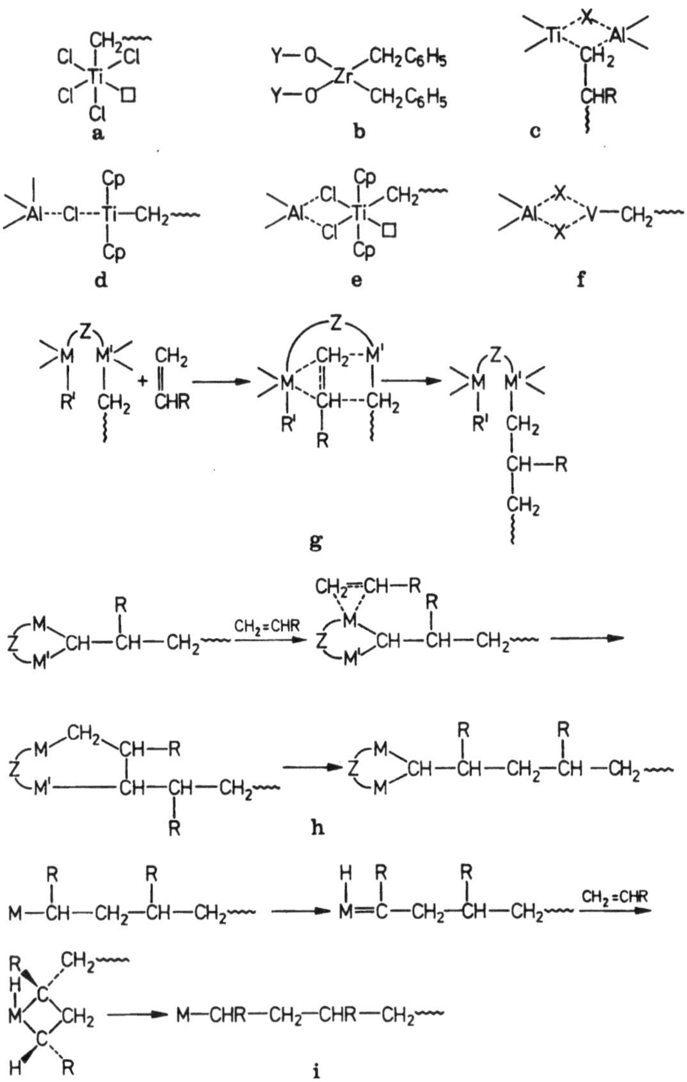

Figure 24. Models of catalytic centers and some proposed mechanism for the α-olefin polymerization. M = M′ or M ≠ M′, X=Hal, Y=Si or Al, Z=org. or inorg. bridging ligands. (Reproduced with permission from ref. [209])

site and another olefin molecule becomes coordinated on Ti free coordinating site with the same prochiral face as the preceding one.

The Cossee-Arlmann model has been analyzed by Corradini *et al.* [397, 401] on the basis of non-bonded interactions at the catalytic site. These authors conclude that lower steric repulsions appear to be implied for a coordination of the polymeric chain on the more hindered octahedral position at a titanium atom (at the (110) surfaces, or at edges, steps and reliefs

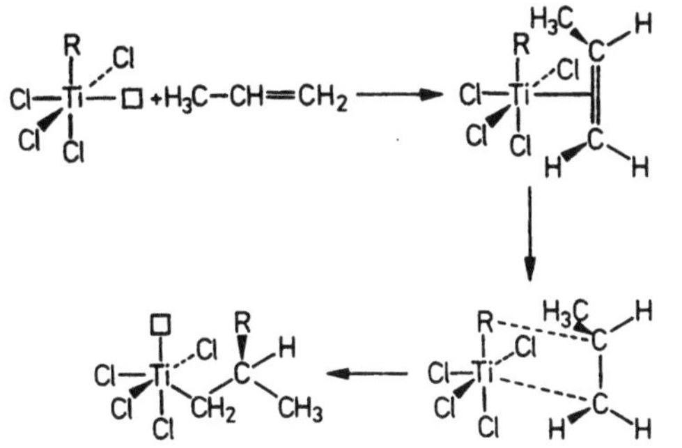

Figure 25. Model of monometallic active site polymerization of olefins proposed by Arlman and Cossee. (Reproduced with permission from ref. [363])

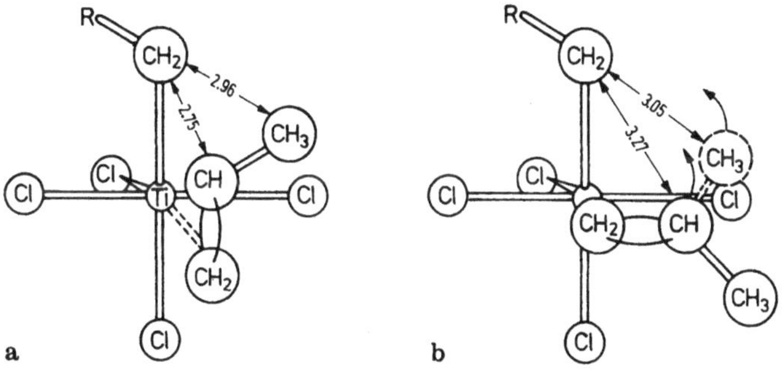

a b

Figure 26. Different modes of olefin π-complexation on an octahedral Ti atom, with the double bond parallel to a Ti—X bond. At left **a**, the mode of complexation assumed by Cossee is shown. At right **b** a *cis-* and *trans*-like mode of complexation are shown: the former corresponds to the methyl group orientation drawn with a dashed line, the latter with a continuous line. Distances are given in angstrom units. (Reproduced with permission from ref. [364])

on (111) surfaces of all the layered modifications of $TiCl_3$) and this, in turn, causes a chiral orientation of the first carbon-carbon bond of the chain: such orientation may be the main factor in determining the isospecificity.

A monoatomic model for the catalytic centre has also been proposed by Ivin *et al.* [402] who suggest the growing polymeric chain to be bound to the transition metal through a double bond (carbene complex) and the insertion reaction to occur through the formation of a metalcyclobutane intermediate (*c.f.* scheme i, Figure 24).

This mechanism has however been objected to by Zambelli *et al.* [367] on the basis of ^{13}C-NMR analysis of isotactic polypropylenes obtained in the presence of catalytic systems prepared starting from ^{13}C enriched organometallic compounds. It is to be observed that carbene complexes are suggested for the catalytic ring opening polymerization of cycloolefin (*c.f.* section 3.F.).

McKinney [403] on the basis of semiempiric calculations of molecular orbitals, suggests that under certain conditions ethylene and propylene polymerization may take place through the formation of a cyclopentane intermediate involving the active site transition metal and two molecules of monomer according to the sequence 1, 3, 4, 5 in Figure 27. The coplanar arrangement of the two monomer molecules and of Ti and the formation of the cyclopentane ring would be particularly favoured for Ti^{II}, whereas they would be scarcely probable for Ti^{IV}.

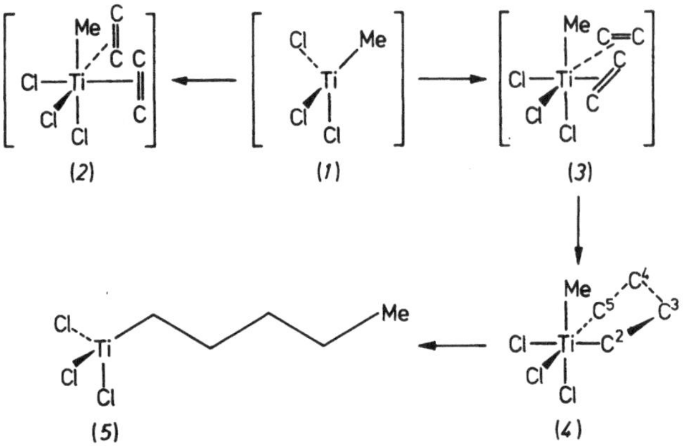

Figure 27. Scheme of the catalytic sequence for Ziegler-Natta polymerization (McKinney). (Reproduced with permission from ref. [403])

Kissin *et al.* [335, 365] propose a model (*c.f.* Figure 28) where the catalytic center contains a transition metal which has two vacant octahedral coordination sites located on faces (001) of violet $TiCl_3$, resulting from the reaction with the organometallic compound.

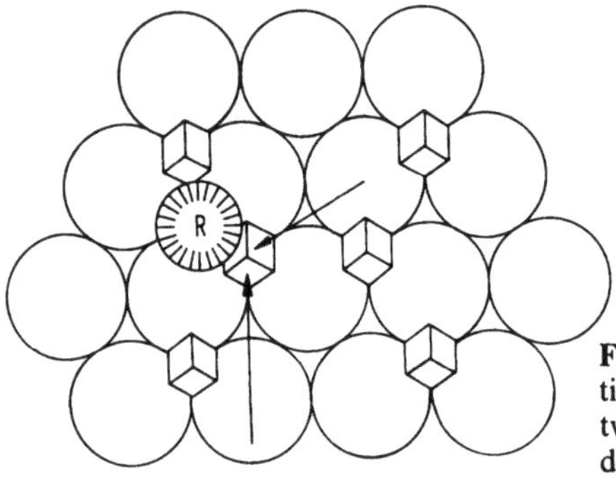

Figure 28. Scheme of the active site of polymerization with an alkyl group of small volume and with two vacancies. The arrows correspond to the directions of monomer approach (Kissin). (Reproduced with permission from ref. [365])

Bimetallic catalytic centres have been proposed by Patat and Sinn [404] who suggest the growing polymeric chain to be bridge linked to both metal atoms (*c.f.* scheme c, Figure 24). Pino *et al.* [372] propose a bimetallic model where the two metal atoms may also be of the same nature and where one of the metal atoms forms a π bond with the olefin whereas the other is bound to the growing chain (*c.f.* scheme g, Figure 24). Henrici-Olivé [405] proposes a model of the catalytic centre containing two halogen bridges between two metal atoms (see scheme e, Figure 24). Ballard [21] hypothesizes the presence of a chlorine atom bridge between two metals atoms (see scheme d, Figure 24). As stated, many experimental data are in closer agreement with the hypothesis of bimetallic catalytic centres, although isospecific monometallic catalysts exist (*c.f.* section 3.A.3). It must be observed, however, that compounds such as $TiCl_3CH_3$ are not catalytically active and become so only in the presence of $TiCl_3$ [159]. Therefore, in this case too, the catalytic centre might have a structure similar to that proposed for bimetallic centres. However, the contrasts between the various hypotheses concerning the structure of the active centres derive mainly from the desire on the part of the various authors to describe with only one model all the catalytic polymerization systems based on transition metals. Actually catalytic centres are likely to have a greater complexity than described by the various models proposed, and considerable research is likely still to be required fully to investigate their structure.

B. Syndiospecific Polymerization of Propylene

1. Foreword

The best catalytic systems for propylene syndiospecific polymerization are those studied by Natta, Zambelli and Pasquon. These systems are based on a vanadium compound and AlR_2Cl, in the presence or in the absence of a weak Lewis base (*e.g.* anisole). They are soluble in hydrocarbon solvents, and are employed at low temperature (*e.g.* 195 K) (*c.f.* section 2.B.5). With some heterogeneous systems based on a Ti compound, small percentages of syndiotactic polypropylene are also obtainable at higher temperatures (*e.g.* 343 K).

Only syndiospecific polymerization with systems of the former type is being considered hereunder.

2. Behaviour of Catalytic Systems — Overall Kinetics

The main characteristics of the relevant systems concerning the influence of the various operating parameters on the course of polymerization may be summarized as below:

(i) By operating with anisole at 195 K, the average molecular weight of the resulting polymer increases almost linearly as a function of time; that is, in the presence of anisole, chain transfer processes are negligible [217].

(ii) In the absence of anisole, the molecular weight of the resulting polymer grows quickly with time to reach then an almost constant value [213].

(iii) By operating in the presence of anisole, the polymerization rate tends to increase slightly with polymerization time to reach then a constant value, which suggests that the catalytic complexes form gradually to reach then a constant concentration [217].

(iv) In the absence of anisole, the reaction rate reaches a constant value quickly [213].

(v) The index of crystallinity due to syndiotactic polymer lowers with polymerization time, probably owing to a greater difficulty in polymer crystallyzation as the molecular weight increases [217].

(vi) The index of crystallinity of syndiotactic type for polymers having the same molecular weight is nearly the same by operating in the presence or in the absence of anisole. This suggests that the stereospecificity of catalytic complexes is not significantly affected by the presence of anisole [217].

(vii) The polymerization rate is considerably affected by the vanadium compound concentration provided reaction is conducted with Al/V and anisole/V ratios within certain ranges. In general, however, the polymerization rate as a function of vanadium compound concentration passes through a maximum [217].

(viii) The polymerization rate is conditioned both by the Al/V ratio and by the nature of the aluminum alkyl compound. This fact, along with those described under the two previous points, suggests that the formation and concentration of the catalytic complexes are affected by equilibrium conditions between the various species (vanadium and aluminum compounds and Lewis base) [217].

(ix) The polymerization rate versus monomer concentration passes through a maximum. This has been interpreted by assuming that polymerization involves two factors: monomer coordination to the catalytic complex, followed by insertion in the growing polymeric chain, and that the concentration of catalytic complexes depends on the monomer concentration [217].

(x) The distribution of the molecular weights of the polymer obtained at temperatures below 233 K is unimodal and quite restricted ($\overline{M}_w/\overline{M}_n \leq 2$), whereas, with polymer obtained at temperatures above 252 K, the distribution is bimodal with $\overline{M}_w/\overline{M}_n > 2$ [406]. This result has been interpreted by admitting the existence of homogeneous and heterogeneous sites. At temperatures below 233 K, only homogeneous sites supplying syndiotactic polymer are present, whereas, at temperatures above 283 K, only heterogeneous sites supplying isotactic polymer are present.

3. Catalytic Complexes: Concentration, Valence of Vanadium

ESR testing and an analysis of kinetic results have led Natta *et al.* [407] to conclude that the catalytic species active in propylene syndiospecific polymerization contain V^{III} species.

The decrements in activity which are observed under certain conditions are due to the reduction of vanadium to V^{II}, which may be reactivated by the addition of weak oxidizing agents [215, 408].

The concentration of active catalytic complexes corresponds to 0.1–0.5% of the vanadium present in the system. This result has been inferred by Natta *et al.* from an examination of kinetic data, and from the molecular weights of the polymers obtained [407], and has been confirmed by Boor and Youngman through quenching with tritiated alcohol [215].

4. Insertion of Monomeric Units — Regiospecificity

Experiments carried out with catalytic systems prepared with ^{14}C labelled $Al(C_2H_5)_2Cl$ [213] and deactivation studies with tritiated alcohol have led to the conclusion that the insertion of monomeric units occurs into a V—C bond [215]. This conclusion is confirmed by NMR analyses performed by Zambelli *et al.* on syndiotactic polymers obtained with 3-^{13}C or 1-^{13}C enriched propylene [409].

Contrary to what was observed for the isospecific polymerization. Takegami and Suzuki [389, 391] have demonstrated that in syndiospecific polymerization propylene insertion occurs into a V—C bond of the secondary (or 2-1) type, namely:

$$V{-}C_2H_5 + C_3H_6 \rightarrow V{-}\underset{}{\overset{\displaystyle CH_3}{\overset{|}{C}H}}{-}CH_2CH_2CH_3$$

As a matter of fact, chiefly *n*-heptane is obtained from the reaction between 1-pentene and syndiospecific catalysts based on V compounds and $AlEt_2Cl$, followed by hydrolysis. The same conclusion has been reached by Zambelli *et al.* on the basis of IR [410] and NMR [411, 413] analysis of ethylene-propylene homopolymers and copolymers obtained with syndiospecific catalysts. These authors point out that the type of insertion is conditioned by steric factors: the secondary insertion of a propylene unit is more frequent when the last monomeric unit of the growing chain is propylene, whereas the primary insertion (or 1–2, namely:

$$V{-}R + C_3H_6 \rightarrow V{-}CH_2{-}\overset{\displaystyle CH_3}{\overset{|}{C}H}{-}R$$

is more frequent when the last monomeric unit of the growing chain is ethylene.

The prevalence of secondary insertion in propylene syndiospecific polymerization has been definitively proved by Zambelli *et al.* through an analysis of the terminal groups of chains interrupted with strong tritiated acids, which have been found to be constituted by *n*-propyl groups instead of by isopropyl ones [392, 414]. It should however be observed that propylene syndiospecific polymerization with soluble catalysts is not thoroughly regiospecific [393, 412, 415, 416], namely the monomer unit enchaiment is not entirely a head-to-tail (h-t)one. As a matter of fact, irregularities of the th-ht type $(\ldots CH(CH_3)CH_2CH_2CH(CH_3)\ldots)$ as well as of the ht-th type $(\ldots CH_2CH(CH_3){-}CH(CH_3)CH_2\ldots)$ result from ^{13}C-NMR analysis in syndiotactic polypropylene. Besides syndiotactic stereoblocks, the polymer contains irregular stereoblocks [392, 393, 412].

The polymerization process may thus be considered a copolymerization with four ht and th stages. A quantitative analysis of such stages has been carried out by Zambelli, Allegra *et al.* [417, 418].

5. *Stereochemistry of the Addition to the Double Bond — Steric Control*

Experiments carried out by Zambelli *et al.* [386, 419] with variously deuterated propylene have shown that, as occurs in isospecific polymerization, syndiospecific polymerization takes place through *cis* addition of the monomer unit to the growing chain. Syndiotacticity is due both to the *cis* addition and monomer unit secondary insertion. Zambelli hypothesizes a four-centre intermediate active complex (*c.f.* Figure 29). Syndiotactic propagation is favoured by the greater stability of complex (a) as compared with that of complex (b) where the methyl groups of the complexed monomer and of the last added monomer unit are in the *cis* position. The secondary insertion on the *trans* complex (*c.f.* Figure 29) is favoured provided that the steric hindrances of substituents bound to the metal of the catalytic complex (growing polymer chain excepted) are not too different. This is the case with a soluble catalyst where the metal-carbon active bond may always be approached by the monomer from the less sterically hindered side. Conversely, in the case of a heterogeneous catalyst, the monomer may always approach the reactive site only from the same side, which preferably produces isospecific polymerization.

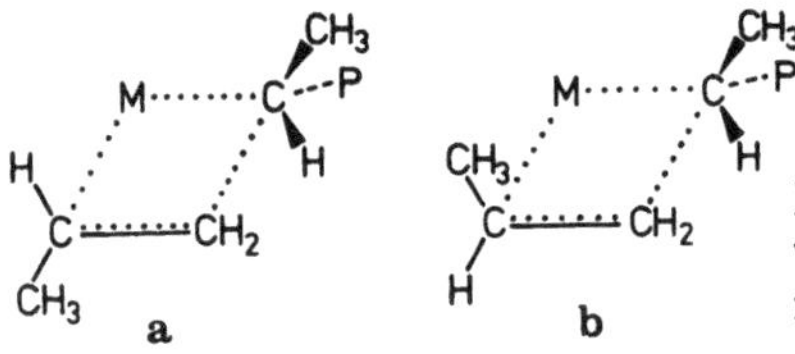

Figure 29. Activated *trans* **a** and *cis* **b** complexes with secondary insertion. (Reproduced with permission from ref. [413])

Syndiospecific polymerization occurs *via cis* ligand migration of the intermediate *trans* complex (*c.f.* Figure 30). In this case, stereoregularity is due to the chirality of the substituted carbon atom of the last monomeric unit of the growing chain [419]. This conclusion is confirmed by the examination of ethylene-propylene copolymers obtained by Zambelli *et al.* [418,

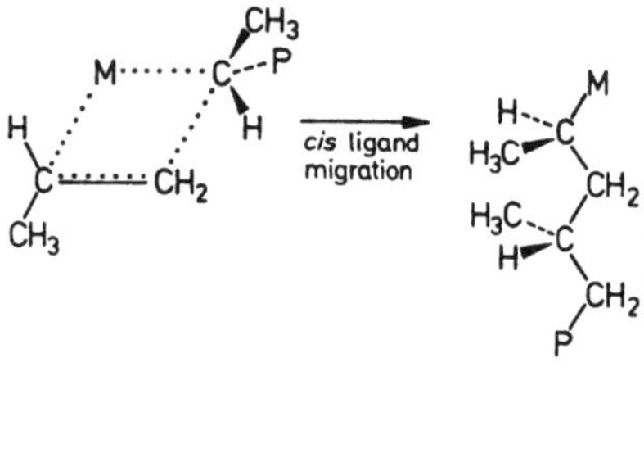

Figure 30. Syndiotactic propagation due to *cis* migration of the *trans* complex ligand. (Reproduced with permission from ref. [413])

419] with syndiospecific catalysts; this results in a product which contains both *meso* and racemic structures when the sequence of propylene units is interrupted by isolated ethylene units (*c.f.* Figure 31).

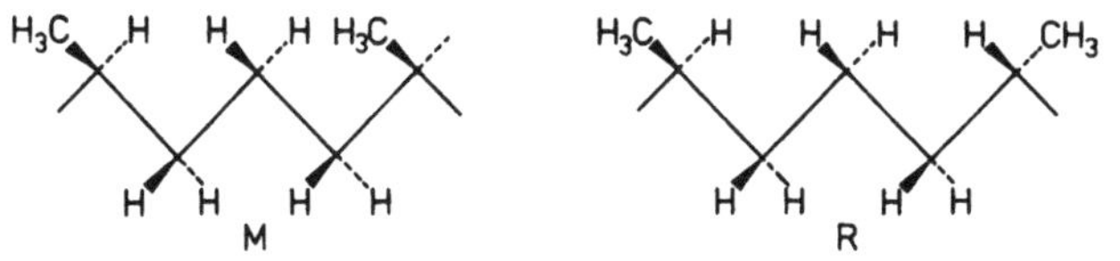

Figure 31. Isolated ethylene unit in a polypropylene chain in meso **M** or racemic **R** environment. (Reproduced with permission from ref. [419])

The existence of a steric interaction between the entering monomer and the last added monomer unit is confirmed by the nature of the ethylene-propylene, ethylene-1-butene and propylene-1-butene copolymers prepared with syndiospecific catalysts [420]. As a matter of fact, a strong tendency to alternate is observed in the case of ethylene copolymers instead of a tendency to the formation of random copolymers.

6. *Model of Catalytic Complexes*

The first hypotheses on the nature of soluble catalytic complexes based on vanadium, active in propylene syndiospecific polymerization, assumed the existence of hexacoordinated complexes having a vacancy for olefin coordination [217]. However, taking into account the type of insertion (secondary) of the monomer units on the metal-carbon bond, it may be inferred that this hypothesis is not acceptable [417].

The latest hypothesis put forward by Zambelli and Allegra [417] proposing catalytic complexes where the V^{III} is pentacoordinated agrees better with the various experimental data so far available concerning both propylene homopolymerization and copolymerization. In this proposal, the V atom is coordinated with three chlorine atoms (two of which bridge bound to aluminium),

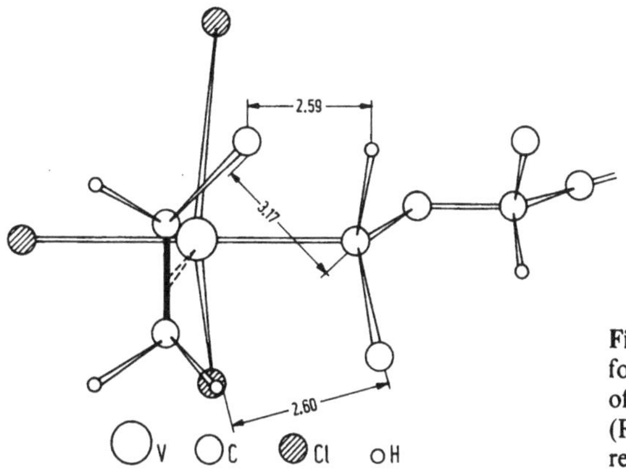

Figure 32. Model of catalytic center for the syndiospecific polymerization of propylene (Zambelli and Allegra). (Reproduced with permission from ref. [417])

with one asymmetric carbon atom (*e.g.* the last atom of the growing polymer chain) and with the coordinated olefin (*c.f.* Figure 32). Prior to olefin complexation and after monomer unit insertion, the catalytic complex is tetracoordinated. Polymerization would take place through equatorial coordination of the monomer to the V atom followed by rotation of the complexed olefin and by its insertion into the V—C bond.

C. Ethylene Polymerization with Ziegler-Natta Catalysts

1. Heterogeneous Catalysts

Consideration concerning the concentration of catalytic complexes and the average lifetime of the growing polyethylenic chains in the polymerization with heterogeneous Ziegler-Natta catalytic systems, has been reported in a previous section (3.A.2). Some other aspects of ethylene polymerization with these catalysts are summarized below.

As pointed out in connection with α-olefin polymerization (*c.f.* section 3.A.1) a different behaviour may be noticed in ethylene polymerization with the various catalytic systems. The greater differences are observed in the dependence of the polymerization rate on the ratio between metal alkyl and transition metal compound. By ethylene polymerization with the halogenides of transition metals at a high valence state and generally soluble in the reacting medium, peak activity may be observed with a given ratio between the two components (*c.f.* Figure 33). A similar behaviour has also been noticed by Böhm [117] for the highly active catalytic system $AlR_3/TiCl_4/Mg(OR)_2$. For $TiCl_3$-based catalytic systems, either constant rate or gradual rate variations may be observed above a minimum threshold of the Al/Ti ratio.

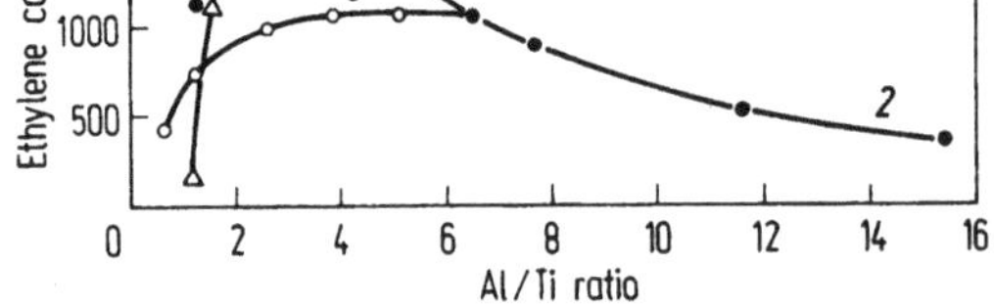

Figure 33. Ethylene consumption during the first 60 min, in dependence in Al/Ti molar ratio, for the polymerization with $TiCl_4$ and different Al alkyls. 1, $[C_2H_4] = 0.024\ mol\ dm^{-3}$, $[TiCl_4] = 6.27\ mmol\ dm^{-3}$, $Al(C_2H_5)_3$, 293 K; 2, As before except 323 K; 3, $[C_2H_4] = 0.045\ mol\ dm^{-3}$ $[TiCl_4] = 5.1\ mmol\ dm^{-3}$, $Al(C_2H_5)_2Cl$, 323 K. (Reproduced with permission from ref. [292])

With regard to the catalytic system $TiCl_3$—$AlEt_2Cl$, Berger and Grievson [76] have not found a dependence of the polymerization rate on the Al/Ti ratio, at least within range 1 to 5.

For Al/Ti ratios >50, Swingler [421] has noticed a lowering of the maximum activity. This result has been explained with a competitive adsorption of the metal alkyl and of the monomer on the catalyst surface. An increase of the polymerization rate with the $AlR_3/TiCl_3$ ratio within the range 1 to 9 has been observed by Schenko *et al.* [293].

The dependence of the rate on the polymerization time is a function of the nature of the catalytic system components as well as of polymerization conditions such as temperature and catalyst and monomer concentrations. For the $AlEt_3$-δ-$TiCl_3$ catalytic system at 313 K, Matsuda and Keii [422] have noticed constancy of polymerization rate during the initial 15 minutes, followed by a gradual lowering to a stationary value, which may be described by the equation

$$\frac{R - R_\infty}{R_0 - R_\infty} = e^{-kt}$$

where R_0 is the initial rate and R_∞ is the rate in steady state conditions.

For the $AlEt_2Cl$—$TiCl_3$ system, the polymerization rate increased gradually with time to a constant value that was maintained for many hours.

Ethylene polymerization with catalysts based on $TiCl_4$ and Al alkyls has been investigated by Schindler [423, 424] and by Chirkov *et al.* [425]. Kinetic curves are characterized by an induction time before maximum rate is attained, followed by a continuous lowering of the polymerization rate with time. The induction time is extremely short in the presence of $AlEt_3$ whereas it may be quite long in the presence of halogenated Al alkyls expecially at low temperature and at low Al/Ti ratios.

The decrease in polymerization rate with time has been ascribed to various causes both chemical and physical. Among the former, may be mentioned a decrease of the number of active centres, often correlated with a lowering of the transition metal valence. Christmas [426], for instance, has studied ethylene polymerization with the catalytic system $VO(O\textit{t}\text{-}Bu)_3$—$Al_2Et_3Cl_3$ at 393 K and has ascertained that the catalyst may be reactivated by addition of trichloroacetic acid esters which probably oxidize the low valence inactive vanadium compounds. The most probable physical cause of the decay in rate with time consists in catalyst encapsulation by the polymer which, being insoluble in the reacting medium, brings about the onset of diffusive barriers between monomer and active centre. Of course, the effect is more conspicuous the higher the polymer yield with respect to the catalyst, and it also depends on how the catalyst primary particles are dispersed in the polymer [427]. For a high yield catalytic system, Böhm [82] has accertained that yields greater than 10^6 g PE (mol Ti)$^{-1}$ must be attained before observing any marked diffusive effect. Under the same polymerization conditions, yields in the order of 10^5 g PE (mol Ti)$^{-1}$ are attainable with $TiCl_3$—$AlEt_3$ catalyst and no lowering of activity due to diffusive phenomena may be observed.

These aspects of polymerization with heterogeneous systems have been reviewed by Chien [204, 428].

For catalysts based on $TiCl_4$ [429, 430] and on $TiCl_3$ [76, 359], the polymerization rate is directly proportional to the concentration of the monomer and of the transition metal salt. A deviation from linearity for monomer concentrations lower than 0.1 mol dm^{-3} has been reported by Böhm [82] for the $AlEt_3$—$TiCl_4/Mg(OR)_2$ catalytic system. Values ranging from 25 to 71 kJ mol^{-1} have been reported for the activation energy of the overall polymerization process with the various catalytic systems [76, 341, 359, 431].

Also insofar as concerns ethylene polymerization, the same polymer chain initiation and termination processes as described (*c.f.* section 3.A.1) for propylene polymerization occur.

2. *Homogeneous Catalysts*

a) Catalysts based on Al alkyls and cyclopentadienyl derivatives of Ti

Among the various homogeneous catalytic systems of the Ziegler-Natta type, those based on cyclopentadienyl derivatives of Ti and Al alkyls have undergone extensive spectroscopic and kinetic examinations in order to infer useful information on the mechanism of Ziegler-Natta catalysis.

One of the greatest difficulties encountered in these studies derives from the fact that, although the said catalytic systems are thoroughly soluble, the polymerization system becomes heterogeneous just after the first minute of polymerization owing to polymer precipitation; discontinuities occur concomitantly in kinetic polymerization curves.

Through a number of experiments at varying ethylene concentration with the $Cp_2TiEtCl$—$AlEtCl_2$ catalytic system Reichert [432] demonstrates that, while the system remains homogeneous, the polymerization rate remains constant and directly proportional to the concentration of the monomer; as soon as the polymer precipitates, a sudden increase in the polymerization rate may be observed up to a maximum, and then the rate continues to decline with time. Such decay has been ascribed to a series of side reactions such as alkyl exchange, hydrogen transfer, and metal reduction with formation of stable, low activity, species [31, 433]. By the use of a plug-flow reactor, Fink [434] has studied the kinetics of the first homogeneous step of polymerization. On the basis of the dependence of the initial rate on the ratio between the components of the catalytic system, according to the concepts developed by Reichert [435, 34], the formation of the active species in two subsequent dynamic equilibria is hypothesized as follows:

$$[Ti] + \frac{1}{2}[Al_2] \overset{K_1}{\rightleftarrows} C1$$

$$C1 + [Al_2] \overset{K_2}{\rightleftarrows} C^* + Al_2'$$

where: [Ti] = Cp_2Ti Propyl Cl, $[Al_2] = Al_2Et_2Cl_4$, Al_2' = a dimeric Al compound of unknown nature, C^* = active species, and C1 is an adduct between [Ti] and $[Al_2]$. The first reaction is quite completely shifted to the right; in fact, the value of K_1 as determined at 240 K *via* NMR spectroscopy [436] ranges between 5×10^3 and 5×10^4 $dm^3\ mol^{-1}$. For the complex Cl, which is considered catalytically inactive, the structure $Cp_2Ti(R)Cl \rightarrow AlEtCl_2$ is proposed. In this structure, both Ti and Al atoms have a tetrahedral coordination and a pair of electrons from a Cl ligand of the Ti atom acts as a Lewis base for a molecule of monomeric Al alkyl. Complexes of this type have already been proposed by other authors for the species considered catalytically active [437, 438].

As regards the second reaction equilibrium, a value of *ca.* 10^{-3} $dm^3\ mol^{-1}$ has been calculated for K_2; therefore equilibrium is shifted towards the formation of active species only at high Al/Ti ratios.

The analysis *via* GPC [434] of the oligomers formed in the first seconds of polymerization has established that for the first insertion step, k_p is dependent on the length of the alkyl group bound to the active centre. At 283 K k_p values are as follows: 140 $dm^3\ mol^{-1}\ s^{-1}$ for ethylene insertion into $Ti{-}C_2H_5$ bond, 90 $dm^3\ mol^{-1}\ s^{-1}$ for insertion into $Ti{-}C_3H_7$ bond and a limit value of 47 $dm^3\ mol^{-1}\ s^{-1}$ for insertion into $Ti{-}C_6H_{13}$ bond.

Agreement has not yet been reached concerning the true structure of the active centre. As a matter of fact, Borisova *et al.* [440] suggest that the true active complex is the primary complex Cl and Hoffmann [441] too believes a coordinatively unsaturated compound of this type represents the best candidate for the active specie. By contrast, Henrici-Olivé hypothesizes the formation of a bimetallic complex with an octahedral structure III where Ti^{IV} is bound to the Al atom *via* a double chlorine bridge and a vacant

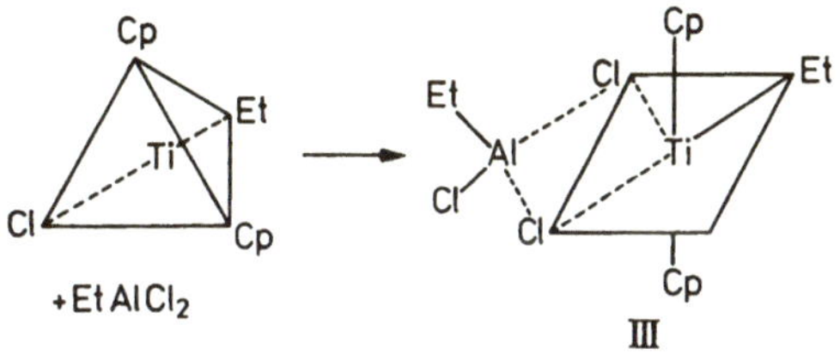

III

coordination site is present for monomer complexation to Ti. The Ti—Et bond would be weakened by the *trans* influence of the bridged Al and by olefin complexation, thus facilitating the insertion reaction.

In a latter study, the same author [442] suggests for the bimetallic complex also a structure of the trigonal bipyramid type which would change to octahedral symmetry only on the approach of a monomer molecule.

Ethylene insertion into a Ti—C bond has been demonstrated by Fink *et al.* [443, 444] for these catalytic systems *via* ^{13}C-NMR spectroscopy by employing ^{13}C enriched ethylene. Fink's experiments have not evidenced the formation of π complexes between olefin and Ti. Henrici-Olivé [445] too has indirectly demonstrated that ethylene insertion occurs into a Ti—Ci bond; as a matter of fact, if small ethylene quantities are fed to the catalytic system $Cp_2Ti(C_4H_9)Cl - Al(CH_3)_3$ and the growing chains are quenched after very short times, only oligomers with an even number of C atoms may be isolated. NMR evidence in favour of olefin insertion in the Lu—CH_3 bond has been brought by Watson [55] for the complex Lu(η^5-C_5Me_5)$_2$$CH_3$-ether which is a soluble catalyst active in ethylene polymerization. The same author points out that olefin insertion in the Lu—CH_3 bond of the alkyl bridged complex Lu(η^5-C_5Me_5)$_2$Al$(CH_3)_4$ occurs at significantly lower rates.

b) Catalysts based on cyclopentadienyl derivatives of Ti, Zr and aluminoxanes

The very high productivity of soluble catalytic systems based on $Cp_2M(CH_3)_2$ (M = Ti or Zr) and polymeric methylaluminoxanes (*c.f.* section 2.A.3a) is justified by Kaminsky [208] on the basis of the assumption that in this case the side reactions causing the quick decay in the activity of the catalysts based on aluminum alkyls and halogenated cyclopentadienyl Ti derivatives are suppressed. After an induction period the polymerization rate reaches a maximum and then keeps constant for very long times (exceeding 20 h), although polymer chain termination reactions also take place during polymerization, evidenced by the low molecular weights obtained at 323–353 K. Clearly, termination appears to be due not to the death of the active centers, but to chain transfer reactions. The catalytic activity increases with aluminoxane concentration up to about 2.6×10^{-2} mol l^{-1}. An increase in activity has also been observed when the degree of oligomerization of the aluminoxane changes from 10 to 14. The productivity relative to Zr(Ti) lowers as the concentration of the transition metal compound rises, the better results being obtained at concentrations lower than 10^{-6} mol dm^{-3}.

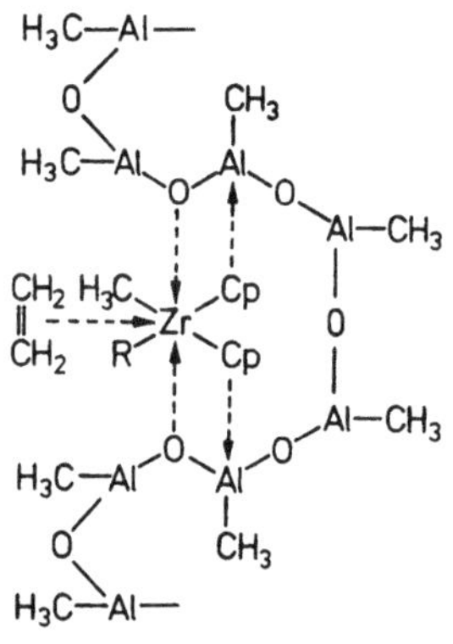

IV

UV spectroscopy has evidenced the formation, in toluene solution, of a complex between catalyst and ethylene, characterized by an absorption at 390 nm, disappearing when ethylene is withdrawn by polymerization [39] Structure IV is hypothesized for the catalyst. In such a structure the interaction between the Al and O atoms of the aluminoxane and, respectively, the cyclopentadienyl rings and the Zr atoms would promote complexation of the olefin and its insertion in the Zr—CH_3 bond.

D. Ethylene Polymerization with Catalysts Supported on Oxides

1. *Phillips Catalysts*

a) Behaviour of catalytic systems — overall kinetics

The behaviour of the catalysts based on chromium oxide supported on SiO_2—Al_2O_3 depends on a number of factors, the most important being the type of support (composition and porosity), chromium concentration, and activation conditions; rationalizing the data reported by the various authors is therefore difficult.

Generally, the polymerization rate changes with time. Yermakov *et al.* [446] have observed that the shape of the kinetic curves is determined by the composition of the catalyst and polymerization conditions such as monomer concentration and temperature. Stationary kinetics have been observed only at very low concentrations of monomer. A maximum in initial polymerization rate as a function of temperature has been found at *ca.* 353 K by Clark [16] and by other authors as well [447, 448], and may be correlated to a variation in the number of active centres with temperature [446].

For the activation energy of the overall polymerization process, values of 29–34 kJ mol^{-1} [16] have been found in the temperature range 233–273 K. Values of 27–67 kJ mol^{-1} have been observed between 303 K and 348 K depending on catalyst composition and activation method; the activation energy of the propagation reaction is 17–21 kJ mol^{-1} [446]. Discordant data concern the dependence of polymerization rate on monomer concentration; some authors [449, 451] have found first order kinetics, whereas deviations from first order have been observed by others [447, 448].

b) Nature and concentration of active centres

The industrial importance of the Phillips catalysts has stimulated many fundamental studies to determine the nature of the active centres and the polymerization mechanism. The degree of oxidation of catalytically active Cr has been studied in particular.

Chromium ions with oxidation number V [447, 449], IV and III ([452] and II [453] have been reported to be active in polymerization. The reason for the discrepancy on the nature of the catalytic sites derives from the fact that a small fraction only of Cr present on the catalyst is catalytically active, and the average oxidation degree of the metal has often been correlated with the catalytic activity.

In a recent study [454], a good correlation has been found by Beck and Lundsford between catalytic activity and concentration of isolated Cr^{III} ions, as determined on the basis of the ability of these ions to give complexes with NO, of the $[Cr(NO)_2]^{+3}$ type. These complexes have been identified through ESR and IR spectroscopy. The authors conclude that the active centres are constituted by mononuclear Cr^{III} ions which are coordinated only to the silica lattice and exhibit at least 2 (and perhaps 3 according to Zecchina *et al.* [455]) vacant coordination sites. The presence of 3 available coordination sites would justify the ability of these catalysts to aromatize acetylene to benzene [13].

As for the Ziegler-Natta catalysis, initiation and propagation reactions are believed to take place through the formation of an organochromium compound, followed by olefin insertion into the metal—carbon bond. The mechanism of formation of the first Cr—C bond, as a result of the interaction of the low valence Cr ion with the olefin, is however obscure and undemonstrated. An oxidative addition of ethylene to coordinatively unsaturated metal ions according to the following scheme [11]:

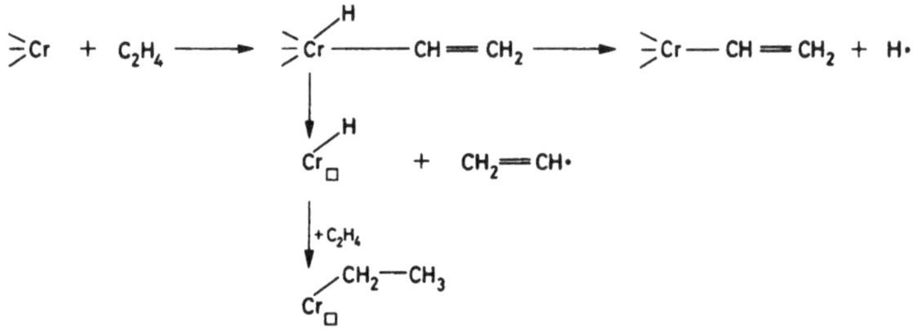

or through transfer of a hydrogen atom from ethylene to an oxygen atom bond to Cr to give the following structure [456]

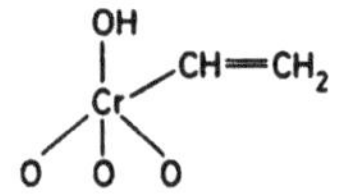

has been hypothesized.

The polymer chains obtained with Phillips catalysts exhibit vinyl and methyl end groups [13]. The termination reaction is therefore believed to take place chiefly by β hydrogen elimination with formation of a terminal vinyl unsaturation:

$$Cr{-}CH_2{-}CH_2{-}P \rightarrow Cr{-}H + CH_2{=}CH{-}P\,.$$

The Cr—H bond might add to an ethylene molecule to regenerate the active centre thus allowing the formation of more polymer molecules on the same catalytic site. The parallelism between the reaction mechanism of the Ziegler-Natta and Phillips catalysts is no longer valid insofar as we consider the results obtained by quenching the polymerization with alcohols. In this case, in fact, radioactivity is present in the polymer when $^{14}CH_3OH$ is used as quenching agent instead of CH_3O^3H as occurs with Ziegler-Natta catalysts. This different behaviour has been explained by admitting the bond between Cr and the polymer chain to be polarized as $Cr^{\delta-} \ldots C^{\delta+}$; consequently, the reaction with alcohol would occur according to the scheme:

$$Cr^{\delta-} \ldots {}^{\delta+}CH_2{-}CH_2{-}P + {}^{14}CH_3OH \rightarrow Cr{-}H +$$
$$+ {}^{14}CH_3O{-}CH_2{-}CH_2{-}P$$

The explanation does not appear fully convincing since, according to the scheme proposed, if added in a stoichiometric quantity, methanol would act as a chain transfer agent and not as active site killer, the Cr—H bond still being capable of regenerating the active centre.

Zecchina *et al.* [457] on the basis of spectroscopic evidences suggest a model for the mechanism of reaction of the Phillips catalyst similar to that proposed by Green for the Ziegler-Natta catalysis. The removal of an α-hydrogen and the formation of an alkylidene compound has been shown; however in this case the α-hydrogen does not form a Cr—H bond but seems to be transferred to an adjacent surface oxygen atom as shown below:

Quenching the reaction with $^{14}CH_3OH$ has been used, along with inhibition methods to provide a correlation between the lowering in the polymerization rate and the quantity of inhibitor added, for the determination of the number of active centres.

According to Hogan [13], Yermakov [22], and Eden [458] for catalysts containing 1% Cr, the number of active centres, as referred to maximum polymerization rate, corresponds to *ca.* 10–15% of total Cr; a value of 1700 $dm^3\ mol^{-1}\ s^{-1}$ has been found for the chain propagation constant at 348 K [22].

2. Organometallic Compounds of Transition Metals Supported on Oxides

As reported under section 2.A.2c, surface complexes form by reaction between organometallic compounds of transition metals and SiO_2 or Al_2O_3. In these complexes, the transition metal is bound both to support through bridge oxygen and to organic groups. It would seem logical to believe polymerization to occur by ethylene insertion in the metal—carbon bonds already present in the catalyst. According to this assumption, radioactive fragments have been found in polystyrene prepared with catalysts based on $Zr(^{14}CH_2{-}C_6H_5)_4$ supported on SiO_2 [349], and moreover the possibility of ethylene insertion into the metal—benzyl bond of the surface complexes of Ti^{III} has been evidenced by ESR analysis [459].

Spectroscopic evidence, however, suggests that the formation of propagation centres may also take place in other ways. No lowering in the intensity of the IR absorption bands attributed to the allyl groups [460] may in fact be observed during the polymerization of ethylene with supported Cr and Zr allyl derivatives. The formation of active centers in these cases is thought to occur either through an oxidative addition to metal ions or through the formation of hydride intermediates [460].

The presence of hydride surface complexes of Ti^{III} as a result of the reaction of hydrogen with the $TiBz_4{-}SiO_2$ catalyst has been ascertained *via* ESR spectroscopy. The analysis of the ESR spectra obtained with these catalysts in the presence of ethylene and hydrogen has shown some fundamental steps of the polymerization reaction such as ethylene coordination to the metal, ethylene insertion into the Ti—H bond and Ti—C bond hydrogenolysis [461]. As regards the $ZrBz_4$—SiO_2 catalyst treated with hydrogen at 473 K, a direct proportionality was observed between catalytic activity and the content of zirconium hydrides in the catalyst [22]. The number of active centres found for catalysts of this type varies according to the transition metal, to its concentration, to polymerization temperature, and to possible activation by hydrogen or UV irradiation [22]. The resulting values are generally lower than 10% of the total transition metal; values ranging from 1000 to 3000 $dm^3\ mol^{-1}\ s^{-1}$ are reported for the chain propagation constant.

E. Alternating Copolymerization of Ethylene with Internal Olefins

The mechanism of the copolymerization reaction of ethylene with cycloolefins or with 2-butene has been discussed in detail by Dall'Asta and Mazzanti [257] who have based their conclusions on the following observations:

(i) Cycloolefins with an even number of carbon atoms do not originate alternating crystalline copolymers with ethylene.
(ii) Conditions being the same, the copolymerization rate for cycloolefins with an even number of carbon atoms is lower than for cycloolefins with an odd number of carbon atoms.
(iii) The differences mentioned under 1 and 2 are observed both with homogeneous and with heterogeneous catalysts.
(iv) Unlike ethylene-*cis*-2-butene copolymerization, the copolymerization of ethylene with cycloolefins having an odd number of carbon atoms gives alternating crystalline copolymers also in the presence of homogeneous catalysts.
(v) In the presence of heterogeneous catalytic systems, a parallelism may be observed between the behaviour of *cis*-2-butene and cycloolefins with an odd number of carbon atoms and that of *trans*-2-butene and cycloolefins with an even number of carbon atoms.

From the foregoing, it may be inferred that copolymerization producing alternating crystalline copolymers is characterized by constancy of steric attack and by *cis*-opening of the double bond.

The authors mentioned above therefore assume cycloolefin reactivity to depend firstly on the tendency of the double bond to undergo a *cis*-addition that depends on the Van der Waals's interactions between the growing chain and the methylene groups of the cycloolefin ring. In cycloolefins having high reactivity, the constant *cis*-opening of the double bond is associated with a constant steric effect as these cycloolefins, in their most favoured energetic configuration, do not exhibit two equivalent faces.

According to these authors, odd rings (C_5 and C_7) differ from even ones (C_6 and C_8) essentially for the following reasons:

(i) Odd rings involve *cis*-addition at the metal carbon bond of the active site because the carbon atoms emerging from the plane of atoms 1,

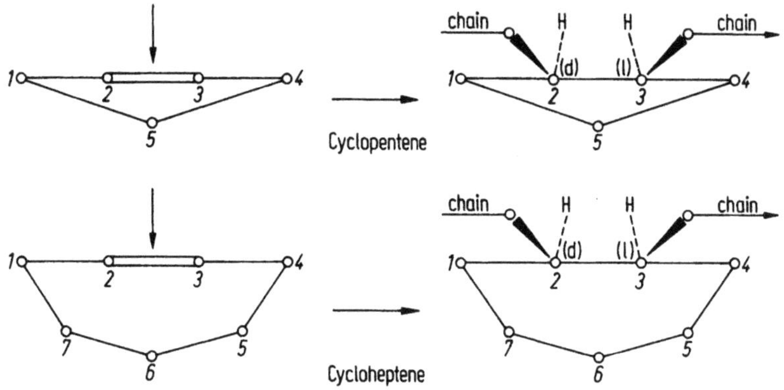

Figure 34. Schematic representation of the *cis*-opening of the double bond in the case of cycloolefins with odd number of ring carbon atoms. (Projection on a plane perpendicular to the plane of the double bond). (Reproduced with permission from ref. [257])

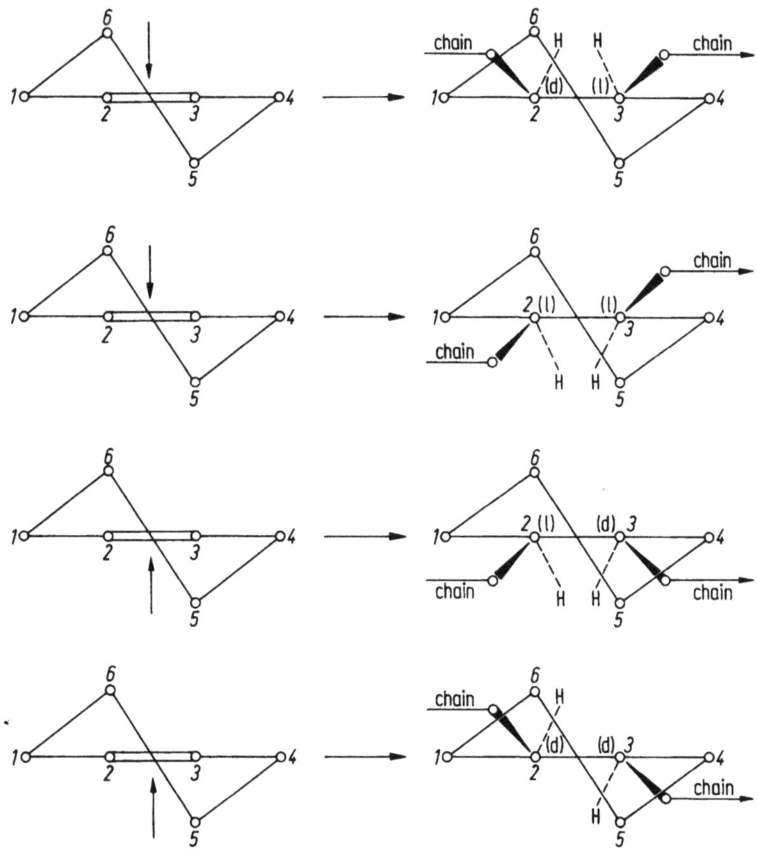

Figure 35. Schematic representation of the opening possibilities of the double bond in the case of cyclohexene (Projection on a plane perpendicular to the plane of the double bond). (Reproduced with permission from ref. [257])

2, 3 and 4 are all from the same side of the plane considered (*c.f.* Figure 34), and consequently the constraints imposed by Van der Waals's distances and the high addition rate are satisfied. Steric interactions (*c.f.* Figure 35) which reduce the *cis*-addition rate occur instead in the case of even rings.

(ii) For the aforesaid reasons, *cis*-addition is always favoured for the odd rings as compared with *trans*-addition, whereas the opposite applies to even rings with binary symmetrical axes, *trans*-addition being favoured in this case.

(iii) As odd cycloolefins possess two non-equivalent faces, it follows that, in the presence of homogeneous catalysts, there is a constant steric attack (namely from the less hyndered face of the ring).

The above assumption agree with *cis*-2-butene behaviour. In this case only a high reactivity is required for the monomer. Constancy of *cis*-opening and steric attack are not required as the two faces of the molecule, defined by the plane containing the carbon atoms, are equivalent. In practice, the reactivity of *cis*-2-butene is comparable to that of C_5 and C_7 cycloolefins and, in the presence of heterogeneous catalysts, *cis*-opening of the double bond occurs. Reactivity is instead lower for *trans*-2-butene as the two types of opening are both disfavoured owing to Van der Waals's interaction of the methyl groups and as the steric attack does not exhibit any favoured position. The structure of the alternating copolymers of this monomer with ethylene is irregular.

F. Ring Opening Polymerization of Cycloolefins

Several hypotheses have been reported concerning the mechanism of ring opening polymerization of cycloolefins [279, 462–464]. Scott and Calderon *et al.* [281] have shown that the ring opening polymerization of cycloolefins having five or more carbon atoms, in the presence of catalysts based on molybdenum or tungsten compounds, is a particular case of the olefin methatesis reaction. It follows that polymerization occurs by scission of the double bond instead of the opening of the α- or β bonds with respect to the double bond as previously thought. The first scheme proposed for the transalkylidenation process assumed a pair-wise scrambling that required an initial bis olefin complex with the transition metal (*e.g.* W) of the catalytic centre:

According to this scheme macrocycles are produced. Low molecular weight macrocyclic oligomers [465] have actually been found in polymers obtained by polymerization of cycloolefins in the presence of the system WCl_6—$Al(C_2H_5)_2Cl$—C_2H_5OH [281, 464]. Yet, high molecular weight polymers are commonly observed during the early stages of cycloolefins polymerization [465]. Dolgoplosh *et al.* [463] suggest that this phenomenon is inconsistent with a pair-wise scrambling mechanism and requires cycloolefins polymerization to propagate by a chain process. Herrison and Chauvin [466] first suggested transalkylidenation to occur through a carbene-metal intermediate, and this mechanism could agree with a chain reaction process [463].

According to this hypothesis the polymerization occurs through a mechanism involving a complex, between the double bond of the cycloolefin and a metal carbene, which undergoes a interchange *via* a metallocyclobutene intermediate:

The formation of carbene active centres could occur through isomerization of an olefin complex [463].

Several authors [462, 463, 467–471] give experimental evidences for this mechanism or support it.

It has to be observed that according to the proposed scheme:

(i) the high polymers obtained by ring opening polymerization of cycloolefins are linear.

(ii) Cyclic oligomers are formed through an intramolecular reaction,

and not through intermolecular condensation of two smaller rings. Wheather this mechanism applies also to ring opening polymerization of cyclobutene monomers with catalysts based on Ru, Os or Ir compounds which fail to polymerize higher cycloolefins, is not clear.

4. Polymerization Processes

A. Polyethylene

Ethylene polymerization processes have undergone a considerable simplification in the last decade thanks to the availability of high yield catalysts which have allowed the elimination of the deashing step. The following processes, all continuously operating, are employed for HDPE and LLDPE synthesis both with Phillips and with Ziegler-Natta catalysts:

1. solution process
2. suspension process
3. gas phase process.

1. Solution Process

At present, the solution process finds a marginal application and is employed both with Phillips catalysts for the synthesis of some types of HDPE and with Ziegler-Natta high yield catalysts for the synthesis of HDPE [472, 473] as well as of LLDPE [474].

In the Phillips solution process cyclohexane, catalyst, ethylene, and, 1-butene if necessary, are continuously fed to a stirred reactor where polymerization takes place at temperatures of 398 K–448 K and at pressures of 2.8–3.5 MPa. The polymer solution is continuously withdrawn and flashed to remove dissolved ethylene. After solvent removal by flashing in devolatilizing extruders, the molten polymer is pelletized.

2. Suspension Process

The suspension process is by far the most widely used both with Phillips and with Ziegler-Natta high yield catalysts. Processes of this type have been developed by Phillips [475], Montedison [476]. Hoechst A.G. [477], Solvay [478, 479], Mitsubishi Chemical Ind. [480] and other companies (*c.f.* Figure 36).

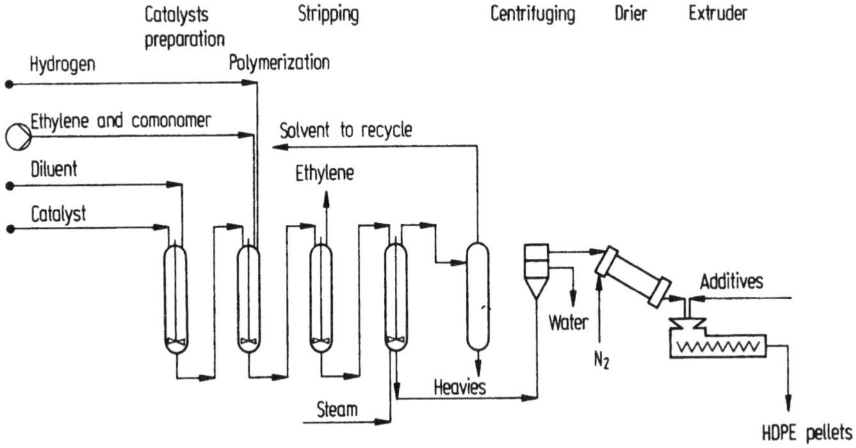

Figure 36. Montedison slurry process for polyethylene. (Reproduced with permission from ref. [481])

In these processes, practically all the polymer is insoluble in the suspending medium which is a poor solvent (isobutane, *n*-heptane). Polymerization is conducted in vessel or loop reactors [475] at temperatures of 353 K–363 K and at pressures of 0.8–1.5 MPa in the presence of hydrogen for molecular weight control and of an α-olefin to adjust polymer density between 0.94 and 0.965 g cm^{-3}.

In some processes the polymerization is carried out in a series of cascade reactors to allow the hydrogen concentration to vary through the operating steps in order to control the distribution of the molecular weights. The slurry, which may contain about 40% by weight polymer, is discharged continuously and may undergo various treatments in the different processes. In some [477], the suspension is centrifugated and 90% of the diluent is recovered and recycled without purification, whereas the balance is conveyed to a solvent treatment unit where low molecular weight polyethylene waxes are also removed. The polymer still impregnated with solvent undergoes steam stripping to remove residual solvent and is then separated from water by centrifugation, dried in a fluidized bed, extruded and pelletized. In other processes [475, 476], the suspension, generally in light hydrocarbons, is directly steam stripped or undergoes simple flashing. These processes are very versatile and allow an easy control of density, molecular weight, and polymer molecular weight distribution.

3. *Gas Phase Process*

The gas phase process [482] developed by Union Carbide for HDPE manufacture has later been extended also to LLDPE [483] synthesis. The process, (*c.f.* Figure 37) which operates continuously, employs a vertical reactor with expanded upper section. Polymerization is done at temperatures below 373 K and at pressures of 1.8–2.5 MPa. A centrifugal single-stage compressor circulates the reaction gas (ethylene and eventually 1-butene) which fluidizes the bed and removes the heat of reaction (about 3.3 kJ g^{-1}). To keep the temperature at values below 373 K, gas conversion is maintained at 2–3% per pass.

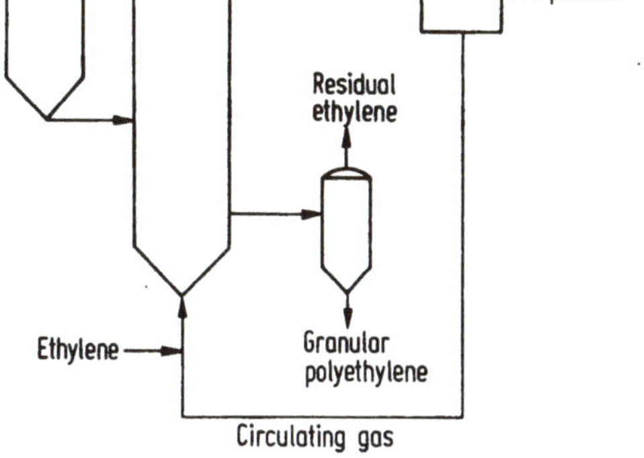

Figure 37. Union Carbide gas-phase process for polyethylene. (Reproduced with permission from ref. [484])

The bed is fluidized at a rate 3 to 6 times higher than that required for minimum fluidizing. The activator (aluminum alkyl) may be introduced along with the catalyst or fed with recycle gas. The polymer is withdrawn periodically from the reactor, separated from the monomer in a degasifier, purified from residual hydrocarbons in a N_2 stream, extruded, and pelletized. The product may also be directly used in the form of free-flowing powder with an average particle size ranging from 0.5 to 1 mm [485].

B. Polypropylene

As in the case of ethylene polymerization, the availability of high yield and high stereospecific catalysts has allowed a considerable simplification of production processes. However, since the availability of these new catalysts dates back only from the second half of the seventies, many processes still utilize traditional catalysts ($TiCl_3$ + $AlEt_2Cl$) and require purification from catalytic residues.

Almost all polypropylene is produced in suspension processes; the suspending medium may be either a hydrocarbon (slurry process) or the same monomer (bulk process). Two plants only operate with the gas phase process [486].

1. Simplified Slurry Process [487]

Solvent (*n*-hexane or *n*-heptane), high yield catalyst, aluminum trialkyl, monomer, and molecular weight regulator (H_2) are fed to the polymerization reactor (*c.f.* Figure 38). Typical reaction conditions are temperatures in the range of 333 K–353 K, and pressures of 0.5 to 1.5 MPa. The diluent keeps

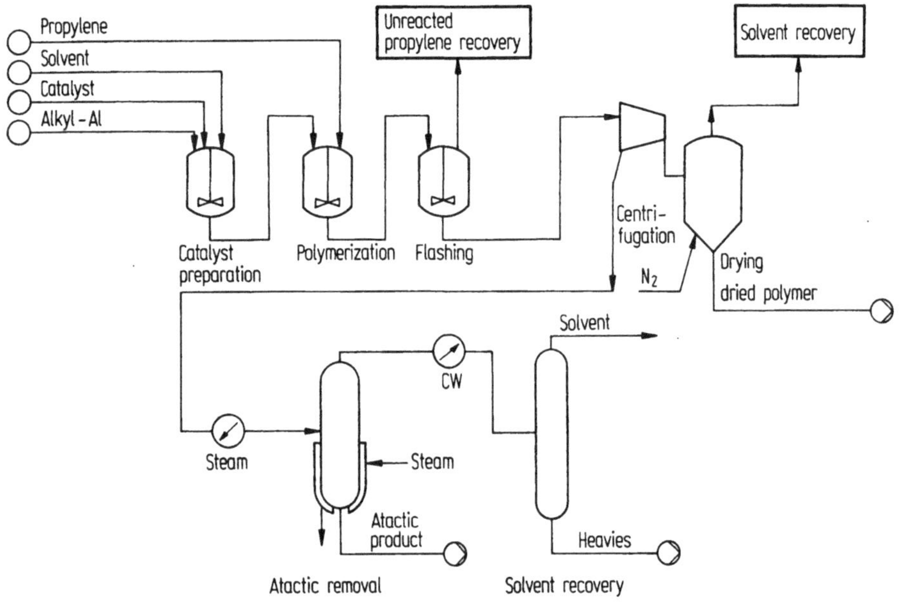

Figure 38. Montedison simplified slurry process for polypropylene (Reproduced with permission from ref. [487])

in suspension the polymer particles and dissolves most of the atactic polymer (2–4%) forming with the isotactic polymer. The polymer slurry containing 40–50% solids is separated from dissolved monomer by flashing and then centrifugated. The polymer, moist with solvent, is dried in a hot nitrogen stream, added with stabilizers and extruded in pellets. The recovery of the solvent from centrifugation is performed by simple distillation to separate the dissolved atactic polymer and heavy products.

As compared with the processes operating with conventional catalysts, in the simplified process the catalyst removal and washing sections do not exist at all; the distillation section is particularly simplified and the waste water treatment is not necessary.

2. Simplified Bulk Process

The polymerization reaction takes place in liquid propylene which performs a dual function of reactant and suspending agent at the same time. Of course, the high monomer concentration favours the polymerization rate. Polymerization is done in vessel or loop reactors [488] (*c.f.* Figure 39). The

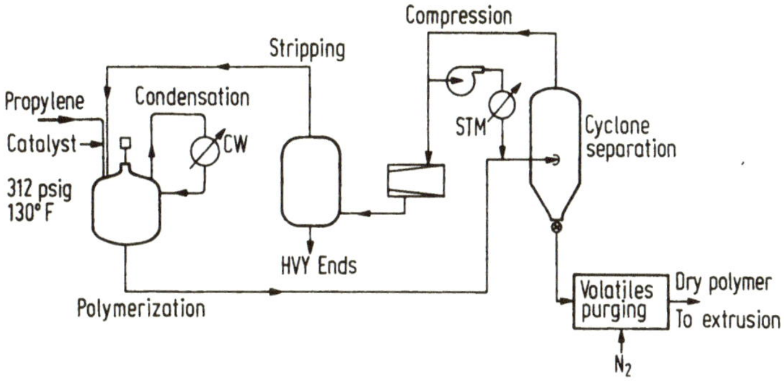

Figure 39. Simplified bulk process for polypropylene. (Reproduced with permission from ref. [488])

heat of reaction may be removed either through an external cooling system or by means of a condenser. In the latter case, about 50% of the polymerization heat is removed by boiling of the monomer that is then condensed and recycled. The slurry containing about 60% by weight polymer is conveyed to a pipeline where most of the monomer undergoes flashing. The polymer then enters a separator cyclone. The powder is separated from the residual monomer by means of a hot N_2 stream, extruded, and pelletized.

3. Gas Phase Process

Polymerization occurs in stirred reactor operating at temperature of 353 K to 363 K at a pressure of *ca.* 3.0 MPa (*c.f.* Figure 40). The heat of reaction is removed almost exclusively by evaporation of recycle liquid propylene that is injected continuously into the reactor. Polymer powder is intermittently withdrawn from the reactor and separated from the monomer in a cyclone.

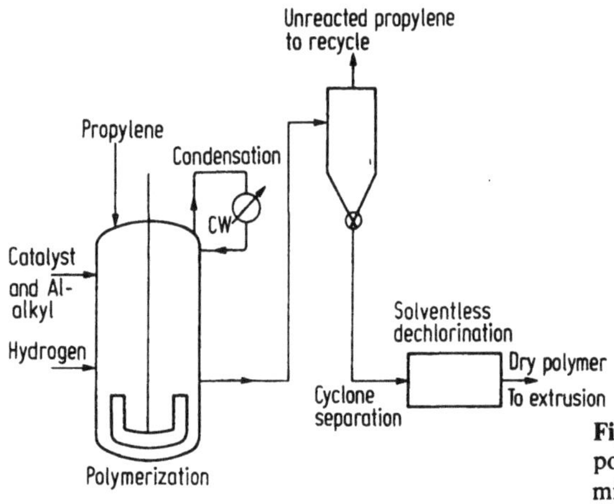

Figure 40. BASF gas-phase process for polypropylene. (Reproduced with permission from ref. [488])

As the process operates with a low yield catalyst, the powder requires a treatment during the extrusion step with a reagent (*e.g.* propylene oxide) which reacts with halogenated catalytic compounds transforming them into metal oxides and volatile chlorinated organic compounds which are removed from the polymer [488].

C. Ethylene — Propylene Copolymers

Olefin elastomers are manufactured by solution and suspension processes, both continuously operating.

1. Suspension Process

This polymerization takes place in liquid propylene which performs a dual function of reactant and suspending agent of the formed copolymer which is insoluble in such system (*c.f.* Figure 41).

Propylene, ethylene, non-conjugated diene monomer as needed, and catalyst are fed to a single polymerization reactor which works at temperatures of

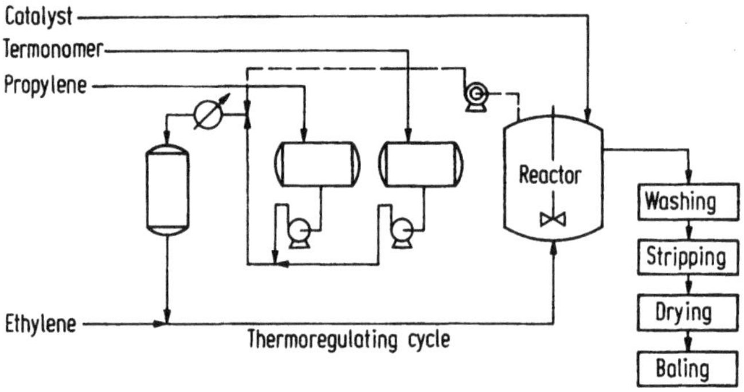

Figure 41. Montedison suspension process for olefin elastomers

253 K–303 K, and at pressures of 0.35 to 1.05 MPa. Reaction heat is removed by vaporizing part of liquified monomers which are condensed and recycled to the reactor. The polymer slurry is continuously withdrawn from the reactor, flashed and steam stripped.

5. Commercial Polymers

A. Polyethylene

The name includes a wide range of products mainly differing as regards the linearity of polymer chains and, consequently, the crystallinity, density, and melting point; characteristics which affect the physico-mechanical as well as the optical properties of the polymer. These products are subdivided into three big families:

1) Low Density Polyethylene (LDPE)
2) Linear Low Density Polyethylene (LLDPE)
3) High Density Polyethylene (HDPE).

The main differences concerning the molecular structure of the various polyethylenes may be roughly visualized as follows:

LDPE = short (mainly-C_4H_9) and long chain branching irregulary distributed along the polymer chains

LLDPE = short ($-C_2H_5$, $-C_4H_9$ or $-C_6H_{13}$) branching regulary distributed along the polymer chains

HDPE = branching practically absent.

1. LDPE

These polymers are obtained by radical polymerization at high temperatures and pressures and, according to polymerization conditions, their density varies between 0.91 and 0.935 g cm^{-3}. Low density is ascribable to branching of the main chain, mainly consisting of ethyl or butyl groups, and due to chain transfer reactions during polymerization. Long branches (above 10 C atoms) are also present, which, although scarcely affecting polymer crystallinity and density, considerably influence the rheological properties.

2. HDPE

Although there are slight differences in the molecular structure of these polymers, which are all prepared with catalysts based on transition metals, all products are characterized by a much lower number of branches, a higher crystallinity, a higher density, and a higher melting point as compared with LDPE.

Some commercial polymers used for particular applications (bottles) are modified by ethylene copolymerization with small quantities of α-olefins in order to enhance the resistance to environmental stress cracking. HDPE is mostly employed in blow moulded and injection moulded applications where rigidity and toughness are required. Other properties which make the polymer very much appreciated are chemical resistance, good gas barrier and dielectric properties. Molecular weight and molecular weight distribution (MWD) affect the rheological properties of the polymer. A blow moulding grade product requires a wide molecular weight distribution ($\bar{M}_w/\bar{M}_n$ above 10), whereas an injection moulding grade product requires a narrow molecular weight distribution ($\bar{M}_w/\bar{M}_n$ below 5). Catalysts and processes must therefore be tailored in order to obtain the desired molecular weight distribution.

3. LLDPE

The family includes polyethylene resins with density ranging from 0.92 to 0.945 g cm^{-3}. These products, which have found widespread application only towards the end of the seventies, are prepared *via* random copolymerization of ethylene with small quantities of α-olefins (1-butene, 1-hexene, 1-octene) in the presence of Phillips or Ziegler-Natta catalysts. The polymer density is controlled by means of the quantity of α-olefin copolymerized with ethylene.

Due to a more linear structure of polymeric chains, LLDPE offers mechanical properties and heat resistance higher than LDPE. However, less branching and higher crystallinity worsen optical properties and involve more difficult processability. The main applications of LLDPE are in the field of films (about 50%) and cable lining.

B. Polypropylene

The commercial polymer consists of isotactic polypropylene with a high degree of steric purity (I.I. above 95%). With the exception of poly-4-methyl-1-pentene, polypropylene is the lightest polymer among commercial products. Among the properties that have underlain the commercial success of the polymer may be mentioned a high melting temperature, good mechanical properties, a very good resistance to environmental stress cracking, excellent dielectric and chemical resistance properties, a low permeability to gas, and a good processability. The presence of tertiary carbon atoms in the polymer chain makes the polymer sensitive to thermal and photo-oxidative degradation. An effective protection with stabilizers is therefore needed. One of the main defects of polypropylene is its brittleness at temperatures below 273 K. This deficiency was solved by blending homopolymer with ethylene-propylene elastomers or by copolymerizing propylene with ethylene after homopolymerization (block copolymers or polyallomers). Polypropylene has found most widespread application both as moulding material and as film- and fibregrade product.

C. Poly-1-Butene

Isotactic polybutene synthesis with Ziegler-Natta catalysts came soon after polypropylene synthesis, but the product has not enjoyed a widespread favour.

One of the main reasons that have restrained the polymer development is the fact that it crystallizes quickly from the melt into an unstable crystalline modification which is slowly converted into a thermodynamically more stable modification. The two forms differ insofar as concerns helical structure and density, therefore the product undergoes deformation with time. The problem has been solved by copolymerizing butene with small quantities of higher α-olefins.

The most important applications of the product are in the fields of commercial piping and film, where its good thermo-mechanical properties and its exceptional creep resistance are appreciated.

D. Poly-4-Methyl-1-Pentene

Among the number of isotactic polymers of α-olefins with branched chain obtained *via* Ziegler-Natta catalysis, only poly-4-methyl-1-pentene has found a modest utilization. The interest for this polymer is connected with its high melting point (508 K), its low density (0.83 $g\,cm^{-3}$). and its extreme clarity. The commercial products are copolymers containing small quantities of linear α-olefins. The presence of the comonomer enhances optical properties as well as other physical properties.

E. Ethylene — Propylene Copolymers

Two different types of olefin elastomer are commercially produced. A saturated (EPM, ethylene-propylene copolymer) and an unsaturated one (EPDM, ethylene-propylene-non conjugated diene terpolymer).

The former requires organic peroxides to effect crosslinking while the latter is cured with sulphur, zinc oxide and accelerator systems of the conventional type. The properties that make olefin elastomers attractive are the excellent resistance to ozone, weathering, and heat and light-aging, in which respect they are superior to natural rubber and most other unsaturated rubbers.

The largest applications of these elastomers are in the automotive field, (except tyre manifacture), and in the electrical industry.

References

1. Natta, G.: Experientia Suppl. **7**, 21 (1957)
2. Pasquon, I.: Stereoregular Linear Polymers, In: Encyclopedia of Polymer Science and Technology, Mark, H. F., Bikales, N. M. (eds.) New York: Wiley (Interscience) 1970, vol. 13 pp. 13–86
3. Standard Oil Co. (Indiana): U.S. Pat. 2.692.258 (Filed April 28, 1951)
4. Friedlander, H. N.: J. Polym. Sci. **38**, 91 (1959)
5. Juweland, O. O., Peters, E. F., Shepard, J. W.: Polymer Preprints **10**, 263 (1969)
6. Tadokoro, Y. T., Kasano, K., Hosai, T.: Kogyo Kogakuasshi **70**, 144 (1967)
7. Phillips Petroleum Company: U.S. Pat. 2.825.721 (Filed Jan, 27, 1953)
8. Guyot, A.: Rev. Caut. Plast., Ed. Plast. **2**, 37 (1965)
9. Kazanski, V. B.: Kinet. Katal. **8**, 1125 (1967)
10. Clark, A.: Catal. Rev. **3**, 145 (1969)
11. Yermakov, Yu. I., Zakharov, V. A.: Adv. Catal. **24**, 173 (1975)
12. Mazakchiev, L. G., Yermakov, Yu. I., Kolovertnov, G. D.: Kinet. Katal. **8**, 188 (1967)

13. Hogan, J.: J. Polym. Sci., Part A-1, **8**, 2637 (1970)
14. Baker, L. M., Carrick, W. L.: J. Org. Chem. **35**, 774 (1970)
15. Clark, A., Finch, J. N., Ashe, B. H.: Proc. Intern. Congr. Catalysis, 3rd, Amsterdam, 1964
16. Clark, A.: Adv. Chem. Ser. **91**, 660 (1969)
17. Clark, A.: Ind. Eng. Chem. **59**, 29 (1967)
18. Phillips Petroleum: U.S. Pat. 3.225.023 (Filed Nov. 30, 1962)
19. Phillips Petroleum: U.S. Pat. 3.622.521 (1971)
20. Union Carbide: U.S. Pat. 4.011.382 (1977)
21. Ballard, D. G. H.: Adv. Catal. **23**, 263 (1973)
22. Zakharov, V. A., Yermakov, Yu. I.: Catal. Rev. Sci. Eng. **19** 67 (1979)
23. Karol, F. J., Karapinka, G. L., Wu Ch Dow, A. W., Johnson, R. N., Carrick, W. L.: J. Polym. Sci., Part A-1 **10**, 2621 (1972)
24. Boehm, H. P.: Adv. Catal. **16**, 179 (1966)
25. Yermakov, Yu. I., Kutznetzov, B. N., Karaktchiev, L. G., Derbeneva, S. N.: Kinet. Katal. **14**, 709 (1973)
26. Ballard, D. G. H.: J. Polym. Sci., Part A-1 **13**, 2191 (1975)
27. Karol, F. J., Johnson, R. N.: J. Polym. Sci., Part A-1 **13**, 1607 (1975)
28. Kazatchiev, L. G., Yermakow, Yu. I.: J. Molec. Catal. **2**, 421 (1977)
29. Chem. Eng. (Nov., 26) 72 (1973)
30. Carrick, W. L., Turbett, R. J., Karol, F. J., Karapinka, G. L., Fox, A. S., Johnson, R. N.: J. Polym. Sci., Part A-1, **10**, 2609 (1972)
31. Natta, G., Pino, P., Mazzanti, G., Giannini, U.: J. Amer. Chem. Soc. **79**, 2975 (1957)
32. Breslow, D. S., Newburg, N. R.: J. Amer. Chem. Soc. **79**, 5073 (1957)
33. Reichert, K. H., Meyer, K. R.: J. Polym. Sci., Part A-1, **10**, 895 (1972)
34. Reichert, K. H., Meyer, K. R.: Makromol. Chem. **169**, 163 (1973)
35. Long, W. P., Breslow, D. S.: Justus Liebigs Ann. Chem. **1975**, 463
36. Andresen, A., Cordes, H. G., Herwing, J., Kaminsky, W., Merck, A., Mottweiler, R., Pein, J., Sinn, H., Vollmer, H.: Angew. Chem. Int. Ed. Engl. **15**, 630 (1976)
37. Sinn, H., Kaminsky, W.: Adv. Organomet. Chem. **18**, 240 (1980)
38. Kaminsky, W., Sinn, H., Woldt, R.: IUPAC Int. Symp. Macromol. Florence 1980 Preprints vol. 2, p. 59
39. Kaminsky, W., Miri, H., Woldt, R.: International Symposium Transition Metal Catalysed Polymerization: Unsolved Problems, Midland, USA, 1981
40. Farina, M., Ragazzini, M.: Chim. Ind. (Milan) **40**, 816 (1958)
41. Bawn, C. E. H., Symox, R.: J. Polym. Sci. **34**, 139 (1959)
42. Tajima, Y., Kunioka, E.: J. Polym. Sci., Part. A-1, **6**, 241 (1968)
43. Hiraki, K., Ivrone, T., Hirai, H.: J. Polym. Sci., Part. A-1, **8**, 2543 (1970)
44. Bawn, C. E. H., North, A. M., Wolker, J. S.: Polymer **5**, 419 (1964)
45. Henrici-Olivé, G., Olivé, S.: Angew. Chem. Int. Ed. Engl. **10**, 776 (1971)
46. Carrick, W. L.: J. Amer. Chem. Soc. **80**, 6455 (1958)
47. Carrick, W. L., Chasar, A. G., Smith, J. J.: J. Amer. Chem. Soc. **82**, 5319 (1960)
48. Karapinka, G. L., Smith, J. J., Carrick, W. L.: J. Polym. Sci. **50**, 143 (1961)
49. Siove, A., Fontanille, M.: Makromol. Chem. **181**, 1815 (1980)
50. Ballard, D. G. H., Van Lenden, PW.: Makromol. Chem. **154**, 177 (1972)
51. Giannini, U., Zucchini, U., Albizzati, E.: J. Polym. Sci., Part B, **8**, 405 (1970)
52. Ballard, D. G. H.: Polym. Prepr. Amer. Chem. Soc. Div. Polym. Chem. **15**, N^{0} 1, 364 (1974)
53. Wilke, G., Bogdanovic, B., Hardt, P., Heimbach, P., Kein, W., Kroner, M., Oberkirch, W., Tanaka, U., Steinrucke, E., Walter, D., Zimmermann, H.: Angew. Chem., Int. Ed. Engl. **5**, 151 (1966)
54. Ballard, D. G. H., Jones, E., Medinger, T., Pioli, J. J. C.: Makromol. Chem. **148**, 175 (1971)
55. Watson, P. L.: J. Amer. Chem. Soc. **104**, 337 (1982)
56. Ballard, D. G. H., Courtis, A., Holton, J., McMeeking, J., Pearce, R.: J. Chem. Soc. Chem. Commun. **1978**, 994
57. Wilke, G.: How nature lost a monopoly. In: Structural Order in Polymers. Ciardelli, F., Giusti, P. (eds.). Oxford, Pergamon Press, 1981, pp. 11–23

58. Ziegler, K., Breil, H., Martin, H., Holzkamp, R.: Ger. Pat. 973626 (Filed Nov. 17, 1953)
59. Ziegler, K., Holzkamp, R., Breil, H., Martin, H.: Angew. Chem. **67**, 426 (1955)
60. Boor, J. Jr. (ed.): Ziegler-Natta Catalysts and Polymerizations. New York—London: Academic Press 1979
61. Gaylord, N. G., Mark, H. F. (eds.): Linear and Stereoregular Addition Polymers. New York: Wiley (Interscience) 1959
62. Reich, L., Schindler, A. (eds.): Polymerization by Organometallic Compounds. New York: Wiley (Interscience) 1966
63. Frankel, M., Rabani, J., Zilkha, A.: J. Polym. Sci. **28**, 387 (1958)
64. Zilkha, A., Calderon, N., Frankel, M.: J. Polym. Sci. **33**, 141 (1958)
65. Zilkha, A., Ottolenghi, A., Frankel, M.: J. Polym. Sci. **39**, 347 (1959)
66. McGowan, J. C., Ford, B. M.: J. Chem. Soc. **1958**, 1149
67. Gilchrist, A.: J. Polym. Sci. **34**, 49 (1959)
68. Kosheskov, K. A., Kargin, V. A., Sheverdina, N. I., Sogolova, T. I., Paleyeva, I. Ye., Paleyev, O. A.: Polym. Sci. USSR (Engl. Transl.) **4**, 1564 (1963)
69. Tanaka, T., Matsuda, T., Kimiginia, K., Okuzumi, I., Akutsu, K.: Kogyo Kagaku Zasshi **73**, 1061 (1970)
70. Bayer, A. G.: Farbenfabriken: Brit. Pat. 856.645 (1961)
71. Cooper, W.: Stereospecific Polymerization In: Progress in High Polymers. Robb, J. C., Peaker, F. W. (eds.). London: Heywood 1960, pp. 288–291
72. Schindler, A.: Mechanism of anionic polymerization In: Crystalline Olefins Polymers. Ruff, R. A., Doak, K. W. (eds.). New York: Wiley (Interscience) 1965, pp. 153–163
73. Ziegler, K., Martin, H., Stedefeder, J.: Tetrahedron Lett. **20**, 12 (1959)
74. Cooper, M. L., Rose, J. B.: J. Chem. Soc. **1959**, 795
75. Friedlander, H. N., Oita, K.: Ind. Eng. Chem. **49**, 1885 (1957)
76. Berger, M. N., Grieveson, B. M.: Makromol. Chem. **83**, 80 (1965)
77. Zakharov, V. A., Bukatov, G. D., Yermakov, Yu. I.: Russian Chemical Rev. **49**, 1097 (1980)
78. Boucher, D. G., Parsons, L. W., Haward, R. N.: Makromol. Chem. **175**, 3461 (1974)
79. Machon, J. P., Hermautaud, R., Houzeaux, J. P.: J. Polym. Sci., Polym. Symposia **52**, 107 (1975)
80. Machon, J. P.: Europ. Polym. J. **12**, 805 (1976)
81. Hsieh, H. L.: Polymer J. **12**, 597 (1980)
82. Böhm, L. L.: Angew. Makromol. Chem. **89**, 1 (1980)
83. Karol, F. J.: Supported catalysts for olefin polymerization. In: Encyclopedia of Polymer Science and Technology, Mark, H. F., Bikales, N. M. (eds.). New York: Interscience 1976. Suppl. Vol. 1, pp. 120–146
84. Diedrich, B.: Appl. Polym. Symp. **26**, 1 (1975)
85. Diedrich, B.: Polym. Prepr. Amer. Chem. Soc. Div. Polym. Chem. **16**, 316 (1975)
86. Weissermel, K., Cherdron, H., Berthold, J., Diedrich, B., Kerl, K. D., Rust, K., Strametz, H., Toth, T.: J. Polym. Sci., Polym. Symp. **51**, 187 (1975)
87. Kashiwa, N.: Polymer J. **9**, 603 (1980)
88. British Petrochemical: Brit. Pat. 841.822 (1955)
89. Cabot Corp: Fr. Pat. 1.349.864 (1964)
90. Baulin, A. A., Rodionov, A. G., Ivanchev, S. S., Domareva, N. M.: Eur. Polym. J. **16**, 937 (1980)
91. Escalona, A. M., Villalba, J.: Polymer **18**, 179 (1977)
92. Maksimov, N. G., Kushnareva, E. G., Zakharov, V. A., Anufrienko, V. F., Zhdan, P. A., Yermakov, Yu. I.: Kinet. Katal. **15**, 738 (1974)
93. Zakharov, Y. A.: Kinet. Katal. **21**, 892 (1980)
94. Solvay Cie: Belg. Pat. 650.679 (1963)
95. Solvay Cie: Belg. Pat. 726.839 (1968)
96. Montedison: Belg. Pat. 728.002 (1968)
97. Mitsui Petrochemical Industries: Jap. Pat. 70-40295 (1967)
98. Eley, D. D., Keiz, D. A., Rudham, R.: J. Chem. Soc. Faraday I **72**, 1685 (1976)
99. Soga, K., Katano, S., Akimoto, Y., Kagiya, T.: Polymer J. **5**, 128 (1973)

100. Eley, D. D., Keir, D. A., Rudham, R.: J. Chem. Soc. Faraday I **73**, 1738 (1977)
101. Baulin, A. A., Semenova, A. S., Stefanovich, L. G., Chirkov, N. M . Stafeyev, A. V.: Vysokomol. Soed. **A16**, 2688 (1974)
102. Baulin, A. A., Sokolov, V. N., Semenova, A. S., Chirkov, N. M., Shalayeva, L. E.: Vysokomol. Soed. **A17**, 46 (1975)
103. Montedison: Brit. Pat. 1286867 (1973)
104. Montedison: Brit. Pat. 1305610 (1973)
105. Giannini, U.: Makromol. Chem. Suppl. **5**, 216 (1981)
106. Bassi, I. W.: Unpublished results
107. Xie Youchang, Gui Linlin, Liu Wauqui, Bu Naiyu, Tang Yauqui and High Activity Polyethylene Catalyst Subgroup: Scientia Sinica **22**, 1045 (1979)
108. Makhtarulin, S. I., Moroz, E. M., Vermel, E. E., Zakharov, V. A.: React. Kinet. Catal. Lett. **9**, 269 (1978)
109. Makhtarulin, S. I., Zakharov, V. A., Moroz, E. M., Maksimov, N. G.: In: Catalysts containing supported complexes. Novosibirsk, Institute of Catalysis **1**, 205 (1980)
110. Zakharov, V. A., Bukatova, Z. K., Makhtarulin, S. I., Chumaevskii, N. B., Yermakov, Ya. I.: Vysokomol. Soed. **A-21**, 466 (1979)
111. Mitsui Petrochemical Industries: U.S. Pat. 3.642.746 (1968)
112. Montedison: Ger. Offen. 2.632.591 (1977)
113. Bart, J. C. J., Bassi, I. W., Calcaterra, M., Albizzati, E., Giannini, U., Parodi, S.: Z. Anorg. Allg. Chem. **482**, 121 (1981)
114. Greco, A., Bertolini, G., Cesca, S.: International Symposium on Macromolecules, Florence Italy, 1980, Vol. 2, p. 36
115. Hoechst A.G.: Belg. Patents 737.778 (1968), 758.994 (1969), 780.530 (1971)
116. Solvay Cie: Belg. Pat. 743.325 (1969)
117. Böhm, L. L.: Polymer **19**, 553 (1978)
118. Haward, R. N., Roper, A. N., Fletcher, K. L.: Polymer **14**, 367 (1973)
119. Duck, E. W., Grant, D., Butcher, A. B., Timms, D. G.: Eur. Polymer Z. **10**, 77 (1974)
120. Damyanov, D., Velikova, M., Petkov, L.: Eur. Polym. J. **15**, 233 (1979)
121. Baulin, A. A., Babaina, E. V., Rodionov, A. G., Kopylov, V. M., Ivanchev, S. S.: Kinet. Katal. **22**, 379 (1981)
122. Radenkov, P., Petrova, T., Petkov, L., Jelyazkova, D.: Eur. Polymer J. **11**, 313 (1975)
123. Radenkov, P., Petkov, L., Karaenev, S., Kyrtcheva, R.: Eur. Polym. J. **12**, 427 (1976)
124. Pekov, L., Kyrtcheva, R., Radenkov, P., Dobreva, D.: Polymer **19**, 567 (1978)
125. Petkov, L., Radenkov, P., Kyrtcheva, R.: Polymer **19**, 570 (1978)
126. Stamicarbon: Belg. Pat. 751315 (1969)
127. Kabanov, V. A., Smetanynk, V. I.: Makromol. Chem. Suppl. **5**, 121 (1981)
128. Mitsubishi Chem. Co.: Jap. Pat. 91797 (1977)
129. Natta, G., Corradini, P.: Atti Accad. Naz. Lincei, Mem. [8] 4, **73** (1955)
130. Corradini, P., Pasquon, I.: Atti Accad. Naz. Lincei, Rend. Cl. Sci. Fis. Mat. Nat. [8], **19**, 453 (1955)
131. Natta, G., Bassi, I. W., Corradini, P.: Makromol. Chem. **18**, 455 (1955)
132. Natta, G., Pasquon, I., Corradini, P., Pegoraro, M., Zambelli, A.: Atti Accad. Naz. Lincei Rend. Cl. Sci., Fis. Mat. Nat. [8] **28**, 539 (1960)
133. Natta, G., Peraldo, M., Allegra, G.: Makromol. Chem. **75**, 215 (1964)
134. Natta, G., Pino, P., Mazzanti, G.: Gazz. Chim. Ital. **87**, 528 (1957)
135. Montecatini: U.S. Pat. 3715344 (Field June 8, 1954)
136. Natta, G., Pino, P., Mazzanti, G.: Ital. Pat. 526101 (1954) to Montecatini
137. Natta, G., Pino, P., Mazzanti, G., Longi, P.: Gazz. Chim. Ital. **87**, 549 (1957)
138. Klem, W., Krose, E.: Z. Anorg. Allg. Chem. **253**, 209 (1947)
139. Reed, J. W., Mac Wood, G. E.: 133rd Meeting ACS, San Francisco April 1958
140. Natta, G.: Chim. Ind. (Milan) **42**, 1207 (1960)
141. Natta, G., Corradini, P., Allegra, G.: J. Polym. Sci. **51**, 399 (1961)
142. Natta, G., Corradini, P., Bassi, I. W., Porri, L.: Atti Accad. Nazl. Lincei Rend. Cl. Sci., Fis. Mat. Nat. [8] **24**, 121 (1958)
143. Natta, G., Corradini, P., Allegra, G.: Atti Accad. Naz. Lincei, Red. Cl. Sci. Fis. Mat. Nat. [8] **26**, 155 (1959)

144. Schumb, W. C., Sundström, J.: J. Amer. Chem. Soc. **55**, 596 (1933)
145. Bock, F., Moser, L.: Monatsch. Chem. 34, 1825 (1913)
146. Wilchinsky, Z. W., Looney, R. W., Tornquist, E. G. M.: J. Catal. **28**, 351 (1973)
147. Stauffer Chemical Co., Speciality Chemical Division, Titanium Trichloride, 1972; Toyo Stauffer Chemical Co., Titanium Trichloride, 1979
148. Ruff, O., Neumann, F.: Z. Anorg. Chem. **128**, 81 (1923)
149. Tornquist, E. G. M., Richardson, J. T., Wilchinsky, Z. W., Looney, R. W.: J. Catal. **8**, 189 (1967)
150. Natta, G., Zambelli, A., Pasquon, I., Giongo, G. M.: Chim. Ind. (Milan) **43**, 307 (1966)
151. Natta, G., Pasquon, I.: Adv. Catal. **11**, 1 (1959)
152. Natta, G., Pasquon, I., Zambelli, A., Gatti, G.: J. Polym. Sci. 51, 387 (1961)
153. Jordan, D. D.: Ziegler-Natta Polymerization Catalysis, Monomers and Polymerization Procedures. In: Stereochemistry of Macromolecules, Vol. 1. Ketley, A. D. (ed.) New York: M. Dekker 1967, pp. 1–45
154. Hoeg, D. F.: Mechanism of Ziegler-Natta Catalysis Experimental Foundations. In: Stereochemistry of Macromolecules, Vol. 1. Ketley, A. D. (ed.). New York: M. Dekker 1967, pp. 47–144
155. Mazzanti, G., Longi, P.: Rend. Ist. Lombardo Sci. Lettere **91**, 755 (1957)
156. Natta, G., Mazzanti, G., Giannini, U., De Luca, D., Bandini, F.: Makromol. Chem. **76**, 54 (1964)
157. Dawans, F., Teyssiè, P.: Bull. Soc. Chim. France **1963**, 2376
158. Natta, G., Pino, P., Mazzanti, G., Lanzo, R.: Chim. Ind. (Milan) 39, 1032 (1957)
159. Beerman, C., Bestian, H.: Angew. Chem. **71**, 618 (1959)
160. Matlak, A. S., Breslow, D. S.: J. Polym. Sci. Part. A.3, 2853 (1965)
161. Benning, C. I., Wszolek, W. R., Werber, F. X.: J. Polym. Sci., Part A-1 **6**, 755 (1968)
162. Boor, J., Jr.: Ind. Eng. Chem. Prod. Res. Develop. **9**, 437 (1970)
163. Boor, J., Jr.: The Nature of the Active Site in the Ziegler-Type Catalyst. In: Macromolecular Reviews. Vol. 2. Peterlin, A., Okamura, S., Goodman, M., Zimm, B. H., Mark, H. F. (eds.). New York—London—Sydney: Interscience Publishers 1967 pp. 115–269
164. Herwig, H., Gumboldt, A. G. M., Weissermal, R. V.: U.S. Pat. 3501415 (1970) to Farbwerke Hoechst A.G.; Idemitsu Kosan Co. Ltd. Brit. Pat. 1184593 (1966)
165. Natta, G., Zambelli, A., Pasquon, I., Giongo, G. M.: Chim. Ind. (Milan) **43**, 1298, 1307 (1966)
166. Vandenberg, E. I.: U.S. Pat. 3,051,690 (1956) to Hercules Powder Co
167. Bua, E., Luciani, L.: Ital. Pat. 554013 (1957) to Montecatini
168. Natta, G., Mazzanti, G., Longi, P., Bernardini, F.: Chim. Ind. (Milan) **41**, 519 (1959)
169. Natta, G., Giachetti, E., Pasquon, I., Pajaro, G.: Chim. Ind. (Milan) **42**, 1091 (1960)
170. Natta, G., Pasquon, I., Giuffrè, L.: Chim. Ind. (Milan) **43**, 871 (1961)
171. Murray, J., Sharp, M. J., Hockey, J. A.: J. Catal. **18**, 52 (1970)
172. Marconi, W., Mazzei, A., Cesca, S., De Maldè, M.: Chim. Ind. (Milan) **51**, 1084 (1969)
173. Cesca, S., Santostasi, M. L., Bertolini, G.: Chim. Ind. (Milan) **51**, 1093 (1969)
174. Mitsubishi Petrochem. Co. Ltd.: Brit. Pat. 1092390 (1966); U.S. Pat. 3622552 (1971)
175. Razuvaev, G. A., Minsker, K. S., Fedoseeva, G. T., By Khovskii, V. K.: Polym. Sci. USSR **2**, 299 (1961)
176. Coover, H. W. Jr., McConnel, R. L., Joyner, F. B.: Relationship of Catalyst Composition to Catalytic Activity for the Polymerization of α-Olefins. In: Macromolecular Reviews. Vol. 1. Peterlin, A., Okamura, S., Goodman, M., Zimm, B. H., Mark, H. F. (eds), New York—London—Sydney: Interscience Publishers 1966, pp. 91 to 118
177. Sivaran, S.: Ind. Eng. Chem. Prod. Res. Dev. **16**, 121 (1977)
178. Farbwerke Hoechst: Germ. Pat. 1940329 (1970)
179. Hagemeyer, H. J., Jr., Edwards, M. B.: Brit. Pat. 943206 (1963) (U.S. Priority, 1959), to Eastman Kodak
180. Edwards, M. B., Park, K. V., Carter, E. H., Jr.: Fr. Pat. 1450785 (1966) (U.S. Priority, 1964) to Eastman Kodak

182. Zambelli, A., Segre, A. L., Marinangeli, A., Gatti, G.: Chim. Ind. (Milan) **48**, 1 (1966)
183. Zambelli, A., Gatti, G., Marinangeli, A., Cabassi, F., Pasquon, I.: Chim. Ind. (Milan) **48**, 333 (1966)
184. Zambelli, A., Giongo, G. M., Segre, A. L.: Chim. Ind. (Milan) **50**, 1185 (1968)
185. Watt, W. R.: J. Polym. Sci. Part. A-1, **7**, 787 (1969)
186. Zambelli, A., Di Pietro, J., Gatti, G.: J. Polym. Sci. Part. A-**1**, 403 (1963)
187. Hermans, J. P., Henrioulle, P.: U.S. Pat. 3769233 (1973) to Solvay Cie
188. Yermakov, Yu. I., Zakharov, V. A.: The number of propagation centers in solid catalysts for olefin polymerization. In: Coordination Polimerization; A Memorial to Karl Ziegler. Chien, J. C. W. (ed.). New York—San Francisco—London: Acad. Press. 1975, pp. 91–133
189. Chien, J. C. W., Hsieh, J. T. T.: J. Polym. Sci. Polym. Chem. Ed. **14**, 1915 (1976)
190. Gardner, K., Parsons, J. W., Haward, R. N.: J. Polym. Sci. Polym. Chem. Ed. **16**, 1683 (1978)
191. Duck, E. W., Grant, D., Kronfli, E.: Eur. Polym. J. **15**, 625 (1979)
192. Kashiwa, N.: International Symposium Transition Metal Catalysed Polymerizations: Unsolved Problems. Midland USA 1981
193. Giannini, U., Cassata, A., Longi, P., Mazzocchi, R.: Belg. Pat. 785332 (1972) to Montedison
194. Giannini, U., Cassata, A., Longi, P., Mazzocchi, R.: Belg. Pat. 785334 (1972) to Montedison
195. Galli, P., Luciani, L., Cecchin, G.: Angew. Makromol. Chem. **94**, 63 (1981)
196. Luciani, L., Giannini, U.: Mater. Plast. Elastomeri **1977**, 417
197. Susuki, E., Tamura, M., Doi, Y., Keii, T.: Makromol. Chem. **180**, 2235 (1979)
198. Soga, K., Terano, M., Ikeda, S.: Polym. Bull. **1**, 856 (1979)
199. Doi, Y., Suzuki, E., Keii, T.: Makromol. Chem. **181**, 293 (1981)
200. Goodall, B. L.: International Symposium Transition Metal Catalysed Polymerizations: Unsolved Problems, Midland USA, 1981
201. Langer, A. W., Burkdardt, T. J., Steger, J.: International Symposium Transition Metal Catalysed Polymerizations: Unsolved Problems, Midland USA, 1981
202. Taylor, T. W., Choi, K. Y., Yuan, H., Ray, H.: International Symposium Transition Metal Catalysed Polymerizations: Unsolved Problems, Midland USA, 1981
203. Kashiwa, N., Yoshitake, J.: Makromol. Chem. Rapid Commun. **3**, 211 (1982)
204. Chien, J. C. W.: New Supported Catalysts for Stereoregular Polymerization of α-olefins. In: Preparation and Properties of Stereoregular Polymers, Lenz, R. W., Ciardelli, F. (eds.). Dordrecht—Boston—London: D. Reidel Publ. Comp. 1980, pp. 113–130
205. Pino, P., Oschwald, A., Ciardelli, F., Carlini, C., Chiellini, F.: Stereoselection and Stereoelection in α-olefins Polymerization, In: Coordination Polymerization, Chien, J. C. W. (ed.) New York: Academic Press, 1975, pp. 25–72
206. Soga, K., Izumi, K., Ikeda, S., Keii, T.: Makromol. Chem. **178**, 337 (1977)
207. Billingham, N. C.: Brit. Polym. J. **6**, 299 (1974)
208. Sinn, H., Kaminsky, W., Vollmer, H. J., Woldt, R.: Angew. Chem. **92**, 396 (1980)
209. Pino, P., Mühlhaupt, R.: Angew. Chem. Int. Ed. Engl. **19**, 857 (1980)
210. Natta, G., Pasquon, I., Corradini, P., Peraldo, M., Pegoraro, M., Zambelli, A.: Atti Accad. Naz. Lincei, Rend. Cl. Sci. Fis. Mat. Nat. [8] **28**, 539 (1960)
211. Natta, G., Pasquon, I., Zambelli, A.: J. Amer. Chem. Soc. **84**, 1488 (1962)
212. Longi, P., Roggero, A.: Ann. Chim. (Rome) **51**, 1013 (1961)
213. Zambelli, A., Natta, G., Pasquon, I., Signorini, R.: J. Polym. Sci. Part C **16**, 2485 (1967)
214. Zambelli, A., Natta, G., Pasquon, I.: J. Polym. Sci. Part C **4**, 411 (1963)
215. Boor, J. Jr., Youngman, E. A.: J. Polymer Sci. Part A-1, **4**, 1861 (1966)
216. Youngman, E. A., Boor, J. Jr.: Makromol. Rev. **2**, 33 (1967)
217. Zambelli, A., Pasquon, I., Signorini, R., Natta, G.: Makromol. Chem. **112**, 160 (1968)
218. Natta, G., Crespi, G., Valvassori, A., Sartori, G.: Rubber Chem. Technol. **36**, (5), 1583 (1963)

219. Pasquon, I., Valvassori, A., Sartori, G.: The Copolymerization of Olefins by Ziegler-Natta Catalysts. In: The Stereochemistry of Macromolecules Vol. 1, Ketley, A. D. (ed.). New York: M. Dekker, 1967, pp. 177–238
220. Crespi, G., Valvassori, A., Flisi, U.: Olefin Copolymers, In The Stereo Rubbers, Saltman, W. M. (ed.), J. Wiley, New York, 1977, pp. 365–431
221. Natta, G., Mazzanti, G., Crespi, G., Moraglio, G.: Chim. Ind. (Milan) **39**, 275 (1957)
222. Natta, G., Mazzanti, G., Valvassori, A., Pajaro, G.: Chim. Ind. (Milan) **39**, 733 (1957)
223. Natta, G., Mazzanti, G., Valvassori, A., Sartori, G., Fiumani, D.: J. Polym. Sci. **51**, 411 (1961)
224. Bier, G.: Angew. Chem. **73**, 186 (1961)
225. Mazzanti, G., Valvassori, A., Pajaro, G.: Chim. Ind. (Milan) **39**, 743 (1957)
226. Mazzanti, G., Valvassori, A., Pajaro, G.: Chim. Ind. (Milan) **39**, 825 (1957)
227. Natta, G., Mazzanti, G., Valvassori, A., Sartori, G.: Chim. Ind. (Milan), **40**, 717 (1958)
228. Marconi, W., Cesca, S., Della Fortuna, G.: Chim. Ind. (Milan) **46**, 1131 (1964)
229. Lukah, C. A., Spurlin, H. M.: Copolymers of α-olefins. In: Copolymerization, Ham, G. (ed.) New York: Wiley Interscience 1964, pp. 115–144
230. Bushick, R. D.: J. Polym. Sci., Part A, **3**, 2047 (1965)
231. Corbelli, L., Milani, F., Fabbri, R.: Intern. Rubber Conference, Nürnberg 1980
232. Corbelli, L.: Ethylene-Propylene Rubbers. In: Developments in Rubber Technology. Whelan, A., Lee, K. S. (eds.), Applied Science Publ., Barking, Essex, England, 1981, pp. 87–129
233. Baldwin, F. P., Borzel, B., Cohen, C. A., Makowski, H. S., Van de Castle, J. F.: Rubber Chem. Techn. **43**, 522 (1970)
234. Gladding, E. K., Fisher, B. S., Collette, J. W.: Ind. Eng. Chem. Prod. Res. Develop. **1**, 65 (1962)
235. German, R., Hank, R., Vaughan, G.: Kaut. Gummi. Kunstst. **19**, 67 (1966)
236. Cunningham, R. E.: J. Polym. Sci., Part A, **3**, 3157 (1967); ibid. Part A, **4**, 1203 (1966)
237. Cesca, S.: J. Polym. Sci., Macromol. Rev. **10**, 1 (1975)
238. Lehr, M. H.: Macromolecules **1**, 178 (1968)
239. Lehr, M. H., Corman, C. J.: Macromolecules **2**, 217 (1969)
240. Seidov, N. M., Dalin, M. A., Absov, A. I.: Dokl. Akad. Nauk, USSR **24**, 46 (1968)
241. U.S. Rubber Co.: U.S. Pat. 3441546 (1969)
242. Stannicarbon: U.S. Pat. 3506629 (1970)
243. Duck, E. W., Cooper, W.: 23rd IUPAC Congress Macromol. Prepr. **2**, 722 (1971)
244. Easterbrook, E. K., Brett, T. J., Loveless, F. C., Matthews, D. N.: 23rd IUPAC Congress Macromol., Prepr. **2**, 712 (1971)
245. Porri, L., Di Corato, A., Natta, G.: Eur. Polym. J. **5**, 1 (1969)
246. Racanelli, P., Porri, L.: Eur. Polym. J. **6**, 751 (1970)
247. Natta, G., Mazzanti, G., Valvassori, A., Sartori, G.: Chim. Ind. (Milan) **40**, 717 (1958)
248. Natta, G., Valvassori, A., Mazzanti, G., Pajaro, G.: Chim. Ind. (Milan) **40**, 896 (1958)
249. Valvassori, A., Sartori, G.: Chim. Ind. (Milan) **44**, 109 (1962)
250. Carbonaro, A., Greco, A.: Chim. Ind. (Milan) **51**, 601 (1969)
251. Suminoe, T., Yamazaki, N., Kambara, S.: Kobushi Kagaku, **20**, 461 (1963)
252. Tsuruta, T., Ishizuka, N.: Autumnal Meeting of Japan Chem. Soc. Sendai, 1968
253. Natta, G., Dall'Asta, G., Mazzanti, G., Pasquon, I., Valvassori, A., Zambelli, A.: J. Amer. Chem. Soc. **83**, 3343 (1961)
254. Natta, G., Dall'Asta, G., Ciampelli, F., Mazzanti, G.: Kolloid Z. **182**, 50 (1962)
255. Natta, G., Dall'Asta, G., Mazzanti, G., Pasquon, I., Valvassori, A., Zambelli, A.: Makromol. Chem. **54**, 95 (1962)
256. Natta, G., Dall'Asta, G., Mazzanti, G.: Chim. Ind. (Milan) **44**, 1212 (1962)
257. Dall'Asta, G., Mazzanti, G.: Makromol. Chem. **61**, 178 (1963)
258. Natta, G., Allegra, G., Bassei, I. W., Corradini, P., Ganis, P.: Makromol. Chem. **58**, 242 (1962)
259. Corradini, P., Ganis, P.: Makromol. Chem. **62**, 97 (1963)

260. Natta, G., Zambelli, A., Pasquon, I., Ciampelli, F.: Makromol. Chem. **79**, 161 (1964)
261. Furukawa, J., Hirai, R., Nakariva, N.: J. Polym. Sci., Part. B, **7**, 671 (1969)
262. Kawasaki, A., Maruyama, I., Taniguchi, M., Hirai, R.: J. Polym. Sci., Part. B, **7**, 613 (1964)
263. Furukawa, J., Amano, H., Hirai, R.: J. Polym. Sci., Part. A-1, **10**, 681 (1972)
264. Furukawa, J.: Angew. Makromol. Chem. 23, 189 (1972)
265. Furukawa, J., Tsuruki, S., Kiji, J.: J. Polym. Sci., Part. A-1, **11**, 2999 (1973)
266. Wieder, W., Witte, J.: IUPAC Int. Symp. Macromol. Florence 1980 Preprint Vol. 2, p. 87
267. Naruzen Petrochemical: Ger. Pat. 1,963,780, 1,964,706 (1970)
268. Bier, G., Gumboldt, A., Lehmann, G.: Plastics Inst. (London) Trans. J. **28**, 98 (1960)
269. Bier, G., Hoffmann, W., Lehmann, G., Seydel, G.: Makromol. Chem. **58**, 1, (1962)
270. Kontos, E. G., Easterbrook, E. K., Gilbert, R. D.: J. Polym. Sci. **61**, 69 (1962)
271. Coover, H. W. Jr., McConnel, R. L., Joyner, F. B., Slonaker, D. F., Combs, R. L.: J. Polym. Sci., Part A, **4**, 2563 (1966)
272. Hagemayer, H. J., Edwards, M. B.: J. Polym. Sci., Part. C, **2**, 731 (1964)
273. Wiberg, K. B., Bartley, W. S.: J. Amer. Chem. Soc. 82, 6375 (1960)
274. Natta, G., Dall'Asta, G., Mazzanti, G.: Angew. Chem. Int. Ed. Engl. **3**, 723 (1964)
275. Natta, G., Dall'Asta, G., Bassi, I. W., Carella, G.: Makromol. Chem. **91**, 87 (1966)
276. Natta, G., Dall'Asta, G., Mazzanti, G., Motroni, G.: Makromol. Chem. **69**, 163 (1963)
277. Dall'Asta, G.: J. Polym. Sci., Part. A-1, **6**, 2397 (1968)
278. Natta, G., Dall'Asta, G., Porri, L.: Makromol. Chem. **81**, 253 (1965)
279. Dall'Asta, G., Ring Opening Polymerizations, In: The Stereo Rubbers, Saltman, W, M. (ed.) J. Wiley, New York, 1977, pp. 285–364
280. Natta, G., Dall'Asta, G., Motroni, G.: J. Polym. Sci. Part. B **2**, 349 (1964)
281. Scott, K. W., Calderon, N., Ofstead, E. A., Judy, W. A., Ward, J. P.: Advances in Chemistry Series, **91**, 399 (1969)
282. Dall'Asta, G., Manetti, R.: Eur. Polym. J. **4**, 145 (1968)
283. Haas, F., Nützel, K., Pampus, G., Theisen, D.: Rubber Chem. Technol. **43**, 1116 (1970)
284. Günther, F., Haas, F., Marwede, G., Nützel, K., Oberkirch, W., Pampus, G., Schön, N., Witte, J.: Angew. Makromol. Chem. **14**, 87 (1970)
285. Calderon, N., Ofstead, E. A., Judy, W. A.: J. Polym. Sci. Part. Part. A-1, **5**, 2209 (1967)
286. Dall'Asta, G., Carella, G.: Brit. Pat. 1,067,367 (Priority, Feb. 11, 1965) to Montecatini
287. Shell Int. Res. Mij.: Brit. Pat. 1118517 (Priority May 20, 1967)
288. Marshall, P. R., Ridgewell, B. J.: Eur. Polym. J., **5**, 29 (1969)
289. Judy, W. A.: Pats. 2.016.360 and 2.016.364 (Priorities Aug. 26, 1968) to Goodyear
290. Keii, T., Sòga, K., Saiki, N.: J. Polym. Sci., Part. C, **16**, 1507 (1967)
291. Schneko, H., Reimöller, M., Weirauch, W., Kern, W.: J. Polym. Sci., Part C, **4**, 71 (1963)
292. Schneko, H., Reimöller, M., Weirauch, W., Bednjiagin, U., Kern, W.: Makromol. Chem. **73**, 154 (1964)
293. Kohara, T., Doi, Y., Shinoyama, M., Keii, T.: Nippon Kagaku Zasshi **1975**, 1664
294. Soga, K., Keii, T.: J. Polym. Sci., Part A-1, **4**, 1971 (1966)
295. Tait, P. J. T.: A Kinetic Model for Hererogeneous Ziegler-Natta Polymerization. In: Coordination Polymerization. Chien, J. C. W. (Ed.). New York: Academic Press 1975, pp. 155–197
296. Zakharov, V. A., Bukatov, G. D., Chumaevskii, N. B., Yermakov, Yu. I.: Makromol. Chem. **178**, 967 (1977)
297. Saltman, W. M.: J. Polym. Sci. **46**, 375 (1960)
298. Vesely, K., Ambroz, J., Vilim, R., Hamrick, O.: J. Polym. Sci. **55**, 25 (1961)
299. Reich, L., Stivala, S. S.: J. Polym. Sci., Part A, **1**, 203 (1963)
300. Pasquon, I.: J. Polym. Sci. **60**, 547 (1962)
301. Rodriguez, L. A. M., Van Looy, H. M., Gabant, J. A.: J. Polym. Sci., Part A, **4**, 1905 (1966)

302. Rodriguez, L. A. M., Van Looy, H. M.: J. Polym. Sci. Part A-1, **4**, 1951 (1966)
303. Van Looy, H. M., Rodriguez, L. A. M., Gabant, J. A.: J. Polym. Sci., Part A-1, **4**, 1927 (1966)
304. Rodriguez, L. A. M., Gabant, J. A.: J. Polym. Sci. **57**, 881 (1962)
305. Ambroz, J., Oseky, P., Mejzlik, J., Hamrik, O.: J. Polym. Sci., Part C, **16**, 423 (1967)
306. Pasquon, I.: J. Pure Appl. Chem. **15**, 465 (1967)
307. Zambelli, A., Pasquon, I., Marinangeli, A., Lanzi, G., Mognaschi, E. R.: Chim. Ind. (Milan) **46**, 1464 (1964)
308. Natta, G., Pasquon, I., Giachetti, E.: Chim. Ind. (Milan) **39**, 993, 1002 (1957)
309. Keii, T.: Kinetics of Ziegler-Natta Polymerization, Tokyo, Kodanska, 1972
310. Bulls, V. N., Higgins, T.: J. Polym. Sci., Part A-1, 8, 1025 (1970)
311. Carradine, W. R., Roase, H. F.: J. Appl. Polym. Sci. **15**, 889 (1971)
312. Munoz-Escalona, A., Parada, A.: Polymer **20**, 474 (1979)
313. Chumaevskii, N. B., Zakharov, V. A., Bukatov, G. D., Yermakov, Yu. I.: Makromol. Chem. **177**, 747 (1976)
314. Burfield, D. R., McKenzie, I. D., Tait, P. J. T.: Polymer **17**, 130 (1976)
315. Vecchi, E., Zannetti, R., Luciani, L., Barbé, P.: Chim. Ind. (Milan) **43**, 741 (1961)
316. Eley, D. D., Keir, D. A., Rudham, R.: J. Chem. Soc. Faraday I, **73**, 1738 (1977)
317. Pasquon, I., Dente, M., Narduzzi, F.: Chim. Ind. (Milan) **41**, 387 (1959)
318. Schmeal, W. R., Stree, J. R.: J. Polym. Sci. Polym. Phys. Ed. **10**, 2173 (1972)
319. Boucheron, B., Levresse, B., Machon, J. P.: Eur. Polym. J. **9**, 1095 (1973)
320. Boucheron, B.: Eur. Polym. J. **11**, 131 (1975)
321. Paja, E., Uhhiat, M., Pajda, A.: Chemik **23**, 86 (1970)
322. Natta, G., Pasquon, I., Svab, J., Zambelli, A.: Chim. Ind. (Milan) **44**, 621 (1962)
323. Cooper, W.: Kinetics of Polymerization initiated by Ziegler Natta and related Catalysts. In: Comprehensive Chemical Kinetics, Bamford, C. H., Tipper, C. F. H. (eds.). Amsterdam—Oxford—New York: Elsevier Sci. Publ. Comp. 1976, pp. 133–249
324. Berger, M. N., Boocock, G., Haword, R. N.: Adv. Catal. **19**, 211 (1969)
325. Carrick, W. L.: Adv. Polym. Sci. **12**, 65 (1973)
326. Tait, P. J. T.: Ziegler-Natta and Related Catalists. In: Developments in Polymerization. Bol. 2. Howard, R. N. (ed.) London: Applied Science Publishers 1979, pp. 81–148
327. Schnecko, H., Jung, K. A., Kern, W.: In: Coordination Polymerization. Chien, J. C. W. (ed.). New York: Academic Press 1975, pp. 73–90
328. Yermakov, Yu. I., Zakharov, V. A.: Russian Chemical Reviews **41**, 203 (1972)
329. Schnecko, H., Kern, W.: Chem. Ztg. **94**, 229 (1970)
330. Tanaka, S., Morikava, H.: J. Polym. Sci., Part A, **3**, 3147 (1965)
331. Schnecko, H., Dost, W., Kern, W.: Makromol. Chem. **121**, 159 (1969)
332. Burfield, D. R.: J. Polym. Sci. Polym. Chem. Ed. **16**, 3301 (1978)
333. Ayrey, G., Mazza, R. S.: Makromol. Chem. **176**, 3353 (1975)
334. Chirkov, N. M.: IUPAC International Macromolecular Symposium, Budapest 1969, p. 297
335. Kissin, Yu. V., Mezhikovsky, S. M., Chirkov, N. M.: Eur. Polym. J. **6**, 267 (1970)
336. Caunt, A. D.: Br. Polym. J. **13**, 22 (1981)
337. Petts, R. W., Waugh, K. C.: Polymer **23**, 897 (1982)
338. Chien, J. C. W.: J. Amer. Chem. Soc. **81**, 86 (1959)
339. Caunt, A. D.: J. Polym. Sci. **64**, 49 (1963)
340. Schnecko, H., Jung, K. A., Grosse, L.: Makromol. Chem. **148**, 67 (1971)
341. Feldman, C. F., Perry, E.: J. Polym. Sci. **46**, 217 (1960)
342. Youngman, E. H., Boor, J. Jr.: J. Polym. Sci., Part B, **4**, 913 (1966)
343. Burfield, D. R., Tait, P. J. T.: Polymer **13**, 315 (1972)
344. Bukatov, G. D., Chumaevskii, N. B., Zakharov, V. A., Zuznetsova, G. I., Yermakov, Yu. I.: Makromol. Chem. **178**, 953 (1977)
345. Jung, K. A., Schnecko, H.: Makromol. Chem. **154**, 227 (1972)
346. Kohn, E., Shurmans, H. J. L., Cavender, J. V., Mendelson, R. A.: J. Polym. Sci. **58**, 681 (1962)

347. Coover, H. W. Jr., Guillet, J. E., Combs, R. L., Joyner, F. B.: J. Polym. Sci., Part A, **4**, 2583 (1966)
348. Childers, C. W.: J. Amer. Chem. Soc. **85**, 229 (1963)
349. Ballard, D. G. H., Jones, E., Wyatt, R. I., Murray, R. T., Robinson, P. A.: Polymer **15**, 169 (1974)
350. Rodriguez, L. A., Van Looy, H. M.: J. Polym. Sci., Part A, **4**, 1971 (1966)
351. Burfield, D. R., Savariaz, C. M.: Macromolecules **12**, 243 (1979)
352. Zakharov, B. A., Bukatov, G. D., Yermakov, Yu. I., Demin, E. A.: Dokl. Akad. Nauk. SSSR, **207**, 857 (1972)
353. Mejtzlik, J., Lesna, M.: Macromol. Chem. **178**, 261 (1977)
354. Varzelhan, V., Burger, T. F., Stein, D. J.: Makromol. Chem. **183**, 489 (1982)
355. Zacharov, V. A., Bukatov, G. D., Yermakov, Yu. I.: Mechanism of Ziegler-Natta Polymerization on the basis of Data on the Number of Active Centers and their Reactivity. In: Polym. Sci. Technol. 1983, 19 (Coord. Polym.) 267–288
356. Chien, J. C. W.: J. Polym. Sci., Part A, **1**, 425 (1963)
357. Tait, P. J. T.: Co-ordination Complex Polymerization. In: Macromolecular Chemistry vol. 1, London Royal Society of Chemistry (1980), pp. 3–21
358. Novokshonova, L. A., Berseneva, G. P., Tsėtkova, V. I., Chirkov, N. M.: Vysokomol. Soėdin. A **9**, 562 (1967)
359. Grievson, B. N.: Makromol. Chem. **84**, 93 (1965)
360. Chumaevski, N. B., Zakharov, V. A., Bukatov, G. D., Yermakov, Yu. I.: Makromol. Chem. **176**, 1959 (1975)
361. Kollar, L., Simon, A., Kallo, D.: Magy. Kem. Poly. **74**, 289 (1968)
362. Guttman, J. Y., Guillet, J. F.: Macromolecules **1**, 461 (1970)
363. Arlman, E. J., Cossee, P.: J. Catal. **3**, 99 (1964)
364. Allegra, G.: Makromol. Chem. **145**, 235 (1971)
365. Kissim, Yu. V., Chirkov, N. M.: Eur. Polym. J. **6**, 525 (1970)
366. Natta, G., Pino, P., Mantica, E., Danusso, F., Mazzanti, G., Peraldo, M.: Chim. Ind. (Milan) **38**, 124 (1956)
367. Zambelli, A., Locatelli, P., Sacchi, M. G., Rigamonti, E.: Macromolecules **13**, 798 (1980)
368. Kohara, T., Shinoyama, M. S., Doi, Y., Keii, T.: Makromol. Chem., **180**, 2139 (1979)
369. Langer, A. W. Jr.: Am. N.Y. Acad. Sci. **295**, 110 (1977)
370. Karol, F. J., Carrick, W. L.: J. Amer. Chem. Soc. **83**, 2654 (1961)
371. Danusso, F., Sianesi, D.: Chim. Ind. (Milan) **44**, 474 (1962)
372. Pino, P.: Adv. Polym. Sci. **4**, 393 (1965)
373. Pino, P., Lorenzi, G. P., Lardicci, L.: Chim. Ind. (Milan) **42**, 712 (1960)
374. Pino, P., Ciardelli, F., Lorenzi, G. P., Natta, G.: J. Amer. Chem. Soc. 84, 1487 (1962)
375. Pino, P., Ciardelli, F., Lorenzi, G. P.: J. Polym. Sci., Part. C **4**, 21 (1963)
376. Pino, P., Guastalla, G., Rotzinger, B., Mulhaupt, R.: International Symposium: Transition Metal Catalyzed Polymerizations: Unsolved Problems, Midland U.S.A., 1981
377. Pino, P., Fochi, G., Piccolo, O., Giannini, U.: to be published
378. Chiellini, E., Marchetti, M.: Makromol. Chem. **169**, 59 (1973)
379. Ciardelli, F., Locatelli, P., Marchetti, M., Zambelli, A.: Makromol. Chem. **175**, 923 (1974)
380. Lazzaroni, R., Salvadori, P., Pino, P.: J. Chem. Soc. Chem. Commun. **18**, 1164 (1970)
381. Chiellini, E., Carlini, C., Botrini, C., Ciardelli, F.: Chim. Ind. (Milan) **51**, 852 (1969)
382. Ciardelli, F., Carlini, C., Montagnoli, C., Lardicci, L., Pino, P.: Chim. Ind. (Milan) **50**, 860 (1968)
383. Pino, P., Guastalla, G., Giannini, U.: to be published
384. Natta, G., Farina, M., Peraldo, M.: Atti Accad. Naz. Lincei, Rend. Cl. Sci. Fis. Mat. Nat. [8] **25**, 424 (1958)
385. Miyazawa, T., Ideguchi, Y.: J. Polym. Sci. Part B **1**, 389 (1963)
386. Zambelli, A., Giongo, M., Natta, G.: Makromol. Chem. **112**, 183 (1968)

387. Natta, G., Corradini, P., Ganis, P., Bassi, I. W., Allegra, G.: Chim. Ind. (Milan) **44**, 532 (1962)
388. Longi, P., Mazzanti. G., Roggero, A., Lachi, M. G.: Makromol. Chem. **61**, 63 (1963)
389. Takegami, Y., Suzuki, T.: Bull. Chem. Soc. Jap. **42**, 848 (1969)
390. Takegami, Y., Suzuki, T., Okazaki, T.: Bull. Chem. Soc. Jap. **42**, 1060 (1969)
391. Suzuki, T., Takegami, Y.: Bull. Chem. Soc. Jap. **43**, 1484 (1970)
392. Zambelli, A., Locatelli, P., Rigamonti, E.: Macromolecules **12**, 156 (1979)
393. Zambelli, A., Bajo, G., Rigamonti, E.: Makromol. Chem. **179**, 1249 (1978)
394. Zambelli, A., Locatelli, P., Sacchi, M. C., Rigamonti, E.: Macromolecules in press
395. Zambelli, A.: Homopolymers and Copolymers of Propylene. In: NMR Basic Principles and Progress, Vol. 4, Berlin—Heidelberg—New York: Springer 1971, pp. 101–108
396. Crains, W. O., Zambelli, A., Roberts, J. D.: Macromolecules **4**, 330 (1971)
397. Corradini, P., Barone, V., Fusco, R., Guerra, G.: Eur. Polym. J. **15**, 1133 (1979)
398. Zambelli, A., Locatelli, P., Zanoni, G., Bovey, F. A.: Macromolecules **11**, 923 (1978)
399. Pino, P., Lorenzi, G. P.: Stereoregular Polymers: Stereochemical Aspects and Synthetic Approaches. In: Preparation and Properties of Stereoregular Polymers. Lenz, R. W., Ciardelli, F. (eds.). Dordrecht—Boston—London: D. Reidel Publ. Comp. 1980, pp. 1–73
400. Cossee, P.: The Mechanism of Ziegler-Natta Polymerization. II. Quantum-Chemical and Crystal-Chemical Aspects. In: the Stereochemistry of Macromolecules. Ketley, A. D., ed. Dekker, M. New York 1967, pp. 145–175
401. Corradini, P., Guerra, G., Fusco, R., Barone, V.: Eur. Polym. J., **16**, 835 (1980)
402. Ivin, K. J., Rooney, J. J., Stewart, C. D., Green, M. L. H., Mahtab, R.: J. Chem. Soc. Chem. Commun. **1978**, 607
403. McKinney, R. S.: J. Chem. Soc. Chem. Commun. **1980**, 490
404. Patat, P., Sinn, H.: Angew. Chem. **70**, 496 (1958)
405. Henrici-Olivé, G., Olivé, S.: Angew. Chem. Int. Ed. Engl. **6**, 790 (1967)
406. Doi, Y., Kinoshiia, J., Morinaga, A., Keii, T.: J. Polym. Sci., Part A-1, **13**, 249 (1975)
407. Natta, S., Zambelli, A., Lanzi, S., Pasquon, I., Mognaschi, E. R., Segre, A. L., Centola, P.: Makromol. Chem. **81**, 161 (1965)
408. Emde, H.: Angew. Makromol. Chem. **60**, 1 (1977); Gumboldt, A., Helberg, J., Schleitzer, G.: Makromol. Chem. **101**, 229 (1967)
409. Zambelli, A., Sacchi, M. C., Locatelli, P.: Macromolecules **12**, 1051 (1979)
410. Zambelli, A., Tosi, C., Sacchi, M. C.: Macromolecules **5**, 649 (1972)
411. Zambelli, A., Wolfgruber, C., Zannoni, G., Bovey, F. A.: Macromolecules **7**, 750 (1974)
412. Bovey, F. A., Sacchi, M. C., Zambelli, A.: Macromolecules **7**, 752 (1974)
413. Zambelli, A., Tosi, C.: Adv. Polym. Sci. **15**, 31 (1974)
414. Zambelli, A., Sacchi, M. C.: Makromol. Chem. **175**, 2213 (1974)
415. Asakura, T., Ando, I., Nishioka, A., Doi, Y., Keii, T.: Makromol. Chem. **178**, 791 (1977)
416. Doi, Y.: Macromolecules **12**, 248 (1979)
417. Zambelli, A., Allegra, G.: Macromolecules **13**, 42 (1980)
418. Locatelli, P., Immirzi, A., Zambelli, A.: Makromol. Chem. **176**, 1121 (1975)
419. Zambelli, A.: A Few Considerations on Stereoregular Propylene Polymerization. In: Coordination Polymerization. Chien, J. C. W. (ed.). New York: Academic Press 1975, pp. 15–24
420. Zambelli, A., Lety, A., Tosi, C., Pasquon, I.: Makromol. Chem. **45**, 73 (1968)
421. Swingler, M. R. M.: Ph. D. Thesis, University of Bradford (1972)
422. Matsuda, T., Keii, T., Kojima, A., Ishida, T.: Kogyo Kagaku Zasshi **71**, 1124, 1128 (1968)
423. Schindler, A.: J. Polym. Sci., Part C, **4**, 81 (1963)
424. Schindler, A.: Makromol. Chem. **73**, 94 (1964)
425. Meshkova, I. N., Kumankova, S. A., Tsvetkova, V. I., Chirkov, N. M.: Polym. Sci. USSR **3**, 1011, 1130 (1963)
426. Christman, D. L.: J. Polym. Sci. Part A-1, **10**, 471 (1972)
427. Singh, D., Merril, R. P.: Macromolecules **4**, 599 (1971)

428. Chien, J. C. W.: J. Polym. Sci. Chem. Ed. **17**, 2555 (1979)
429. Ludlum, D. B., Anderson, A. W., Ashby, C. E.: J. Amer. Chem. Soc. **80**, 1380 (1958)
430. Feilchenfeld, H., Jeselson, M.: J. Phys. Chem. **63**, 720 (1959)
431. Zakharov, V. A., Chumaevskii, N. B., Bukatov, G. D., Yermakov, Yu. I.: Makromol. Chem. **177**, 763 (1976)
432. Reichert, K. H.: Angew. Makromol. Chem. **94**, 1 (1981)
433. Henrici-Olivé, G., Olivé, S.: Adv. Polym. Sci. **6**, 421 (1969)
434. Fink, G., Zoller, W.: Makromol. Chem. **182**, 3265 (1981)
435. Meyer, K., Reichert, K. H.: Angew. Makromol. Chem. **12**, 175 (1970)
436. Fink, G., Rottler, R., Kreiter, C. G.: Angew. Makromol. Chem. **96**, 1 (1981)
437. Long, W. P.: J. Amer. Chem. Soc. **81**, 5312 (1959)
438. Long, W. P., Breslow, D. S.: J. Amer. Chem. Soc. **82**, 1953 (1960)
439. Schnell, D.: Dissertation, TU München (1973)
440. Borisova, L. F., Fuschman, A. E., Vizen, E. I., Chirkov, N. M.: Eur. Polym. J. **9**, 953 (1973)
441. Laucher, J. W., Hoffmann, R.: J. Amer. Chem. Soc. **98**, 1729 (1976)
442. Henrici-Olivé, G., Olivé, S.: Chemtech. **1981**, 746
443. Fink, G., Rottler, R.: Angew. Makromol. Chem. **94**, 25 (1981)
444. Fink, G., Rottler, R., Schnell, D., Zoller, W.: J. Appl. Polym. Sci. **20**, 2779 (1976)
445. Henrici-Olivé, G., Olivé, S.: J. Polym. Sci. Part B **8**, 271 (1970)
446. Zakharov, V. A., Yermakov, Yu. I.: J. Polym. Sci. Part A 1, **9**, 3129 (1971)
447. Kazansky, V. B., Turkevich, J.: J. Catal. **8**, 231 (1967)
448. Ayscough, P. B., Eden, C. E., Steiner, H.: J. Catal. **4**, 278 (1965)
449. Eley, D. D., Rochester, C. H., Scurrel, M. S.: Proc. R. Soc. London, A, **329**, 361 (1972)
450. Vuillaume, G., Revillon, A., Spitz, R., Guyot, A.: J. Macromol. Sci. Chem. Ser. A **5**, 559 (1971)
451. Guyot, A., Daniel, J. C., Durrien, M., Ptack, M.: J. Polym. Sci. Part A **3**, 1765 (1965)
452. Przhevalskaya, L. K., Shvets, V. A., Kazansky, V. B.: J. Catal. **39**, 363 (1975)
453. Krauss, H. L., Mums, L. Z.: Z. Naturforsch. B **34**, 1628 (1979)
454. Beck, D. D., Lunsford, J. H.: J. Catal. **68**, 121 (1981)
455. Zecchina, A., Garrone, E., Morterra, C., Coluccia, S.: J. Phys. Chem. **79**, 978 (1975)
456. Pecherskaya, Yu. I., Kazansky, V. B., Voevodsky, V. V.: Act. Congr. Int. Catal. 2nd, 1961, Paper 108
457. Ghiotti, G., Garrone, E., Coluccia, S., Morterra, C., Zecchina, A.: J. Chem. Soc. Chem. Commun. **1979**, 1032
458. Eden, C., Feilchenfeld, H. F., Haas, Y.: J. Catal.: **11**, 263 (1968)
459. Zakharov, V. A., Dudchenko, V. K., Maksimov, N. G., Yermakov, Yu. I.: React. Kinet. Catal. Lett. **7**, 419 (1977)
460. Zakharov, V. A., Dudchenko, V. K., Paikshtis, E. A., Karakchiev, L. G., Yermakov, Yu. I.: J. Molec. Catal. **2**, 421 (1977)
461. Zakharov, V. A., Maksimov, N. G., Nesterov, G. A., Anufrienko, V. F., Stachastenev, V. K., Yermakov, Yu. I.: J. Molec. Catal. **4**, 167 (1978)
462. Calderon, N., Ofstead, E. A., Judy, W. A.: Angew. Chem. Int. Ed. Engl. **15**, 401 (1976)
463. Dolgoplosk, B. A., Makovetsky, K. L., Gdenko, T. G., Korshak, Yu. V., Tinyakova, E. I.: Eur. Polym. J., **10**, 901 (1974)
464. Wasserman, E., Ben-Efraim, D. A., Volawski, K.: J. Amer. Chem. Soc. **90**, 3286 (1968)
465. Calderon, N., Ofstead, E. A., Judy, W. A.: Central Regional Meeting of the American Chemical Society, Akron, Ohio, May 1968
466. Herisson, J. L., Chauvin, Y.: Makromol. Chem. **141**, 161 (1970)
467. Katz, T. J., McGinnis, J.: J. Am. Chem. Soc. **97**, 3265 (1975)
468. Grubbs, R. H., Bunk, P. L., Carr, D. C.: J. Amer. Chem. Soc. **97**, 3265 (1975)
469. Casey, C. P., Burkhardt, T. J.: J. Amer. Chem. Soc. **95**, 5833 (1973); **96**, 7808 (1974)
470. Casey, C. P., Tuinstra, H. E., Saeman, M. C.: J. Amer. Chem. Soc. **98**, 608 (1976)

471. Chauvin, Y.: Olefin Metathesis: in Giulio Natta, Present Significance of his Scientific Contribution, Carrà, S., Parisi, F., Pasquon, I., Pino, P. (Eds.), Editrici di Chimica, Milano, 1982, pp. 214–230
472. Parker Forsman, J.: Hydrocarbon Processing **51**, 130 (1972)
473. De Bree, S. D.: Hydrocarbon Processing **53**, [2] 115 (1974)
474. Dow Chemical: U.S. Pat. 4120820 (1978)
475. Phillips Petroleum Company: U.S. Pat. 3.248.179; 3.257.362 (1966)
476. Heath, A.: Chem. Eng. (Apr. 3), 66 (1972)
477. Kreutzer, H., Diedrich, B.: Chem. Eng. (Aug. 5), 62 (1974)
478. Stevens, J.: Hydrocarbon Processing **4** [11], 179 (1970)
479. Chem. Eng. News (May 10), 38 (1971)
480. Mishitani, K.: Chem. Econ. and Eng. Rev. **7**, 37, 59 (1971)
481. Hydrocarbon Processing, **58** [11], 224 (1979)
482. Rasmussen, D. M.: Chem. Eng. (Sept. 18), 104 (1972)
483. Chemical Age, (Nov. 18), 20 (1977)
484. Chem. Eng. News (Dec. 5), 21 (1977)
485. Hydrocarbon Processing, **58** [11], 225 (1979)
486. Wisseroth, K.: Chemiker Zeitung **101**, 271 (1977)
487. Giannini, U., Luciani, L., Collina, A., Barbé, P. C.: High Yield Polypropylene Process, Catalyst Performance and Polymer Characteristics, paper presented at the 4th Nat.l and 1st. Latin-American Congress of Petrochemistry, San Carlos de Bariloche, Nov. 14–20, 1976
488. Brockmeier, N. F.: Latest Commercial Technology for Propylene Polymerization; International Symposium — Transition Metal Catalyzed Polymerization — Unsolved Problems, Midland, U.S.A. 1981

Chapter 3

Metal Catalysed Skeletal Reactions of Hydrocarbons

G. L. C. Maire and F. G. Garin

Institut de Chimie
Université Louis Pasteur de Strasbourg
4, rue Blaise Pascal
F-67008 Strasbourg, France

Contents

1. Introduction

In this review article we shall deal with metal catalysed skeletal reactions of hydrocarbons. The aim is to show how it is possible to recognize and demonstrate the reaction mechanisms and to correlate the changes of activity and selectivity with changes in electronic properties or changes in geometry of the metal aggregates. In spite of their great importance in petrochemistry, the mechanisms of hydrocarbons reactions on metal catalysts are much less understood [1, 2] than in homogeneous catalysis, although the latter field has been developed much more recently [3].

Under the reaction conditions, some of the reactants, products, intermediates or other species are adsorbed on the catalyst surface, and the properties of the catalyst surface sites are considerably influenced by the presence of such adsorbed species. Moreover, it is extremely difficult to characterize a reaction intermediate. Even when an adsorbed species has been characterized by physical techniques, one never can prove that this species is a reactive one and is involved in the catalytic reaction.

Catalysis is a rate phenomenon and catalytic kinetics are essentially a multiparameter problem. In metal catalysed skeletal reactions of hydrocarbons the catalyst surface may be covered by many different species a majority of which is not really involved in the critical pathway of the catalytic act but are located in dead-end side lines, whilst only a few are really reaction intermediates. A consequence of such complicated systems, is that an overall kinetic study of the reaction does not give any information on the intimate reaction mechanism. Obviously, the first task when studying such catalytic systems, is to distinguish among the various possible pathways, and this is best achieved by tracer techniques.

Another consequence of surface heterogeneity is the possibility of changing the reaction mechanism by modifying the catalyst surface. For example, some specific sites may be selectively poisoned, so that one reaction pathway becomes greatly predominant.

In some cases, such as catalysis by alloys, the composition of the catalyst at the surface is different from that of the bulk. Under these circumstances, the properties of the catalysts *in the working state* is one of the most important parameters to measure in order to elucidate the mechanisms of skeletal reactions of hydrocarbons. In such a manner Tamaru *et al.* [4] have successfully elucidated the mechanisms of a number of catalytic reactions; for instance, the hydrogenation of olefins, the water gas-shift reaction, and the H_2—D_2 exchange reaction over zinc oxide.

The recent development of metallocarbene and metallocarbyne chemistry [5, 6], and the unambiguous proofs which were given for their participation in some reactions such as metathesis [7] (olefin-carbene addition) and dimerization (dicarbene recombination) [8] allow one now to reconsider for example the mechanisms of skeletal isomerization on metals. The exhaustive review published in 1982 by Gault [9] provides an excellent discussion on the variety of hydrocarbon reactions involving carbon—carbon bond rupture and formation, a number of which may be accounted for by a limited number of adsorbed species and elementary steps. As discussed by Gault [9] besides metallacyclobutanes and metallacarbenes, simpler species such as π-alkyl and π-allylic species may also initiate skeletal rearrangements, especially on poor carbene-promoter metals like palladium [10]. The corresponding reactions, 1-2-alkyl shift and alkyl-alkene insertion, also have analogues in coordination chemistry.

We shall review in this chapter a number of experimental facts, *i.e.* selectivity, structural effects, and kinetic data which could not previously be satisfactorily explained. They receive now a straightforward interpretation when assuming that metallacarbene and metallacarbyne intermediates are involved in skeletal rearrangements of hydrocarbons. Although the present state of theoretical physics and chemistry does not allow yet a quantitative theoretical treatment of a metal surface in the presence of hydrocarbon adsorbates, we believe that the qualitative description of the mechanisms of the skeletal reactions of hydrocarbons on metals, made in this review, may provide useful models and fill the gap between homogeneous and heterogeneous catalysis.

An attempt will be made to compare the catalytic activity and/or selectivity of supported metal clusters of known particle size (prepared by the impregnation and decomposition of metal cluster compounds on supports) with that displayed by catalysts containing the typical crystallite sizes available from conventional methods of heterogeneous catalyst preparation. Based on recent conclusions [11, 12, 13], metal cluster complexes represent a unique class of chemical compounds in that they involve two, three, or many more atoms incorporated into a single molecular aggregate stabilized by attendant ligands. New approaches to the preparation of supported metallic catalysts by the decomposition of surface organometallic complexes have made it possible to prepare highly dispersed particles of transition metals with improved activity and selectivity for skeletal reactions of hydrocarbons.

2. Reactions of Alkanes and Cyclanes on Metals

Almost twenty five years ago P. H. Emmett edited the series Catalysis, volume VI of which was devoted to hydrocarbon catalysis and was concerned with the isomerization, cracking and hydroreforming of hydrocarbon molecules over the (then) new "dual-function catalysts" [14]. In a survey of isomerization catalysts and their conditions of operation, Condon [18] proposed that "all inorganic catalysts may be broadly described as acidic (electron pair-accepting), basic (electron pair--donating), or mono-electronic (oxidative-reductive), hydrogen active". It was pointed out that representative of electronic catalysts active for isomerization are various forms of pure or supported nickel, palladium or platinum, and activated charcoal. With such catalysts isomerization appeared to be limited to shifts in the position of double bonds, interconversion of geometrical isomers of olefins, and inversion of configuration at a saturated carbon. The catalysts most generally effective for isomerization of hydrocarbons were recognized as being of the "acidic" type. In fact the earliest investigations had shown that isomerization of hydrocarbons occurred on acidic catalyst (*e.g.* $AlCl_3$, H_2SO_4, $SiO_2-Al_2O_3$). On this type of catalyst the following classical reactions have been reported,

(a) interconversion between isomers with at least one tertiary carbon as for isohexanes and dimethylcyclohexanes [19, 20, 21],

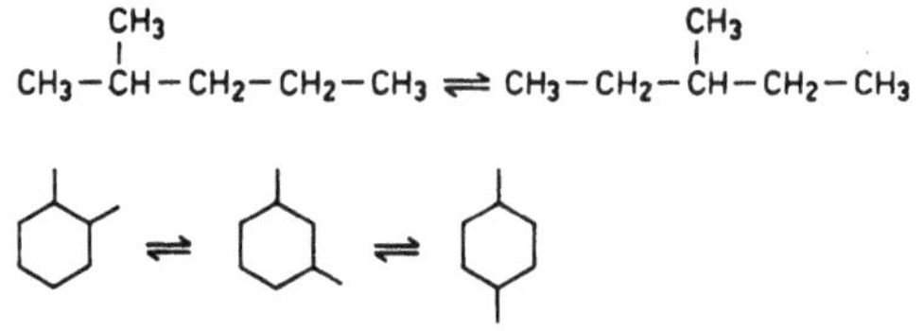

(b) isomerization of hydrocarbons without tertiary carbon as for isomerization of *n*-butane to isobutane, or cyclohexane to methylcyclopentane [18].

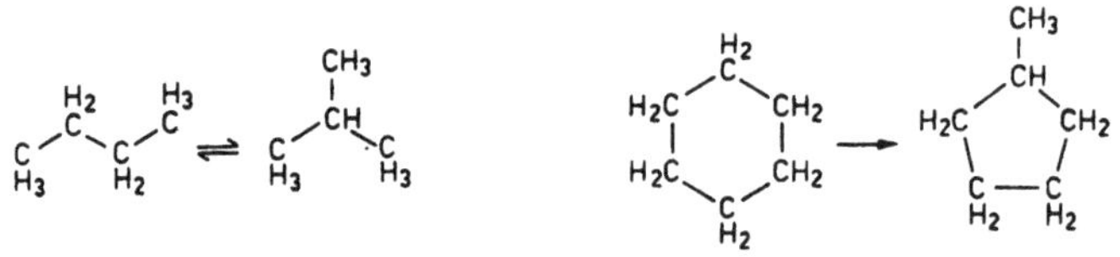

Type (a) reactions are easier than type (b) reactions. Thus in sulfuric acid at 298 K [18], or on silica-alumina at 423 K, type (a) reactions occur but not type (b) reactions. On $AlCl_3$ both types of reactions take place, but type (a) reactions are more rapid than type (b) reactions. Such acidic catalysts at higher temperatures lead also to cracking of alkanes and cycloalkanes. Thus *n*-hexadecane is cracked over silica-alumina at 773 K [24, 25]

$$C_nH_{2n+2} \rightarrow C_mH_{2m} + C_{n-m}H_{2n+2-2m}$$

All these reactions have been interpreted according to mechanisms which include carbonium ions as reaction intermediates. Thus the isomerizations of *n*-butane and isohexanes are described in the following manner by carbonium ion rearrangement [26]

$$CH_3-\overset{+}{C}H-CH_2-CH_3 \rightarrow CH_3-\underset{\underset{\displaystyle CH_3}{|}}{C}H-\overset{+}{C}H_2 \rightarrow CH_3-\underset{\underset{\displaystyle CH_3}{|}}{\overset{+}{C}}-CH_3$$

$$CH_3-\underset{\underset{\displaystyle CH_3}{|}}{C}H-\overset{+}{C}H-CH_2-CH_3 \rightarrow CH_3-\overset{+}{C}H-\underset{\underset{\displaystyle CH_3}{|}}{C}H-CH_2-CH_3$$

It was found that addition of a transition metal to an acidic catalyst increased considerably the rates of isomerization reactions. Among the different transition metals it was rapidly established that platinum was the most efficient cocatalyst.

Catalytic reforming is one of the basic petroleum refining processes. Catalysts for reforming reactions are small crystallites of Pt or Pt alloys supported on acidic alumina or silica-alumina. The catalyst is said to be bifunctional, since both the metal and the oxide components play active roles. Furthermore the presence of platinum allows the occurrence hydrogenation and dehydrogenation reactions not permitted on the acidic support alone. The importance of catalytic reforming over bifunctional catalysts helped to focus attention on the catalytic chemistry of hydrocarbons at metal surfaces.

A. Reactions Occurring on Metals

The skeletal reactions of hydrocarbons on metal represent an important class of reactions in heterogeneous catalysis both from a theoretical and an industrial point of view, in spite of some uncertainty concerning the role of the metal and its dispersion on the support [15]. The main reactions can be summarized as follows,

(a) Dehydrogenation of alkanes to olefins

$$C_nH_{2n+2} \rightarrow C_nH_{2n} + H_2$$

(b) Isomerization of alkanes *e.g.* isomerization of *n*-hexane to isohexanes

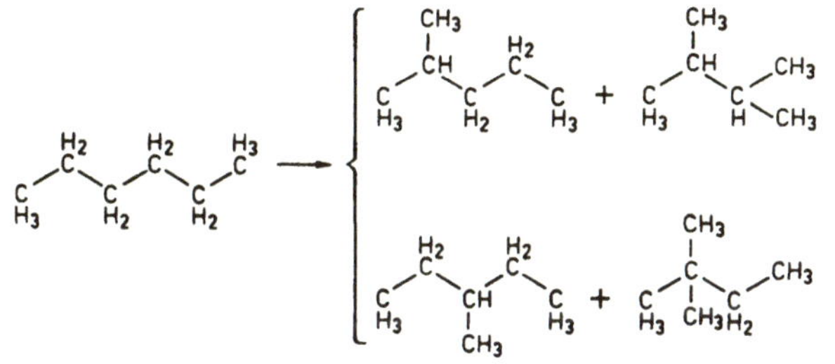

(c) Isomerization and aromatisation of alkylcyclanes

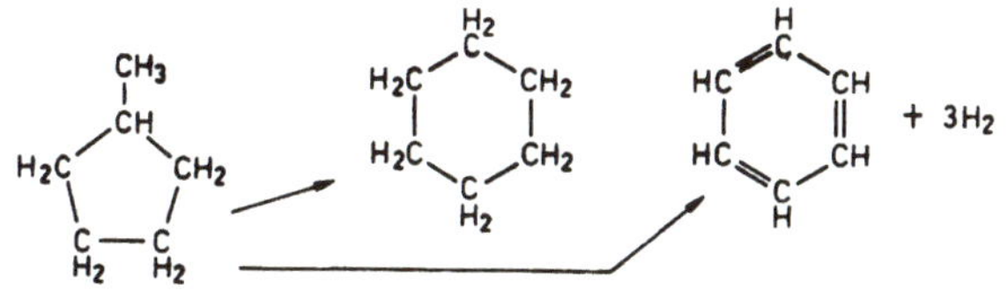

(d) Isomerization of substituted cyclanes or aromatics

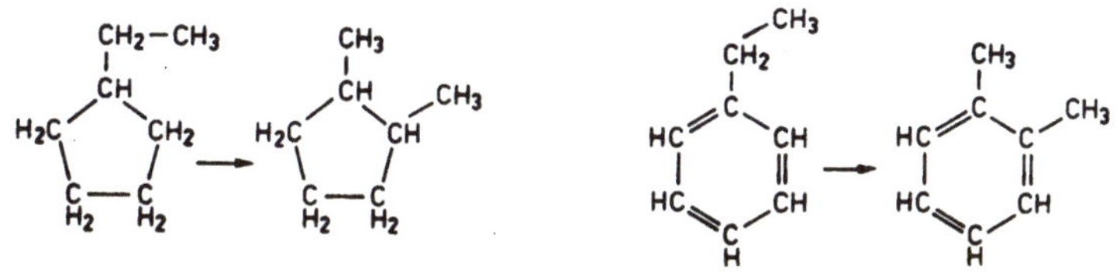

(e) Dehydrocyclization of alkanes and olefins to cyclanes or aromatics

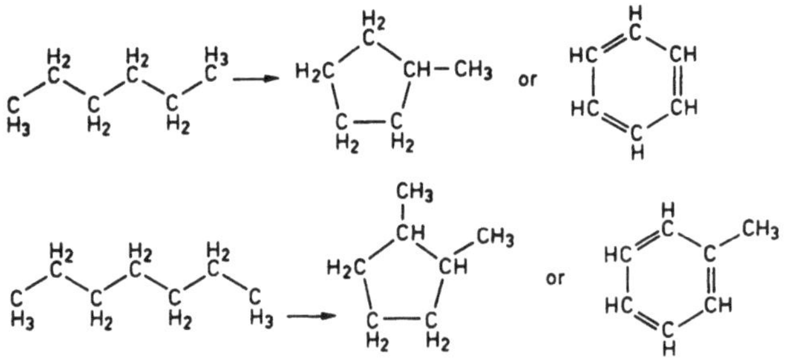

(f) Hydrocracking reactions of alkanes, hydrodesulfurization and hydrodenitrogenation

C_nH_{2n+2} → lower molecular weight alkanes

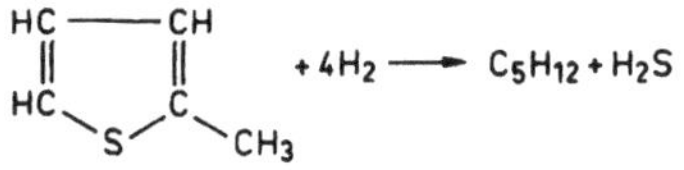

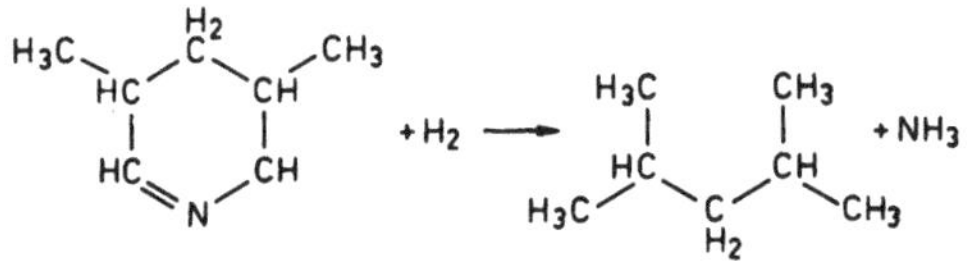

It is generally accepted that on reforming catalysts — platinum dispersed on an acidic support — skeletal isomerization necessarily occurrs according to a bifunctional mechanism which includes three consecutive steps: dehydrogenation on the metal, isomerization of the resulting olefin on the acidic support by a carbonium ion mechanism, and re-hydrogenation of the isomerized olefin. These processes may be represented thus [15], for a bifunctional Pt/Al_2O_3 catalyst.

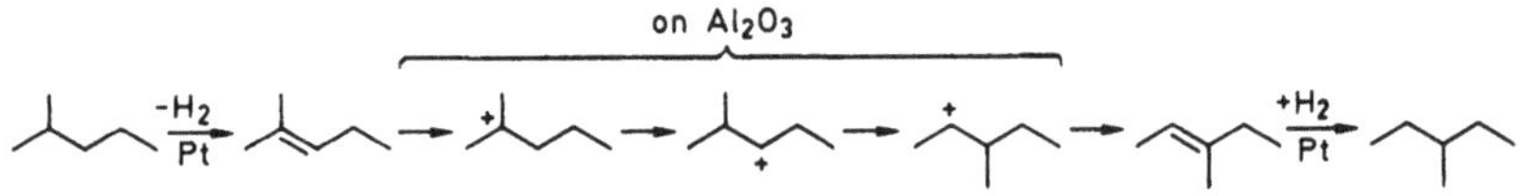

Nevertheless, the fact remains that in the absence of an acidic function, platinum alone is able to effect skeletal rearrangements of alkanes or cycloalkanes. The main lines of evidence for this conclusion are summarized below, but it needs to be pointed out that, *under reforming conditions*, such platinum — only catalysed skeletal reactions probably make only a minor contribution to the overall reaction which occurs on bifunctional catalyst systems since the presence of an acidic function is essential in order to obtain the necessary activity and selectivity for the reforming conversion.

The evidence for platinum — only catalysed skeletal rearrangement reactions comes from work with unsupported metal (evaporated metal films [16]) and with metal dispersed on inert supports such as Pt/charcoal or Pt/pumice [17, 27–29].

Using non-acidic charcoal as a support, it has been shown [27–29] that platinum catalyses reactions such as the hydrogenolysis of polymethylcyclopentanes, the dehydrocyclization of alkanes, and the aromatization of substituted alkanes via a cyclopentane-type intermediate

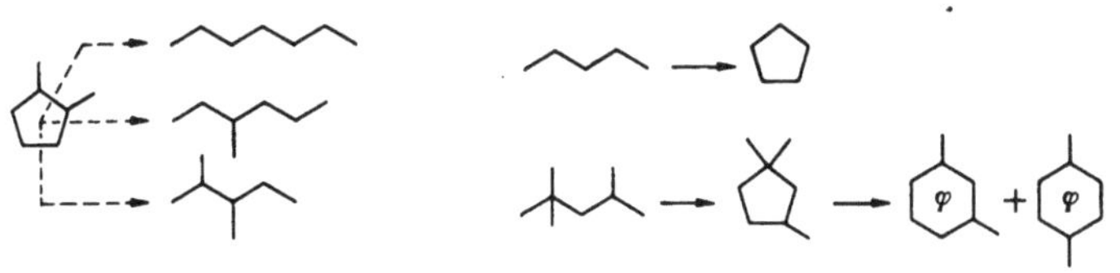

In addition, skeletal rearrangements of alkanes have been demonstrated [1, 16, 30] on evaporated platinum films under conditions such that the glass substrate is catalytically inactive. Isobutane and neopentane are isomerized respectively to *n*-butane and isopentane

The various metals act specifically for these reactions of isomerization since platinum isomerizes neopentane to isopentane whereas palladium does not [31, 32] except under special experimental conditions (oriented films of palladium and rather low partial pressure of hydrocarbon [33]). On the other group VIII transition metals the hydrocracking reactions are favoured.

B. Activity and Selectivity versus the Character of the Metal

As already pointed out, the reactions of isomerization take place on platinum and palladium for small molecules like isobutane and neopentane. Moreover nickel and especially cobalt lead to an extensive cracking to methane.

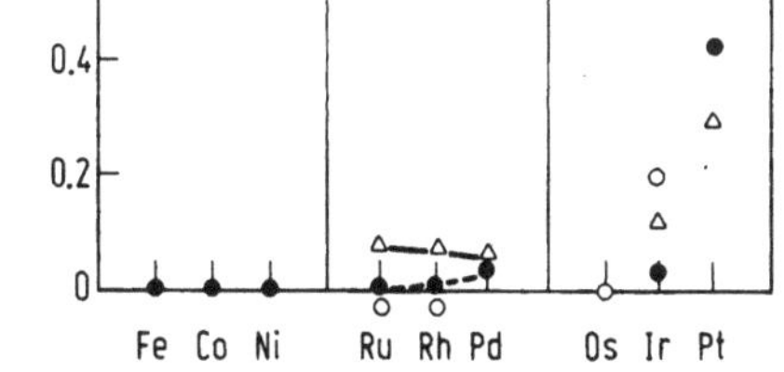

Figure 1. Ratios isomerization/hydrogenolysis (△, ○), and dehydrocyclization/hydrogenolysis (●), obtained on the transition metals of Group VIII. (Reprinted with permission from Concept de Coordinence en Catalyse Hétérogène Appliquée (paper R. Montarnal), copyright by C.N.R.S. 1975, reproduced by courtesy of Dr. G. Martino. Ref. [35])

Montarnal *et al.* [35] has summarized the selectivity for isomerization to hydrogenolysis in the manner represented in Figure 1. The variation of the selectivity for isomerization versus the character of the metal is quite considerable, but it depends also very much on the experimental conditions and on the nature of the hydrocarbon. The activity for cyclization and dehydrocyclization is important on the metals favouring isomerization whithout there being any straight parallelism between these different reactions.

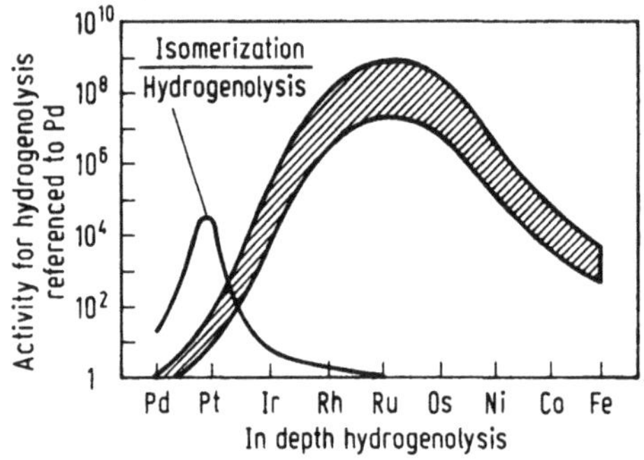

Figure 2. Schematic evolution of the activity in hydrogenolysis and of the ratio isomerization/hydrogenolysis in function of the In depth hydrogenolysis. (Reprinted with permission from Concept de Coordinence en Catalyse Hététogène Appliquée (papier R. Montarnal), copyright by C.N.R.S. 1975, reproduced by courtesy of Dr. G. Martino. Ref. [35])

Montarnal *et al.* [35] correlated the activity for hydrogenolysis and the ratio of isomerization to hydrogenolysis with the character of the metal and to the "in depth hydrogenolysis" as seen in Figure 2. As the "in depth hydrogenolysis" seems to be closely related to the strength of chemisorption of the alkanes, these authors proposed a classification of the metals as function of an increasing coefficient of chemisorption.

However, to study the specificity of the metals one might look, for example, at the different reactions given by a more complicated molecule like 1,1,3-trimethyl-cyclopentane. For such a molecule the main skeletal rearrangement reactions are the following [36]:

(i) aromatization to xylenes corresponding to a cycle enlargement at the quaternary carbon atom,

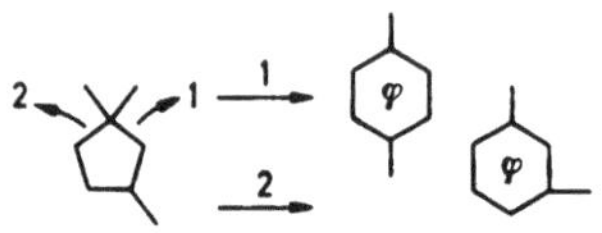

(ii) aromatization to toluene corresponding to a cycle enlargement at the tertiary carbon atom followed by demethylation,

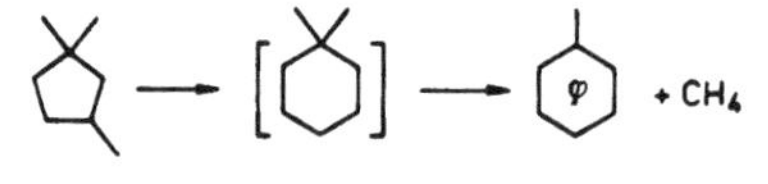

(iii) demethylation leading to the gemdimethylcyclopentane,

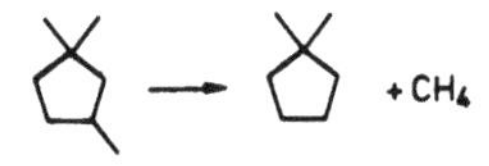

(iv) total degradation into methane,

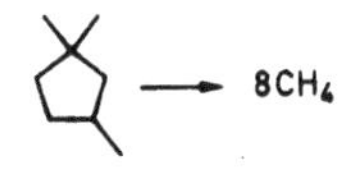

(v) demethylation followed by ring opening leading to heptanes,

(vi) demethylation followed by ring opening and degradation leading to a series of alkanes from C_2 to C_6,

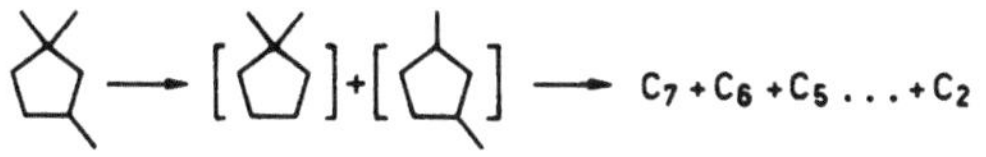

Reaction (i) takes place mostly on platinum where 50% of the initial products are xylenes. Reaction (ii) operates mostly on palladium (35% of toluene). One finds here again the difference between platinum and palladium for the isomerization of isobutane and neopentane, the reaction with ring enlargement being in fact a partical case of bond shift. Reaction (iv) (total degradation into methane) is the only one which occurs on cobalt over all range of temperatures. On iron and tungsten at low temperature (470 K) the demethylation reaction takes place, but at higher temperature (570 K) total degradation into methane occurs. Both nickel and rhodium allow ring opening, but rhodium leads only to the formation of acyclic hydrocarbons, while on nickel the ring opening is followed by further degradation to all members of alkanes from C_1 to C_7.

It is worthwhile to note at this stage that the catalysts are modified during the reaction. On a tungsten film at high temperature the reaction is identical to the one given by cobalt, that is complete conversion of trimethylcyclopentane to methane. This reaction is quickly poisoned and is followed by a low reaction rate giving mostly xylenes [36]: "Superficial tungsten carbide" seems to provide the same catalytic selectivity as platinum [37].

Although skeletal isomerization of alkanes on metals has been known for a long time, the reaction mechanism is still a matter of controversy due to the difficulty in identifying the reactive adsorbed intermediates, and the question is addressed in subsequent Sections.

3. Skeletal Reaction Mechanisms

The initial act of chemisorption of an alkane molecule at a metal surface is dissociative, and considerable insight into this process has been obtained by the use of deuterium exchange reactions particularly when the initial distribution of deutero-isomers is determined [39]. Nevertheless, it should be noted that such deuterium exchange reactions usually begin at temperatures much below those for skeletal reactions. Thus, although in principle the measurement of initial product distributions from deuterium exchange may be considered as a method for obtaining information about adsorbed hydrocarbon species under skeletal reaction conditions, in practice this is usually difficult to achieve, and it has only been successful in some special cases.

A. Small Molecules: The Bond-Shift Mechanism

Anderson and Baker [30] have shown that over evaporated platinum *films* at about 570 K the butanes underwent isomerization (accompanied by hydrocracking) and that platinum was the sole seat of catalytic activity. To assess the nature of the adsorbed intermediate, we note that for hydrocarbons of more than two carbon atoms, particulary over palladium, an intermediate π-bonded to the surface has been suggested by Kemball *et al.* [41] for deuterium exchange and *cis-trans* isomerization. The essential feature of such a π-bonded intermediate is that the molecule should have not less than three adjacent non-quaternary carbon atoms. Thus, isobutane would be able to form a π-bonded intermediate, but neopentane would not. The fact that Anderson and Avery [40] observed isomerization of neopentane makes it certain that a π-bonded intermediate is not necessary for isomerization. Furthermore, the fact that neopentane and isobutane react over platinum at comparable rates and temperatures suggests that a π-bonded intermediate is not involved in the reaction of isobutane. In further confirmation of this, we note also that it was concluded that π-bonded intermediates formed much more readily on palladium than on platinum [41].

In the absence of a π-bonded intermediate for these reactions, one must presume that the adsorbed intermediates are bonded to the surface by σ bonds.

To elucidate the mechanism of the skeletal isomerization of saturated hydrocarbons Anderson and Avery [31] studied the influence of hydrocarbon geometry on the reaction and they have compared the reaction of ethane, *n*-butane, isobutane and neopentane. Below 543 K neopentane reacted with an activation energy and frequency factor very similar to those for the butanes, and since neopentane cannot form a 1-2 diadsorbed intermediate without prior fragmentation, it was concluded that a 1-2 diadsorbed intermediate is not primarily involved in the butane reactions either.

This assumption is in agreement with the observation that ethane, which can only be diadsorbed in the 1-2 mode reacted with a much higher activation energy than the other hydrocarbons. It was concluded that the surface precursor to the present reaction with the butanes is 1-3 diadsorbed. (The term diadsorbed means bonding to the surface at two carbon atoms).

As a consequence of this work, Anderson and Avery [42] proposed a mechanism of isomerization of aliphatic hydrocarbons at a platinum surface. The essential features are as follows. Initially the 1 and 3 carbon atoms were probably singly bonded to the surface by dissociative adsorption through conventional σ bonds. One of these adsorbed carbon atoms will change its hybridization to sp^2 by the loss of a further hydrogen atom, then the intermediate formed (A) resembles that proposed by Zimmerman and Zweig [43], (Figure 3).

The model proposes that the surface precursor is 1,3 diadsorbed with a double bond to the surface at one carbon atom (Structure A), and that isomerization occurs by the transformation of this to a bridged structure (Structure B).

The likelihood of isomerization was discussed in terms of the energy difference between the precursor and the corresponding bridged intermediate and the energies were calculated by a simple Hückel MO method. One objection which may arises with this mechanism concerns its non reversibility in the isopentane-neopentane reactions. The sp^3 hybridization of the bridged carbon atom C_2 (Figure 3) does not restrict the isomerization of neopentane,

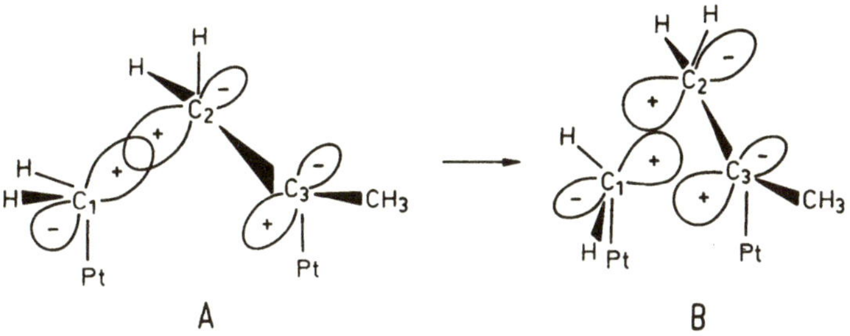

Figure 3. Anderson's mechanism proposes that the surface precursor is 1, 3 diadsorbed with a double bond to the surface at one carbon atom (structure **A**), and that isomerization occurs by the transformation of this to a bridged structure (structure **B**). (Reprinted with permission from Journal of Catalysis, copyright by Academic Press, reproduced by courtesy of Prof. J. R. Anderson. Ref. [42])

whereas the sp^2 nature of the carbon atom to which the migrating group is eventually bound does exclude the possibility of neopentane being a product from isopentane isomerizations. In fact, small amounts of neopentane are formed on platinum/alumina catalysts at 535 K [44].

B. Larger Molecules: Bond-Shift and Cyclic Mechanisms

A comparative study of the hydrogenolysis of methylcyclopentane and isomerization of hexane isomers on platinum catalysts showed the identity of the product distributions from these two reactions, the almost complete absence of 2,3-dimethylbutane and the formation of methylcyclopentane during isomerization. These results strongly suggested that a common intermediate with a five membered ring structure is responsible for these reactions [17].

Scheme 1 summarizes the relationships between isomerization, hydrogenolysis and dehydrocyclization.

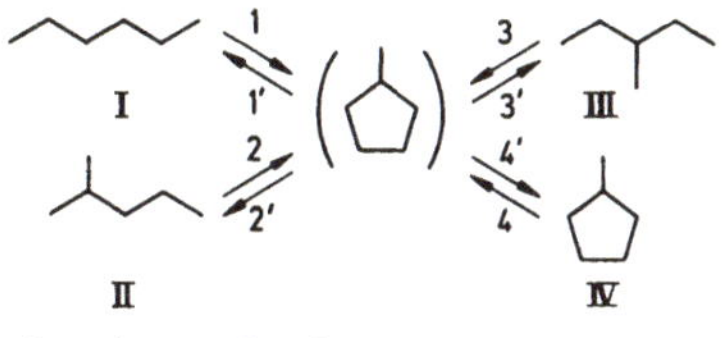

Scheme 1

Isomerization of *n*-hexane, for instance, is represented by the consecutive elementary steps 1-3′ or 1-2′, its dehydrocylization by the reactions 1-4′; hydrogenolysis of methylcyclopentane by any of the steps 4-1′, 4-2′ and 4-3′.

Further investigations have shown that at high temperatures on platinum and palladium films, and on supported catalysts with low metal content, a typical distribution is obtained with any cyclic hydrocarbons, corresponding to an equal chance of breaking the different cyclic bonds. π-allylic triadsorbed species were thought to be responsible for this type of hydrogenolysis called *non-selective*.

On 10% platinum/alumina, as on platinum films, at low temperatures, and on nickel, only the disecondary CH_2—CH_2 cyclic bonds of cyclobutanes and cyclopentanes are broken. $\alpha,\alpha,\beta,\beta$, tetraadsorbed species could be associated with this *selective* hydrogenolysis.

The analysis of the product distributions obtained from methylcyclopentane on Pt/Al_2O_3 catalysts suggest that a third mechanism takes place on this metal. It is *partly selective* and involves the break of disecondary bonds mainly, but also of secondary-tertiary bonds [45].

Among these mechanisms, one corresponds to an equal chance of rupturing the five cyclic bonds of methylcyclopentane. Since ditertiary CH—CH bonds were opened according to this mechanism, and since *cis-trans* equilibrium of disubstituted cyclopentanes was reached during hydrogenolysis[1] it was assumed that the species responsible for isomerization and dehydrocyclization of hexanes were α,β,γ triadsorbed species [34]. The first elementary step of isomerization could thus be described by the following reactions.

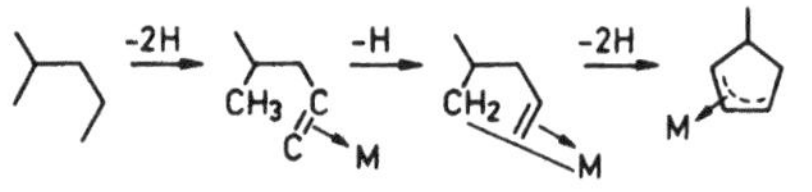

Such a scheme is similar to that proposed by Shephard and Rooney to explain the dehydrocyclization of *o*-methylethylbenzene to indane and it should be noted that the results obtained by these authors were a very good confirmation for the proposed cyclic mechanism of isomerization [47], thus

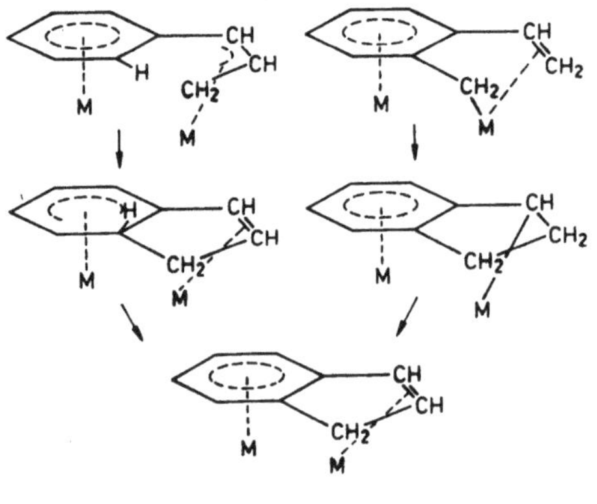

Scheme 2

[1] Such species π-bonded to the metal were shown to cause inversion of the carbon atom [46, 41].

In 1968 an adsorbed cyclopropane intermediate was proposed by Muller and Gault [36] to explain the aromatization of 1,1,3-trimethylcyclopentane to xylenes. This reaction, corresponding to a ring enlargement at the quaternary carbon atom,

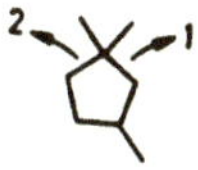

1: *p*-xylene
2: *m*-xylene

is very similar to the reaction of isomerization of neopentane to isopentane discovered by Anderson and Baker [16, 30]. The mechanism proposed can account for the reaction isopentane→neopentane which was not the case for the bond shift mechanism proposed by Anderson [42].

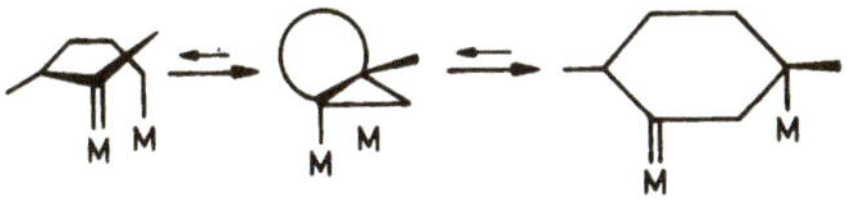

This mechanism involving an $\alpha\alpha\gamma$ triadsorbed species and *adsorbed cyclopropane intermediate*, may act reversibly and explain the formation of a quaternary carbon atom. In order to understand the mechanism of this reaction, the deuterolysis of *gem*dimethylcyclopropane was studied on the same metal as used with dimethylcyclopentane. Neopentane represents always more than 99% of the reaction products except on platinum at 573 K where 4.2% of isopentane was formed initially. On any metal *except* platinum, 1, 3-d_2 neopentane was obtained. On platinum, large amounts of 1, 1, 3-d_3 and 1, 1, 3, 3-d_4 neopentanes were also formed. These observations suggest the presence of $\alpha\alpha\gamma$ triadsorbed species. Anderson arrived to similar conclusions by quite different experiments and reasonings [16, 30, 31].

In summary, we have to keep in mind that there are at least two different mechanisms of isomerization operating on platinum catalysts. One of these involves a cyclic intermediate; the other one involves an alkyl shift. Aromatization also occurs under the same conditions provided the reactant molecule is sufficiently large.

C. The ^{13}C Tracer Technique

We saw before that two basic mechanisms are involved in skeletal rearrangements of hydrocarbons in metal catalysis. To investigate in more details these mechanisms, use has been made of a ^{13}C tracer technique which allows all the elementary steps involved in the reactions to be distinguished.

Whilst in the case of very small molecules, bond shift is the only possible isomerization reaction, in the case of larger molecules with five carbons at least in the chain, a cyclic mechanism predominates in the case of extremely dispersed catalysts. On most catalysts — films, bulk metals, supported catalysts of medium and low dispersion — both isomerization mechanisms, bond-shift and a cyclic mechanism compete, and the first task is to evaluate the contribution of each of them. As an example, consider the isomerization of 2-methylpentane-2-^{13}C. This reaction yields two different 3-methylpentane isomers, 3-^{13}C and 2-^{13}C, according to whether the reaction mechanism is cyclic or bond shift (scheme 3)

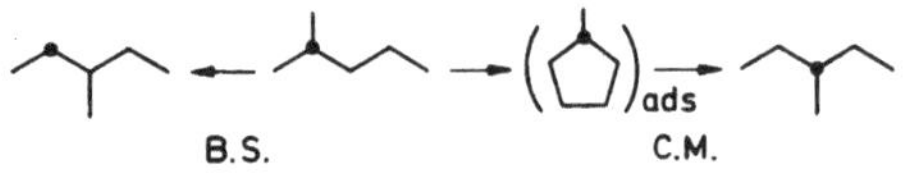

Scheme 3

More generally, when more than two parallel pathways are possible to yield the same molecule, several reacting hydrocarbons labeled in various positions have to be used. For instance isomerization of 2-methylpentane to *n*-hexane may occur in 3 different ways either by cyclic mechanism (C), or by methyl shift (A) or propyl shift (B). 2-methylpentane-2-^{13}C in this case allows only an estimate of the contribution of bond-shift A, and another molecule, 2-methylpentane-4-^{13}C is required to estimate the contribution of the cyclic mechanism [48] (schemes 4, 5)

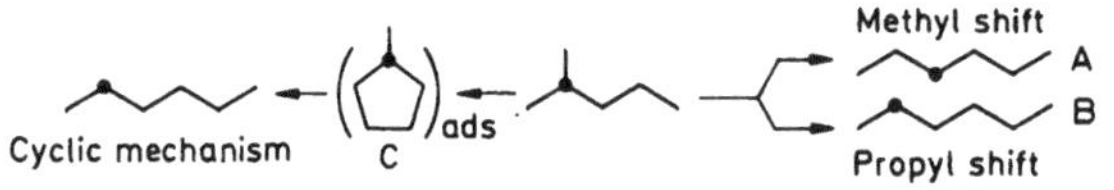

Scheme 4

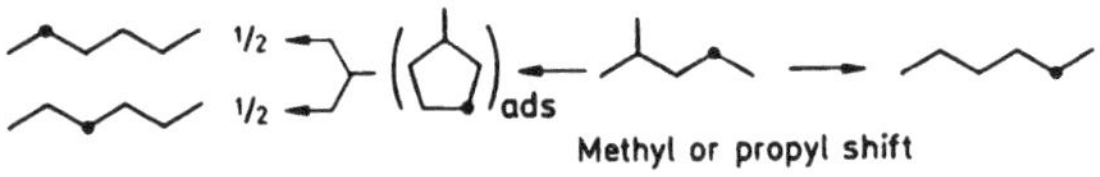

Scheme 5

The use of labeled molecules allows also the determination of the amounts of self-isomerized products, *i.e.* of molecules obtained by positional isomerization but having the same skeletal structure as the reactant. Formation of such molecules may contribute an important part to the overall reaction. For instance, on a Pt catalyst of low dispersion, self-isomerized 2,3-dimethylpentane is the major reaction product [49], and may be traced by labeling the reactant on carbon 2, (scheme 6)

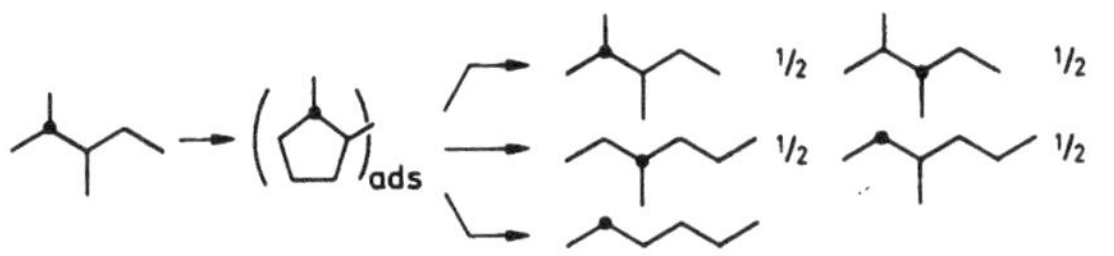

Scheme 6

Similarly, methyl shift in isopentane is a reaction ten times faster than chain lengthening and can be observed only with the appropriate labeled hydrocarbon [44], (scheme 7)

Scheme 7

For *n*-pentane on a highly dispersed Pt/Al_2O_3 catalyst, the major reaction is self-isomerization by a cyclic mechanism and is traced as a label-scrambling reaction when using *n*-pentane-2-^{13}C as a reactant, (scheme 8)

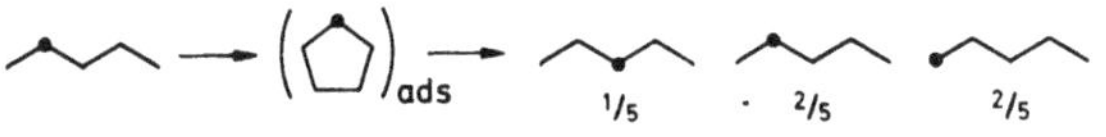

Scheme 8

During the past decade, isomerization of hexanes or pentanes have very often been taken as a test reaction to study particle size or alloying effect. In these studies, the product distribution and especially the percentage of cyclic molecules (cyclopentane or methylcyclopentane) in the reaction products have often been used to estimate the contribution of the cyclic mechanism. Such a procedure may be misleading. First, self-isomerization reactions, which may be predominant, are overlooked when not using labeled molecules. Secondly, the percentage of cyclic products is highly dependent upon experimental conditions [44], increasing with decreasing hydrogen pressure or increasing hydrocarbon pressure or temperature. Comparison between two catalysts *are only valid then when using strictly identical experimental conditions*. Moreover, a third less obvious and more fundamental reason to question the product distribution approach results from the mechanism itself: as shown for 3-methylhexane isomerization [50], the ratio ϱ between cyclic and acyclic products varies widely with conversion, even at very small conversion when the distribution of the acyclic isomers remains practically constant. This can be easily accounted for by considering the simplified kinetic scheme 9, where reactant A_1 yields only a single acyclic isomer A_2 and one cyclic molecule A_3 *via* a common cyclic intermediate C.

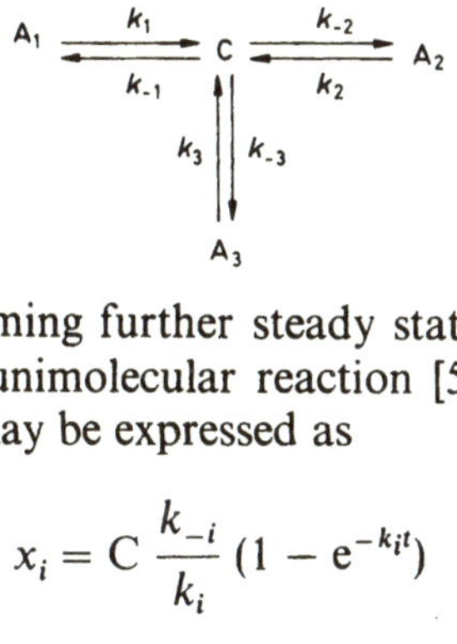

Scheme 9

Assuming further steady state concentration C for this intermediate and pseudounimolecular reaction [51], the mole fractions of reactant and products may be expressed as

$$x_i = C\frac{k_{-i}}{k_i}(1 - e^{-k_i t}) \tag{1}$$

The ratio ϱ between the amounts of cyclic and acyclic products is

$$\varrho = \frac{k_{-3}/k_3(1 - e^{-k_3 t})}{k_{-2}/k_2(1 - e^{-k_2 t})} \tag{2}$$

where k_2, k_3, k_{-2} and k_{-3} represent pseudounimolecular rate constants. Since the "adsorption" rate constant k_3 is always much larger (by 2 to 3 orders of magnitude) than the dehydrocyclization rates constant k_2 even at very small conversion, $1 - e^{-k_2 t}$ may be approximated to $k_2 t$ and $1 - e^{-k_3 t}$ is already close to unity. Therefore, the ratio ϱ decreases sharply with contact time. The only correct way then to compare the contribution of the cyclic mechanism for various catalysts is to determine the percentage of cyclic products at various contact times and to extrapolate to zero conversion. Even in this case, such a comparison is valid only for a given set of experimental conditions and may be misleading, since the desorption rate constant k_{-3} may vary relative to the hydrogenolysis rate constant, k_{-2}, from catalyst to catalyst. From the above discussion, it ensues that ^{13}C tracer techniques are the most suitable for estimating the contributions of the various mechanisms. However this method provides information only if skeletal rearrangement and not adsorption-desorption is rate-determining [44, 52].

D. Kinetics

By tracing the displacement of the various carbon atoms during the isomerization process, carbon-13 labeling allows one to calculate precisely, after having eliminated all the side reaction products, the relative contributions of the various mechanisms. We will take, as a kinetic example pentane isomerization on Pt/Al_2O_3 catalysts [44, 105].

1. Isomerization of n-Pentane-2-C^{13} on 10% Pt/Al_2O_3

Consider the Reaction *n*-pentane → *n*-pentane *via* a cyclic mechanism

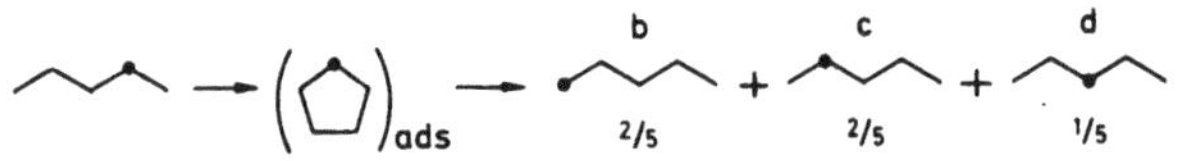

Reaction 1

This reaction can only be observed when using labeled hydrocarbons. The location of carbon-13 in the various molecules is determined by comparing the mole fractions of the labeled isomers to the ones obtained for the various pure isotopic species. These values are obtained from mass spectrometric analysis after correction for the naturally occurring isotopes and C—H fragmentation [53, 54].

From the mole fractions *b* and *d* of *n*-pentane-1-^{13}C and 3-^{13}C it is ascertained that *isomerization via a cyclic mechanism is the major scrambling process* and it is possible to deduce the conversion $\alpha' = 2b + d$ for this scrambling reaction 1, and to compare with the conversion α for the *n*-pentane → isopentane isomerization. Both conversions α and α' are similar on 10% Pt/Al_2O_3 but an increase in temperature or a decrease in hydrogen pressure favors the scrambling reaction *versus* chain shortening [44].

In the case of isopentanes, the determination of the concentrations of the various isotopic varieties is much more accurate than for the *n*-pentanes, which allows one to calculate precisely, after having eliminated all the side-reaction products, the relative contributions of *the two possible bond shift isomerization reactions A and B* (*c.f.* Reaction 2)

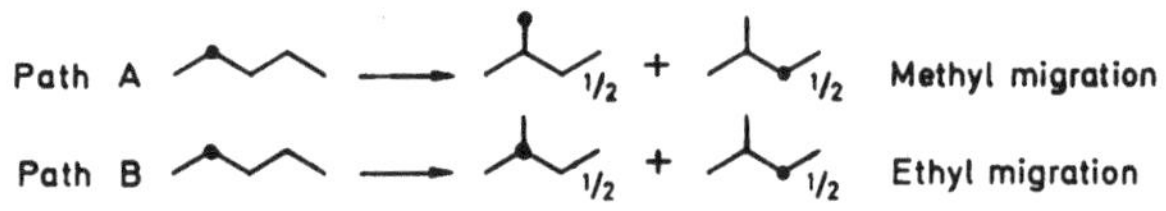

Reaction 2

Scheme 10 shows the various consecutive and parallel reactions which lead to isopentane. Besides the molecules formed directly from *n*-pentane-2-

^{13}C (e_r, f_r, g_r), isopentanes may appear as the results of consecutive reactions, either from the primary isopentanes by methyl shift, reaction 3, or by chain shortening from *n*-pentanes obtained from a cyclic mechanism, reactions 1 + 2′.

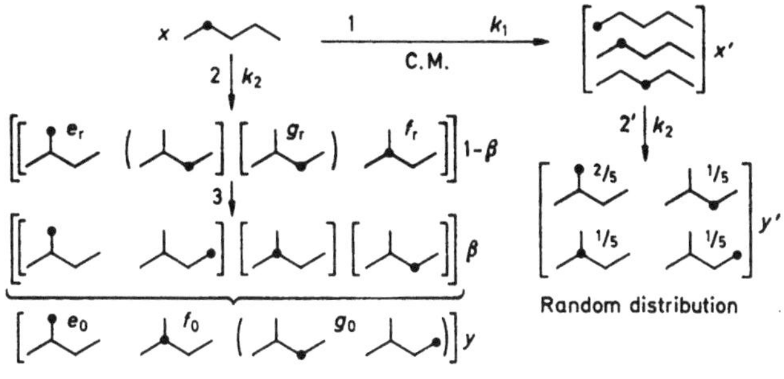

Scheme 10

In this scheme, β is defined as the mole fraction of isopentanes converted according to reaction 3.

In the latter case, the carbon-13 is distributed randomly among all the carbon atoms of the various *n*-pentane and isopentane molecules. From the concentrations of the isomerized *n*-pentanes (x') and isopentanes ($y + y'$), it is easy, assuming that each reaction 1, 2 and 2′ is first order, to deduce the concentration y and the distribution e_0, f_0, g_0 of the isotopic varieties for all the isopentanes, primary and secondary, obtained from *n*-pentane-2-^{13}C by chain shortening.

The results show that methyl shift (Reaction 3) is a very fast reaction, when compared with both the cyclic *n*-pentane→*n*-pentane isomerization, and the *n*-pentane→isopentane isomerization [44].

On the other hand, the percentage of bond shift A (*c.f.* reaction 2) remains fairly constant (26 ± 2%). The experimental conditions, temperature and hydrogen pressure, do not seem to greatly affect the relative importance of the two bond shift mechanisms A and B.

2. *Isomerization of 2-methylbutane-2-^{13}C on 10% Pt/Al_2O_3*

In order to investigate the methyl shift and to compare with the isopentane → *n*-pentane isomerization, the reactions of 2-methylbutane-2-^{13}C (2MB-2-^{13}C) on 10% Pt/Al_2O_3 have been studied under various experimental conditions. The labeling of the molecule allows one to distinguish between three elementary reactions: methyl shift (reaction 3)

Reaction 3

and chain lengthening according to bond shift (A and B, reaction 4)

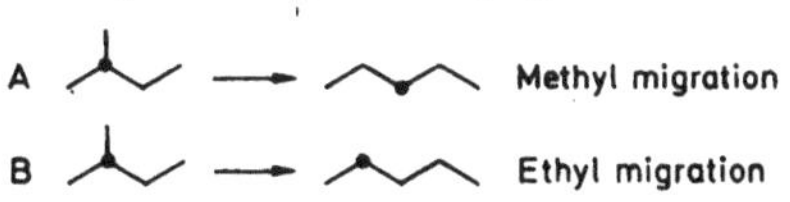

Reaction 4

The conversion α' for reaction 3, measured by the amount of 2-MB-3-^{13}C, is always larger, by a factor of 3 to 6, than the conversion α for the isopentane

→ *n*-pentane isomerization. The ratio α'/α depends on temperature and hydrogen pressure, but not on hydrocarbon pressure. Since methyl shift is fast when compared with isomerization to *n*-pentane, an accurate estimate of mechanisms A and B requires corrections taking account of this fast reaction. Similarly, the consecutive isomerization *via* a cyclic intermediate, proven by the presence of small but definite amounts of *n*-pentane-1-^{13}C in the reaction products, should also be considered. The various consecutive and parallel reactions are summarized in scheme 11. All reactions are assumed to be first order *versus* hydrocarbon and the *n*-pentane → isopentane isomerization is neglected. It has been assumed that the isotope effects were the same for steps 1 and 1′ and 2 and 2′.

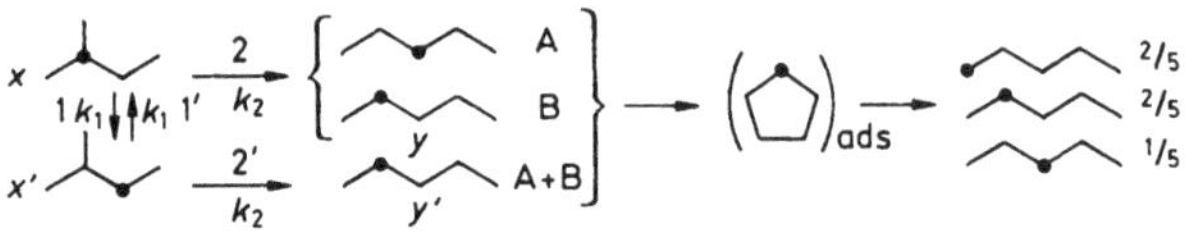

Scheme 11

With these assumptions, the initial distribution of the isotopic varieties of pentanes are calculated from the observed ones first by taking into account the scrambling process and then by eliminating the *n*-pentanes formed by the successive reaction 2′. The value found for the contribution A is 26 ± 2%, and this is approximately the same as the value of A found for the reverse reaction (*n*-pentane → isopentane isomerization).

3. Isomerization and Hydrogenolysis: Kinetic Data on 10% and 0.2% Pt/Al$_2$O$_3$ Catalysts

When *n*-pentane or isopentane is reacted over a Pt/Al_2O_3 catalyst, a number of reactions take place: chain lengthening or shortening, methyl shift, demethylation, internal fission, or *n*-pentane isomerization. Some of these reactions can only be recognized by using labeled hydrocarbons. In order to determine those reactions which involve a common intermediate, a knowledge of kinetic data such as apparent activation energies and kinetic orders with respect to hydrogen is required.

i) Activation Energies. The apparent activation energies, E_a, have been determined for the various reactions of pentane-2-^{13}C and 2-methylbutane-2-^{13}C; they are reported in Table 1, together with the corresponding frequency factors, *A*, and the kinetic orders with respect to hydrogen, *b*. According to Table 1, the activation energies allow the classification of the different reactions into four groups on the 10% Pt/Al_2O_3 catalysts: (a) *n*-pentane isomerization according to a cyclic mechanism (*ca.* 294 kJ mol^{-1}); (b) the isomerization of *n*-pentane to isopentane and the reverse reaction (*ca.* 231 kJ mol^{-1}); (c) a group of reactions including methyl shift, isomerization of isopentane to neopentane, and hydrocracking of an internal C—C bond (*ca.* 189 kJ mol^{-1}); (d) demethylation (*ca.* 147 kJ mol^{-1}).

An interesting feature which appears from Table 1 is the correlation which exists between the apparent activation energies and the kinetic orders for hydrogen for all the reactions of *n*-pentane and isopentane on low or

Table 1. Kinetic data for isomerization and hydrogenolysis reactions on 10% and 0.2% Pt/Al_2O_3 catalysts

0.2% Pt/Al_2O_3 catalyst			Reactions	10% Pt/Al_2O_3 catalyst		
Activation energy/ kJ mol^{-1}	$\log_{10}$ A[a]	b*		Activation energy/ kJ mol^{-1}	$\log_{10}$ A[a]	b*
323 ± 4	27.6 ± 1	−2.1 ± 0.3		299 ± 6	21.0 ± 0.5	−3.4 ± 0.1
293 ± 4	25.0 ± 0.5	−1.6 ± 0.2		232 ± 6	14.7 ± 0.5	−1.8 ± 0.1
237 ± 6	20.4 ± 1	−1.15 ± 0.25		188 ± 12	10.5 ± 1	−1.3 ± 0.2
226 ± 4	18.9 ± 1	−1.05 ± 0.2		161 ± 12	7.5 ± 1	−0.6 ± 0.2
293 ± 4	24.9 ± 0.5	−1.9 ± 0.3		228 ± 6	14.4 ± 0.5	−2.3 ± 0.15
293 ± 4	25.9 ± 1	−1.7 ± 0.3		190 ± 6	9.9 ± 0.5	−1.9 ± 0.2
260 ± 8	20.7 ± 1	−1.2 ± 0.2		189 ± 12	10.4 ± 1	−1.65 ± 0.15
287 ± 6	23.9 ± 1	−1.5 ± 0.3		189 ± 12	10.8 ± 1	−1.65 ± 0.2
228 ± 6	19.3 ± 1	−0.95 ± 0.15		147 ± 12	2.5 ± 1	−0.7 ± 0.1

[a] A, frequency factor/molec s^{-1} (g catal)$^{-1}$
* Kinetic order with respect to hydrogen

high dispersed catalysts. The more negative is the order, the higher is the activation energy.

In Table 2 are collected the apparent activation energies and kinetic orders with respect to hydrogen for all the various skeletal reactions over both types of platinum catalysts. One of the most striking results is the very high negative orders with respect to hydrogen found for all the hydrogenolysis and isomerization reactions.

Table 2. Kinetic data for isomerization and hydrogenolysis reactions on 10% and 0.2% Pt/Al_2O_3 catalysts

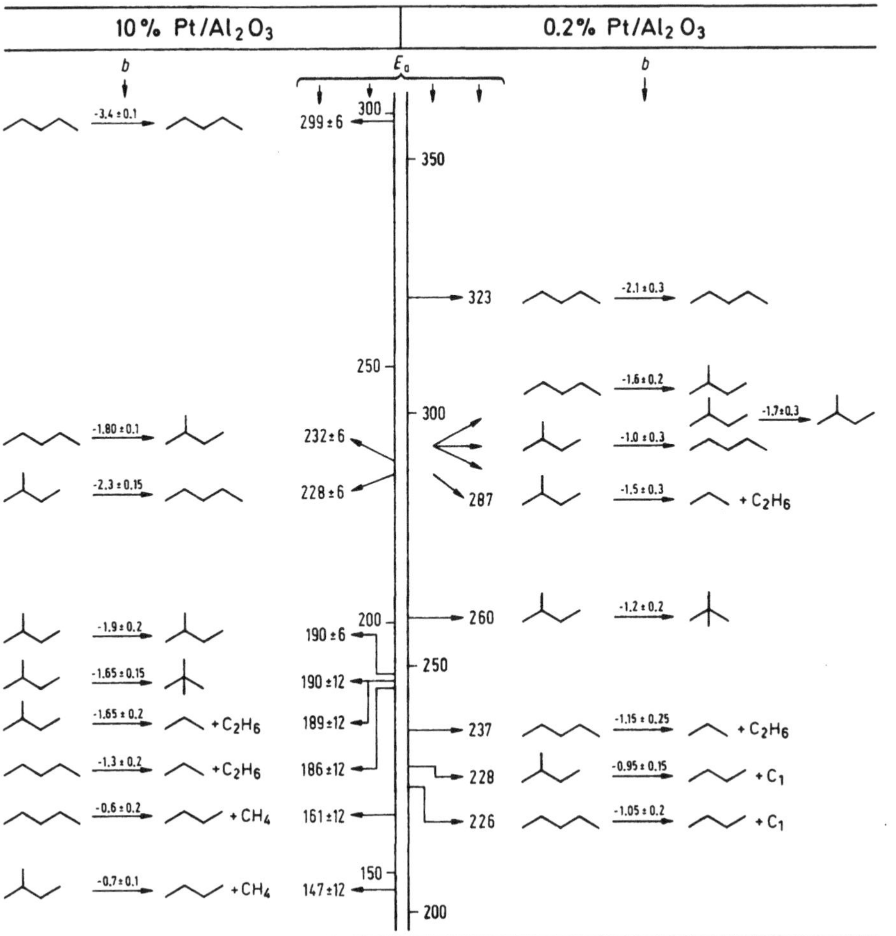

E_a activation energy / kJ mol^{-1} ; b kinetic order with respect to hydrogen

ii) Kinetic Models. Of the two schemes that have already been proposed for pentane rearrangements [44], both involve a series of dehydrogenation steps, all equilibrated, followed by the rate determining step, which is the skeletal rearrangement of a highly dehydrogenated species. The two schemes differ in the nature of the skeletal rearrangement step.

The first treatment of this sort was proposed by Cimino, Boudart and Taylor [55] for the case where the skeletal rearrangement is carbon-carbon bond rupture. This treatment implies a progressive dehydrogenation of the adsorbed molecule, forming a series of species in equilibrium, one being involved in the rupture of the carbon-carbon bond,

$$C_2H_6 \rightleftarrows (C_2H_5)_{ads} + H_{ads} \qquad \textit{Reaction 5a}$$

$$C_2H_{5\,ads} + H_{ads} \rightleftarrows (C_2H_x)_{ads} + aH_2 \qquad \textit{Reaction 5b}$$

$$C_2H_x + H_2 \rightarrow CH_y + CH_z \qquad \textit{Reaction 5c}$$

Provided this rupture is the rate determining step and equilibrium is completed for the previous reactions, the rate may be expressed as

$$r = k\,\frac{(Kp_{C_2H_6}/p_{H_2}^a)\,p_{H_2}}{1 + Kp_{C_2H_6}/p_{H_2}^a}$$

where $a = (6 - x)/2$. This expression approximates to

$$r = k' p_{C_2H_6} p_{H_2}^{1-na} \qquad (3)$$

A negative order for hydrogen of $-b$ would then correspond to the removal of a minimum of $2a = 2 + 2b$ hydrogen atoms from the molecule before the hydrogenolysis takes place. A modification of this treatment has been proposed by Sinfelt and associates [56], who assumed that on some metals such as cobalt and rhenium, where C—H and C—C bond ruptures (or formations) are of comparable rates, equilibrium *5a* was not attained. Provided a stationary state is reached for ethyl radicals, the order for hydrogen should vary in this case between 0 and $1 - \dot{a}$, according to temperature and hydrogen pressure. Here again $(2 + 2b)$ represents the minimum number of hydrogens removed from the molecule to form the reactive species. This figure is lowered to $2b$ if the bimolecular cracking reaction *5c* is replaced by a unimolecular decomposition *5d*

$$C_2H_x \rightarrow CH_y + CH_z \qquad \textit{Reaction 5d}$$

The isomerization of pentanes could be interpreted in a similar way. In the scheme shown below it is assumed that each dehydrogenation step is in equilibrium, and that the rate determining step is the isomerization reaction.

$$C_5H_{12} \rightleftarrows (C_5H_{11})_{ads} + H_{ads}$$

$$(C_5H_{11})_{ads} \rightleftarrows (C_5H_{10})_{ads} + H_{ads}$$

$$(C_5H_{12-x})_{ads} \rightarrow (i\text{-}C_5H_{12-x})_{ads}$$

$$2\,H_a \rightleftarrows H_2 \qquad \textit{Reaction 6}$$

Accordingly a negative order of $-b$ means that $2b$ hydrogen atoms at least have been removed from the molecule before the skeletal rearrangement takes place. The high negative orders found for the various reactions would imply then that *four* to *seven* hydrogen atoms have been removed when forming the precursor for bond shift or cyclic type isomerization. This seems rather unlikely. In the case of ethane hydrogenolysis, the values found for $-b$ leads to the conclusion that the species C_2H_x were extremely poor in hydrogen: however this hydrogen deficiency could in some way be considered as the driving force for the rupture of the carbon-carbon bond. Such a consideration does not hold in the case of skeletal isomerization where the product is desorbed as a complete molecule and not as small C_1 fragments. The classical mechanism represented in the above scheme does not therefore seem adequate to explain the isomerization reactions.

An alternative scheme uses the concept of reactive adsorption introduced by Frennet and coworkers [57]. This has the advantage of accounting for very high negative orders for hydrogen without introducing too extensively dehydrogenated species.

In a recent paper [58] the large variation in kinetic orders for hydrogen were explained by assuming that not one but several sites were involved in hydrocarbon adsorption. The main feature of the competitive adsorption model lies in the association of two factors:

- both gases (here H_2 and the hydrocarbon) may be adsorbed on the same sites and there is thus a competition for adsorption, and
- the hydrocarbon molecule needs a large number Z of free neighbouring sites to be adsorbed. As before, dehydrogenative adsorption proceeds stepwise, leading to a partly dehydrogenated adsorbed species from which skeletal rearrangement occurs. The dehydrogenative adsorption steps are assumed to be equilibrated, and the rearrangement reaction is rate controlling.

Since the z sites required for hydrocarbon adsorption are subject to competition from hydrogen adsorption, a large negative kinetic order with respect to hydrogen arises naturally from the model. A detailed analysis [58] leads to the result that the overall kinetic order for hydrogen for the rearrangement reaction, γ, may be expressed as a linear combination of m, the number of hydrogen atoms removed from the reactant hydrocarbon molecule in forming the adsorbed reactive species, and z, the number of metal atoms required for hydrocarbon adsorption,

$$\gamma = m(\beta + \alpha - 1) + z(\beta + \alpha') \qquad (4)$$

where α, α' and β are the formal orders with respect to hydrogen (*i.e.* the exponent in the dependence on hydrogen pressure) for the terms θ_H^o, $(1 - \theta_H^o)$ and $(1 - \theta_C)$: here θ_H^0 is the hydrogen coverage which the surface would have at the prevailing hydrogen pressure in the absence of adsorbed hydrocarbon, and θ_C is the coverage by adsorbed hydrocarbon.

Although this model can clearly easily account for numerically large kinetic exponents for hydrogen, the number of disposable parameters is large, and there seems little chance of using the theory to evaluate *m* from experimental data.

E. Discussion of Various Mechanisms

1. Bond Shift Mechanisms

Several reaction mechanisms have been proposed for bond shift isomerization. Although each of them account for some specific peculiarity of the reaction, depending upon the structure of the reactant, no conclusive proof can be offered for any of them.

In order to explain the isomerization of neopentane on platinum, Anderson and Avery [31, 42] proposed a mechanism involving as precursor a $\alpha\alpha\gamma$ triadsorbed species and as an intermediate state a π-complex of the Dewar type attached to the surface by two carbon metal bonds.

Scheme 12

A modification of this mechanism involves a dissociatively adsorbed cyclopropane intermediate [36]. Stabilization of this intermediate by methyl substitutents explains the predominance of ethyl shift (path B) over methyl shift (path A) in isomerization of isopentane to *n*-pentane [44, 59].

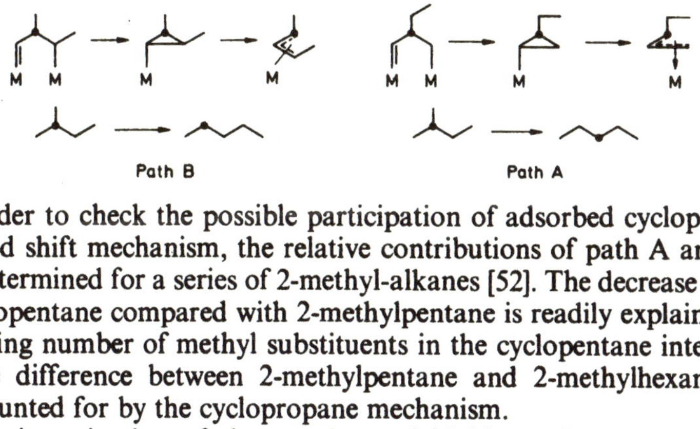

Scheme 13

In order to check the possible participation of adsorbed cyclopropane in the bond shift mechanism, the relative contributions of path A and path B were determined for a series of 2-methyl-alkanes [52]. The decrease of path B from isopentane compared with 2-methylpentane is readily explained by the decreasing number of methyl substituents in the cyclopentane intermediate, but the difference between 2-methylpentane and 2-methylhexane cannot be accounted for by the cyclopropane mechanism.

In an investigation of the reactions of highly strained cage molecules, Rooney and coworkers [60] found that these polycycloalkanes were easily isomerized under mild conditions on Pd and Pt supported catalysts. To explain the rearrangement of such molecules, it seems difficult to invoke either Anderson's or the cyclopropane mechanism: formation of $\alpha\alpha\gamma$ triadsorbed species is hardly conceivable on stereochemical grounds, and even less likely is the formation of a π allylic adsorbed species which is the normal final product in both mechanisms. Moreover, when the reactions are effected in the presence of D_2, the deuterium distribution of adamantene dimers and its isomers were binominal [60a], showing that only one hydrogen.

atom was removed at one time during both the exchange and the isomerization processes which occur at comparable rates when reacted in a large excess of deuterium at 429 K on 2% Pd/pumice catalyst. The results clearly reveal that, in certain cases, a surface alkyl is sufficient for skeletal rearrangement [61]. McKervey *et al.* proposed a mechanism for isomerization where the transient chemisorbed species, I, has partial double bond character with the shifting group symmetrically bridged [62]. The organic part of the complex can first be considered separately as a three-center free radical as in the non-classical carbonium ion (scheme 14)

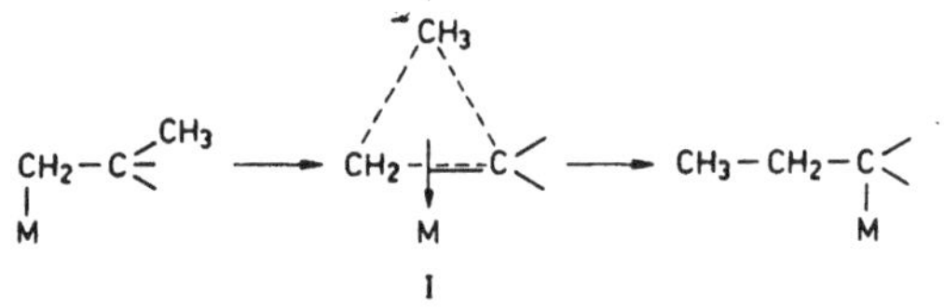

Scheme 14

Although Rooney's mechanism explains nicely the particularities of the isomerization of cage molecules on Pt and Pd catalysts, its extension to more simple molecules is not completely obvious. As a matter of fact, none of the three mechanisms discussed above account in their original form for the differences of behaviour between Pt and Pd in skeletal rearrangement. The easiest way to explain these differences is to assume that in the case of Pd the precursor species is attached to the metal by three adjacent carbon atoms, and only by two in $\alpha\gamma$ positions in the case of Pt. This assumption is based on the known propensity of Pd to form π-allyl complexes [63].

Muller and Gault [64] have suggested an alternative type of intermediate for isomerization on palladium because they have noted that 1,1,3-trimethylcyclopentane is largely converted to *p*- and *m*-xylenes by ring expansion at the quaternary center on Pt, but ring expansion at the tertiary center is preferred on palladium giving an adsorbed 1,1-dimethylcyclohexane that demethylates and is finally desorbed as toluene.

On the basis of the above facts and the ease with which palladium interconverts π-bonded olefin to allylic complexes, these authors suggest the mechanism illustrated for isobutane in scheme 15 [64]

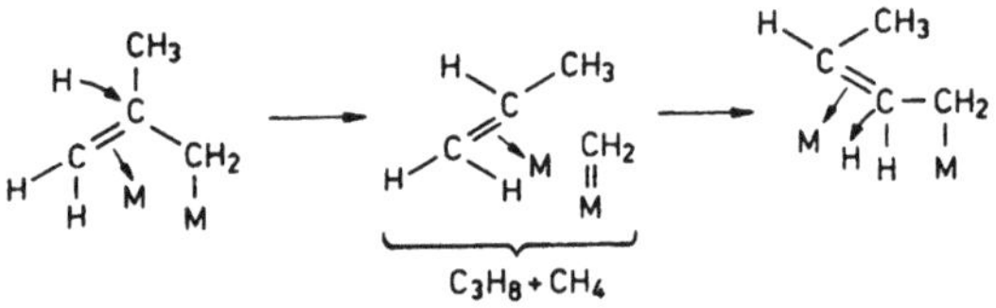

Scheme 15

At this stage we may note that each step of this mechanism proposed in 1972 has been invoked in well known reactions of organometallic chemistry. In 1976 Ephritikhine *et al.* proposed that a nucleophilic attack by H^- occurs on the central carbon of the η^3-allylic group giving the metallacyclobutane [65]. These findings may be related to the first step in scheme 15. In 1970 Herisson and Chauvin [7] proposed for the mechanism of metathesis the dismutation of metallacyclobutane into metallacarbene and adsorbed olefin. This reaction is formally analogous to the second step of the scheme 15.

Finally, it has been observed, for example, with propene [66] that the β hydrogen atoms of coordinated olefins may migrate to the metal giving η^3 allylic hydrides. This observation, when taken together with the fact that η^3 allylic ligands may undergo addition to the central carbon atom led Green *et al.* [65] to suggest the mechanism in scheme 16, for the formation of metallacyclobutane derivates

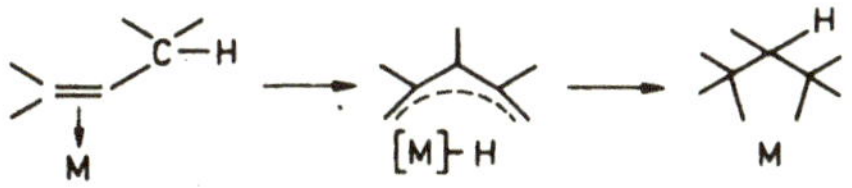

Scheme 16

Gault in a recent review [9] proposed a mechanism for reaction on Pd

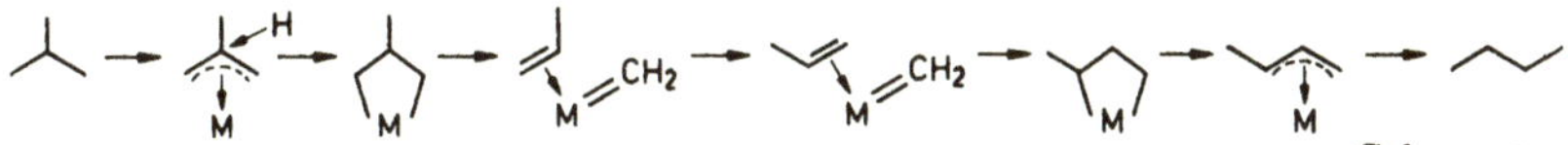

Scheme 17

which provides a straightforward explanation for bond shift isomerization on platinum. In fact since Tipper's finding on the ring opening of cyclopropane with dichloroplatinum [67], many platinacyclobutanes have been characterized [68] and the adsorbed metallacyclobutane is similar to the trimethylene di-σ-complex of platinum [69]. Thus a *direct formation of metallacyclobutane* in the case of platinum would explain the unique propensity of platinum to promote isomerization of neopentane to isopentane

Scheme 18

[44] with regard to the mechanism of rearrangement of platinacyclobutanes, Puddephatt *et al.* [70] have shown how easy is the isomerization of such complexes involving a *concerted process* in which complete C—C bond cleavage to give a carbene complex does not occur (scheme 19)

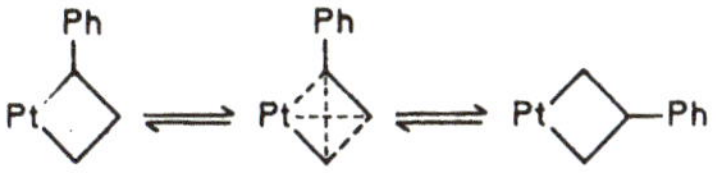

Scheme 19

In 1979, Casey *et al.* [71] pointed out that during the rearrangement of platinacyclobutanes no cross-over products were observed and that the reaction proceeds with complete retention of stereochemistry. They suggested that two opposite rotations will occur in the transition state where the plane of the alkene and the plane of the carbene ligand become perpendicular as in I (scheme 20)

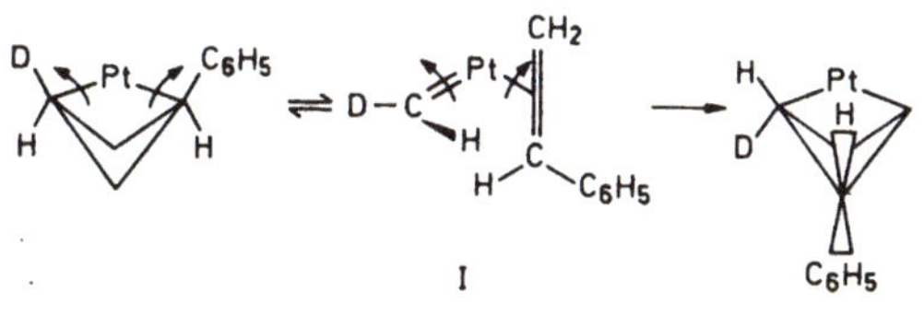

Scheme 20

It is worthwhile to note that platinacyclobutanes rearrangements have been very nicely studied in homogeneous catalysis: but all these, transient species are attached to one metal. In order to account for the possibility that not one but several metal atoms are involved in the bond shift, Garin and Gault [44] proposed the concerted mechanism in scheme 21

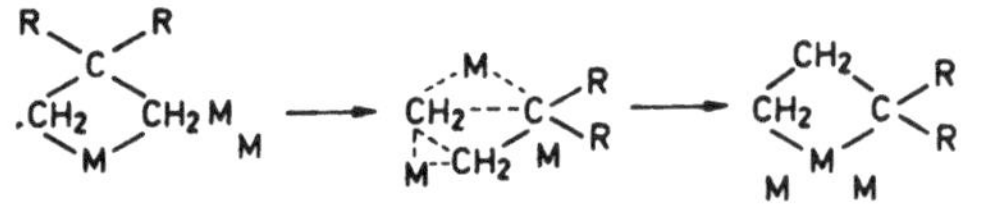

Scheme 21

Finally Clarke and Rooney in their review [63] suggested two mechanisms to account for the difference of behaviour between Pd and Pt in bond shift isomerization. In view of the known propensity of Pt to promote $\alpha\gamma$ exchange with deuterium of paraffins [41, 36] they refocused attention on the $\alpha\gamma$ species diadsorbed on one metal atom as the precursor for bond shift in simple alkanes. The following mechanism for neopentane isomerization on Pt is then feasible (scheme 22)

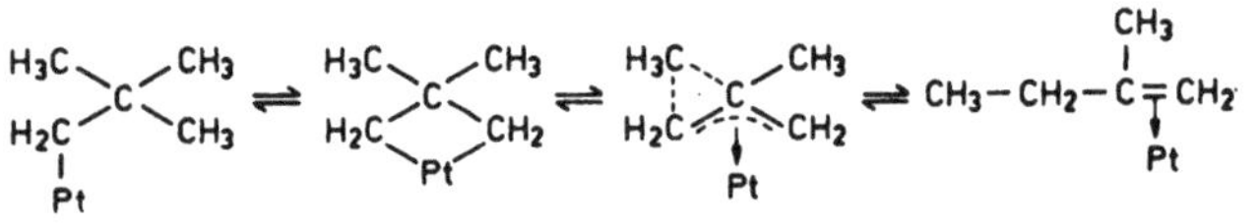

Scheme 22

As in the previous mechanism it is assumed that on platinum the metallacyclobutane is directly formed, while in the case of Pd the indirect route to the $\alpha\gamma$ diadsorbed species would result from a 1,2-hydrogen shift, which is relatively easy [72] *via* transient species of π-allylic character (scheme 23)

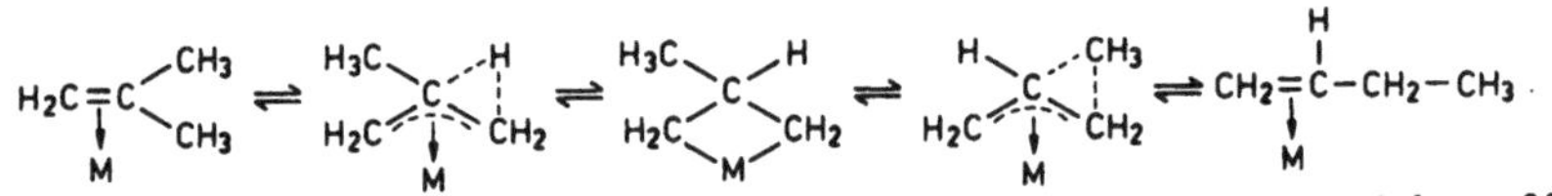

Scheme 23

It now remains to try to make a choice between these proposed mechanisms.

a) Existence of two bond shift mechanisms

From the kinetic parameters determined for all the reactions of *n*-pentane over the 10% Pt/Al_2O_3 catalysts of low dispersion, it is notable that the bond shift reaction is divided into two groups (Table 2) with activation energies differing by 40 kJ mol^{-1}. Interesting also is the distinction among the hydrocracking reactions between internal fission and demethylation.

Two mechanisms are possible for the bond shift reaction with the lower activation energy. Anderson's mechanisms with $\alpha\alpha\gamma$ triadsorbed precursor (Figure 3) and the mechanism involving metallacyclobutane could account for these reactions where isomerization *and* hydrogenolysis are associated. The fact that all these reactions are almost negligible on palladium (internal fission, methyl shift and neopentane isomerization [33]) and that Anderson's mechanism does not discriminate between the various isomerization reactions were arguments proposed by Garin and Gault [44] in favour of the metallacyclobutane dismutation mechanism.

The mechanism, shown in scheme 24 consists of a reversible dismutation of metallacyclobutane into metallacarbene and π-adsorbed olefin. It was chosen because it accounts for a number of facts.

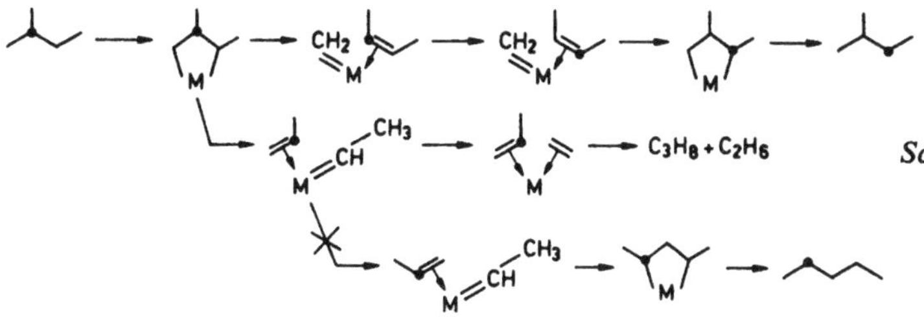

Scheme 24

First the rotation of the adsorbed olefin accounts for the observed bond shift reaction. Secondly, when attempting to apply this mechanism to the isomerization of isopentane to *n*-pentane, an adsorbed ethylidene would be formed, and it is well known from carbene and metallacarbene chemistry that such a species is rapidly isomerized by hydrogen shift to an adsorbed olefin [6]. The metallacyclobutane then cannot be reformed and isomerization is replaced by hydrogenolysis of the internal C—C bond. Thus at the same time it is explained why internal fission has the same activation energy as methyl shift and why chain lengthening does not take place by this mechanism but requires a more activated process. Moreover, this mechanism assumes rotation of the adsorbed olefin, a process which requires an activation energy of 40–60 kJ mol^{-1} [66]. Replacing the methyl group in the π-adsorbed olefin by more bulky substituents would increase the rotation barrier and therefore decrease significantly the contribution of the bond shift mechanism. Thus we can explain the fact that the percentage of bond shift mechanism decreases on going from pentanes to hexanes and from hexanes to heptanes. Similarly, for the same molecule, some bond shift reactions could be disfavored relative to others. This possibility is illustrated in scheme 25, where two methyl shifts for the 3-methylhexane molecules are presented [50]

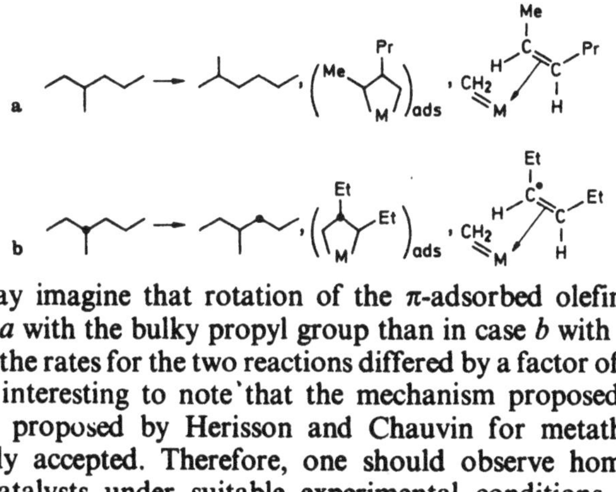

Scheme 25

One may imagine that rotation of the π-adsorbed olefin is more difficult in case *a* with the bulky propyl group than in case *b* with ethyl substituents. In fact, the rates for the two reactions differed by a factor of 6 to 8.

It is interesting to note that the mechanism proposed here is the same as that proposed by Herisson and Chauvin for metathesis [7] and now currently accepted. Therefore, one should observe homologation on the same catalysts under suitable experimental conditions. These conditions were indeed found, and homologation actually takes place, under very low hydrogen pressure, but to a very small extent because reformation of the metallacyclobutane is always much faster than methylene migration [73].

While some good evidence may be provided in favour of a metallacyclobutane mechanism for the reactions of isopentane to isopentane (^{13}C positional isomerization), to neopentane, and to internal fissions, the mechanism for the bond shift reactions of higher activation energy (*e.g.* reaction of isopentane to *n*-pentane and the reverse) is not elucidated yet. Since no hydrogenolysis reactions belong to the latter reactions, one should favour for these reactions either the cyclopropane mechanism (scheme 13) or, better, a concerted mechanism such as Rooney's mechanism in its original (scheme 14) or modified form (scheme 22). We have seen that the cyclopropane mechanism could not account, in the isomerization of a series of 2 methyl-alkanes to *n*-alkane, for the increase of methyl shift (A) relative to alkyl shift (B) with increasing chain length. In contrast, Rooney's mechanism would explain this progressive increase if one assumes that the rate of formation or the transient species with three center orbitals decreases with increasing the size of the migrating alkyl group. In its original form however Rooney's mechanism does not explain the predominance in each case of path B over path A. This objection could be rejected if one assumes, as Clarke and Rooney [63] do, that vinyl or alkylvinyl and not alkyl groups are the migrating entities.

This assumption is based on molecular orbital theory [74] where a vinyl shift is predicted to be much easier than a methyl or ethyl shift (scheme 26)

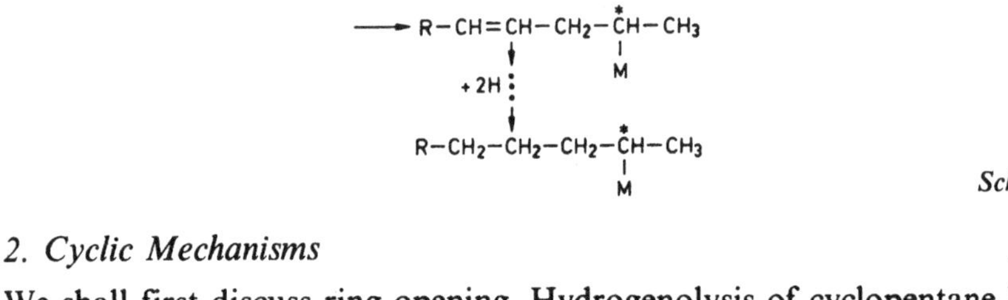

Scheme 26

2. Cyclic Mechanisms

We shall first discuss ring opening. Hydrogenolysis of cyclopentane hydrocarbons on platinum has been extensively studied by Kazanskii and his coworkers and the results obtained are very well summarized by Gault [9] and Newham [19]. In a study of the hydrogenolysis of methylcyclopentane and 1,3-dimethylcyclopentane on a series of Pt/Al_2O_3 catalysts with different metal loading from 0.15 to 20%. Gault in 1957 [75, 76] found that the distribution of the reaction products changed substantially with changing the percentage of platinum on the carrier. An almost *selective* rupture of the

CH_2-CH_2 bonds was found on the more concentrated catalysts (more than 2% of Pt), while on the less loaded ones (less than 1% of Pt) the rupture probabilities of the five C—C bonds of the ring were approximately equal. (scheme 27).

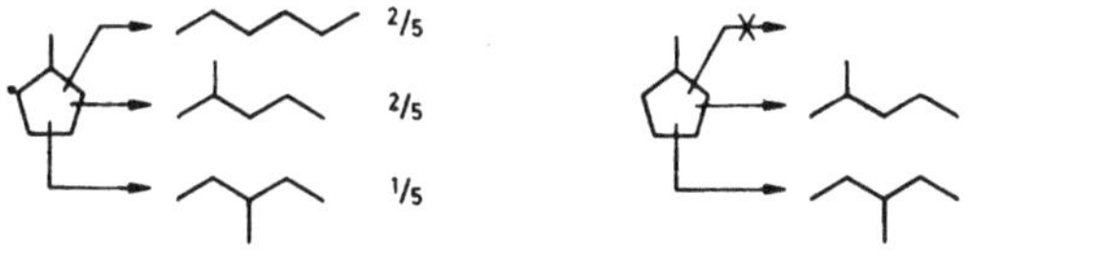

non-selective hydrogenolysis on high dispersed catalysts

selective hydrogenolysis on low dispersed catalysts

Scheme 27

A more careful study of methylcyclopentane hydrogenolysis, made with two catalysts of extreme dispersions, showed that on 0.2% Pt/Al_2O_3 the product distribution did not vary with temperature between 493–593 K, while it varied significantly on the catalyst of low dispersion [34].

On this 10% Pt/Al_2O_3 catalyst all the observed distributions appeared as linear combinations of two limit distribution. One of these included only 2- and 3-methylpentanes and corresponded to a completely selective hydrogenolysis of the di-secondary C—C bonds. The second limit distribution

Table 3. Product distributions corresponding to the non-selective ring-opening of cyclic hydrocarbons on 0.2% Pt/Al_2O_3

	Reaction product	Obtained/%	Calculated/%
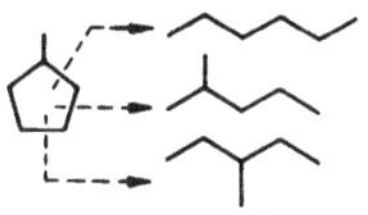	*n*-Hexane	38	40
	2-Methylpentane	41	40
	3-Methylpentane	21	20
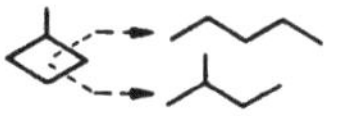	*n*-Pentane	53	50
	Isopentane	47	50
a) 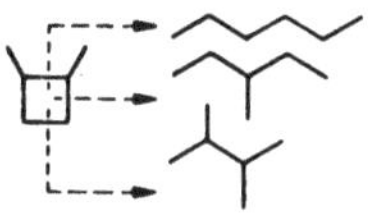	*n*-Hexane	23.5	25
	3-Methylpentane	46	50
	2,2-Dimethylbutane	30.5	25
a) 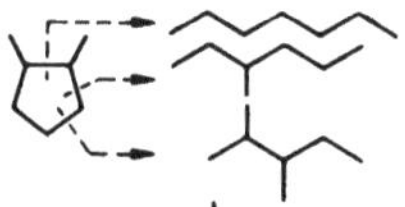	*n*-Heptane	26	20
	3-Methylhexane	31	40
	2,3-Dimethylpentane	43	40
b) 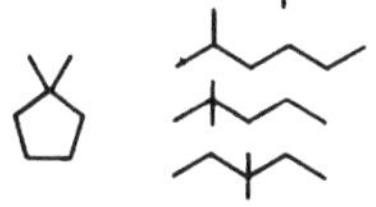	2-Methylhexane	0	0
	2,2-Dimethylpentane	75	75
	3,3-Dimethylpentane	25	25

[a] these two products were equilibrated mixtures of *cis* and *trans* isomers

[b] results obtained by Raik [80] and Kazanskiĭ [78]

contained *n*-hexane, but was different from the one obtained on 0.2% Pt/Al_2O_3 catalyst. Gault in 1965 proposed then that hydrogenolysis of cyclopentanes could occur according to three distinct mechanisms [45]: One, the non selective, mechanism A, corresponds to an equal chance of breaking any cyclic bond (Table 3). Two other mechanisms, the completely selective one, B, where only di-secondary bonds are broken, and a partly selective one, C, involving the breaking of di-secondary bonds mainly, but also of secondary-tertiary bonds. These two mechanisms B and C compete on catalysts of low dispersion. Moreover, it was found that hydrogenolysis of methylcyclopentane on iridium catalysts was completely selective, whatever the temperature and whatever the dispersion of the metal on the carrier [77]. Mechanism B therefore, like mechanism A is real, and the temperature dependency of the product distribution on 10% Pt/Al_2O_3 shows also the existence of a third mechanism C.

Mechanism B: it is tempting to correlate this mechanism with an adsorbed species obtained by losing four hydrogens on two carbon atoms. On the other hand, as di-secondary, secondary-tertiary and di-tertiary (Table 3) C—C bonds are broken in the non-selective hydrogenolysis (mechanism A) an $\alpha\beta$ diadsorbed species (adsorbed π-olefin) has been suggested [45] which can account for the inertness of quaternary-secondary [52] and quaternary-tertiary C—C bonds [78, 79] (*c.f.* scheme 28).

Although most of the product distributions in hydrogenolysis of cyclopentanes and cyclobutanes are accounted for either by mechanisms A or B, a number of anomalies remain which, we believe, should be attributed to a third mechanism, C, analogous to the bond shift *via* a metallacyclobutane mechanism. Since this mechanism accounts not only for bond shift but also for C—C bond rupture, hydrogenolysis of cyclic hydrocarbons might also occur accordingly. To avoid a too severely strained metallabicycloalkane, a methyl substituent will be involved in the process.

Experimental evidence in favor of this mechanism is:

(i) From *trans*-1,2-dimethylcyclobutane a small but significant amount of *n*-hexane is formed

(ii) From 1,2-dimethylcyclopentane on 10% Pt/Al_2O_3 a large amount of *n*-heptane is obtained (the selective mechanism B would have been largely predominent over the non-selective mechanism A)

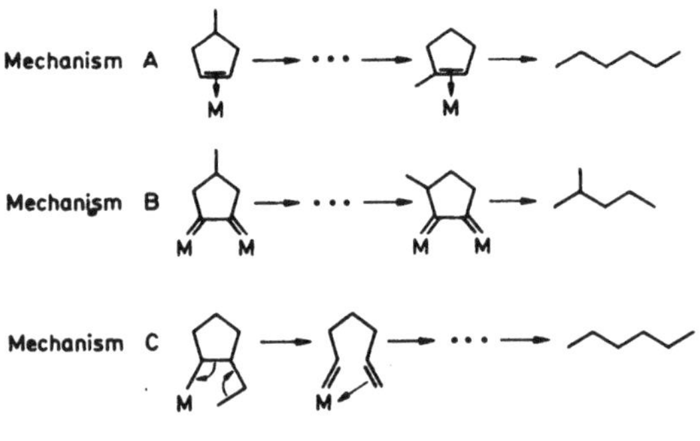

Scheme 28

(iii) The *n*-hexane which appears at higher temperatures on the catalyst of low dispersion in methylcyclopentane hydrogenolysis could also be accounted for by the metallacyclobutane mechanism.

Having characterized the three competing hydrogenolysis reactions (ring opening) by their precursor species — $\alpha\alpha\beta\beta$ tetraadsorbed species (or 1,2-dicarbene), (B); $\alpha\beta$ diadsorbed species (or π-adsorbed olefin), (A); and $\alpha\gamma$ diadsorbed species (or metallacyclobutane), (C); we consider now the process of isomerization according to a cyclic mechanism. In general terms, the following steps occur:

(i) Dehydrocyclization to a cyclopentane intermediate
(ii) Interconversions between adsorbed cyclopentane intermediates
(iii) Ring opening and desorption of the acyclic product.

A more complete description of the mechanisms requires knowledge of the precursor species. For mechanism C, identical to one for the bond shift mechanism already discussed, the precursor is a 1,1-carbene-4,5-π-adsorbed olefin, and dehydrocyclization consists in a carbene-olefin addition. On account of the strain in the metallabicycloalkane, it could be a transition complex between the diadsorbed carbene-olefin and a π-allylic complex,

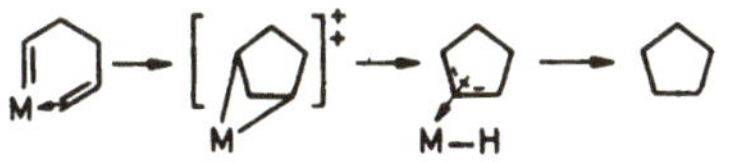

Scheme 29

In favor of the mechanism, Levisalles and his coworkers synthesized a mononuclear complex of tungsten having an alkylidene group and a double bond at the same time [81]. An X-ray structure of such a complex shows that the alkylidene function and the coordinated double bond are perpendicular to each other and the carbenic carbon is near the double bond (Figure 4),

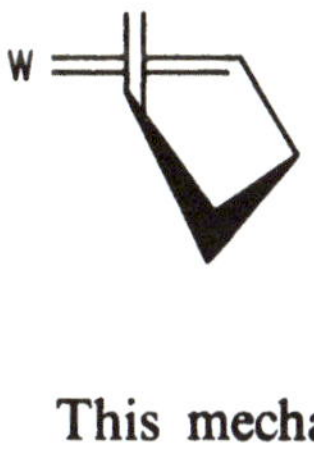

Figure 4. Schematic representation of the complex $W(CO)_4C(OET)(CH_2)_2CHCH_2$ depicted in ref. [81] showing that the alkylidene function and the coordinated double bond are perpendicular to each other. (Reprinted with permission from J. of Organometallic Chemistry, copyright by Elsevier Scientific Publ. Co., reproduced by courtesy of H. Rudler)

This mechanism has some analogies with the mechanism proposed by Shephard and Rooney [47] where an alkene-alkyl insertion occurs with a 1,2-π-5-σ-triadsorbed precursor which is very similar to the one invoked by Herington and Rideal [82] for aromatization of alkanes by 1–6 ring closure. A slightly different version was also proposed to account for cyclic type isomerization of hexanes and non-selective hydrogenolysis of methylcyclopentane [34] (scheme 30),

Scheme 30

A common feature in schemes 29 and 30 is that they require the preliminary formation of a π-adsorbed olefin. Therefore the reactions of three substituted pentanes, *v.i.z.* 2,2,4-trimethyl (I), 2,2,3-trimethyl (II), and 2,2,4,4-tetramethylpentanes (III) have been studied over platinum and palladium films at 570 K [64]. On platinum, the cyclopentane obtained by a simple dehydrocyclization process is the main reaction product and its rate of formation is practically independent of the structure of the reacting molecule. On palladium the dehydrocyclization rate is much larger for II than for I and III. These results suggest that different intermediates are responsible for dehydrocyclization on Pt and Pd.

On palladium the dehydrocyclization of II but not III can only be accounted for if the intermediate is an adsorbed olefin. Adsorbed olefins were proposed as intermediate in 1–6 dehydrocyclization on chromium oxide [82–84]. They were also thought to be responsible for 1–5 ring closure on oxides and metals, and were described by Shephard and Rooney in the case of metal as species A, 1,2-π-bonded and 5-σ-bonded to a single metal atom [47] (scheme 31),

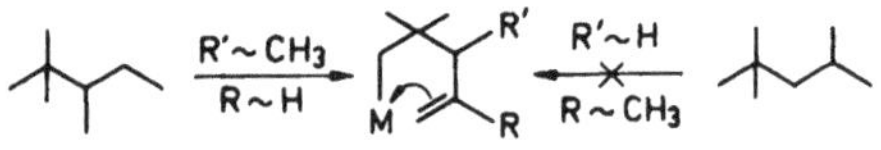

(II) Species A (I)

Scheme 31

species such as A (scheme 31) are very sensitive to steric factors [82].

On platinum, the precursor for the dehydrocyclization process is necessarily attached to the metal by two carbon atoms, only 1 and 5, and not by three carbon atoms 1, 2 and 5 as suggested by Barron *et al.* [34] and Shephard and Rooney [47]. The precursor species does not require olefin formation and the metallacyclobutane mechanism (alkene-alkyl insertion mechanism) should be rejected in this case. A 1,1,3-triadsorbed precursor such as B (scheme 32) was suggested to be responsible for the bond shift rearrangement on platinum [31, 34, 42]. Gault and Muller [64] suggested that the precursor for dehydrocyclization was a 1,1,5-triadsorbed species, C, (scheme 32),

Species B Species C

Scheme 32

An alternative route, possibly more energetically favoured, could be the transient formation of an intermediate E, (scheme 33), in which the two *p* orbitals of carbon atoms 1 and 5 are coupled together with a *d* metal orbital

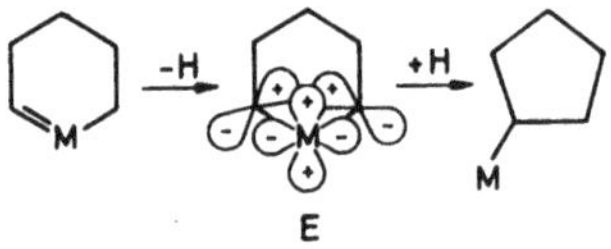

E

Scheme 33

The carbene-alkyl insertion mechanism does not account for the selectivity in hydrogenolysis and 1,5 ring closure. Thus statistical carbon-carbon ring rupture of cyclopentanes are observed with a number of cyclopentanes on highly dispersed platinum/alumina catalysts but, under the same conditions, quaternary-secondary C—C bonds are not ruptured in hydrogenolysis of cyclopentanes, and 2-methylhexane is not isomerized to 3,3-dimethylpentane. The *mechanism A* therefore involves, for 1–5 ring closure, only methylenic and methylic carbon atoms and, for ring opening only carbon atoms bearing at least one hydrogen atom. 1–5 dehydrocyclization in this case is best described then as a dicarbene recombination to form a π-adsorbed olefin [85]

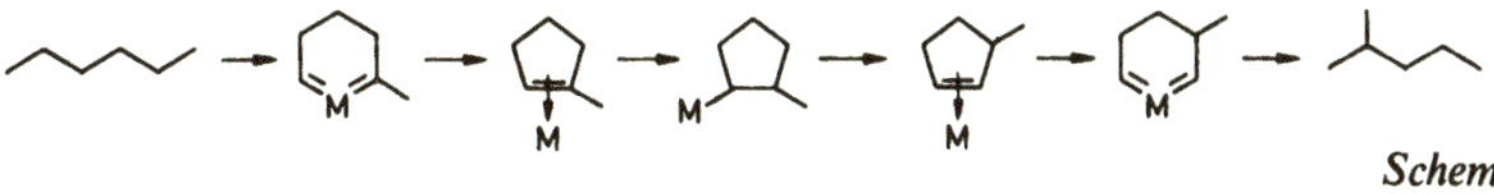

Scheme 34

On the other hand, another mechanism taking place on iridium [77], does not allow the isomerization of *n*-hexane but interconverts very rapidly 2-methylpentane and 3-methylpentane *via* a methylcyclopentane intermediate. Conversely, on iridium only the CH_2—CH_2 bonds are ruptured during the hydrogenolysis of substituted cyclopentanes. This selective hydrogenolysis had already been observed on platinum catalysts of very low dispersion at low temperatures [45]. Since two methyl groups in 1–5 positions are required for dehydrocyclization, and two adjacent methylene groups in cyclopentane hydrogenolysis, Gault *et al.* suggested that this *selective cyclic mechanism B* consists in a dicarbyne recombination to form a dicarbene species [85]. All intermediate species are necessarily attached to two metal atoms and it is possible that the reaction involves as a first step addition of two hydrogen atoms to form an acyclic dicarbene (scheme 35),

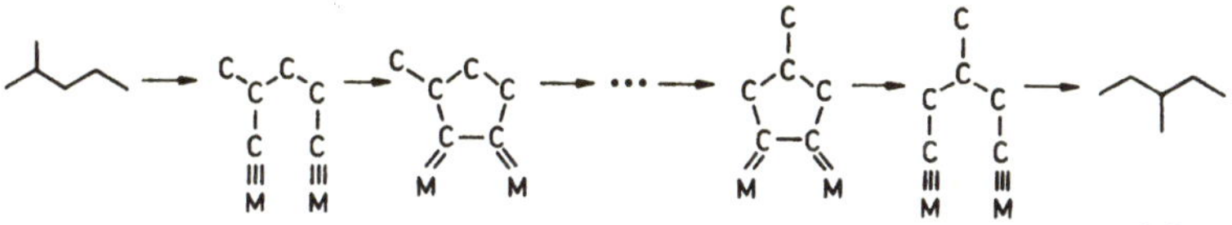

Scheme 35

A common precursor for both dicarbene and dicarbyne mechanisms is adsorbed metallacarbene. Carbene-metal generation from σ-bonded alkyl groups was first proposed by Muetterties [86, 87] when he observed CH_4 generation by reacting $(CH_3)_2Zn$ with WCl_6,

$$WCl_6 + (CH_3)_2Zn \xrightarrow{-ZnCl_2} (Cl_4)W\begin{matrix}\diagup CH_3 \\ \diagdown CH_3\end{matrix} \rightleftarrows (Cl_4)\overset{\overset{\displaystyle H}{|}}{\underset{\underset{\displaystyle CH_3}{|}}{W}}{=}CH_2$$

$$\rightarrow (Cl_4)W{=}CH_2 + CH_4$$

The CH_3—W $\rightleftarrows$ CH_2=W—H equilibrium, proceeding by α hydrogen migration from the σ alkyl to the metal, has been demonstrated by Green [88]. In view of the possibility of α-hydrogen transfer to metal, it remains possible that the carbene complex has structure II rather than III as suggested by Calderon [89] (scheme 36),

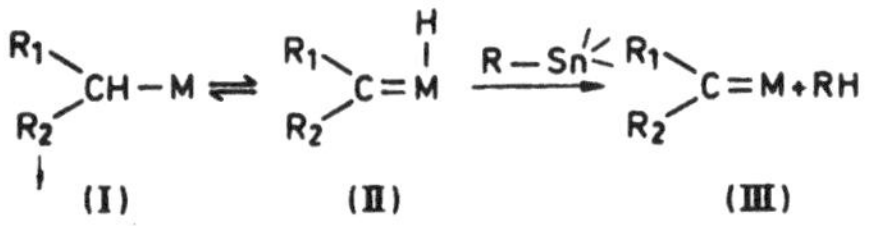

Scheme 36

Strong arguments in favor of hydride-metal involvement in the generation of carbene-metal complexes are comprehensively presented in Rooney's monograph on olefin metathesis [90].

Spectral evidence has been presented for the existence of M—H groups in systems involving Ir, Ru or Os which metathesize highly strained cyclic substrates [91]. On these metals further dehydrogenation to metallacarbyne is also easy [45, 92] and that explains why the selective mechanism B for dehydrocyclization (or hydrogenolysis) predominates over the non-selective mechanism A [77, 93]. On platinum, besides the α-hydrogen elimination route, obvious in the case of 2,2,4,4-tetramethylpentane, another possible path could be the isomerization of π-adsorbed olefin *via* a 1,2-hydride shift [94],

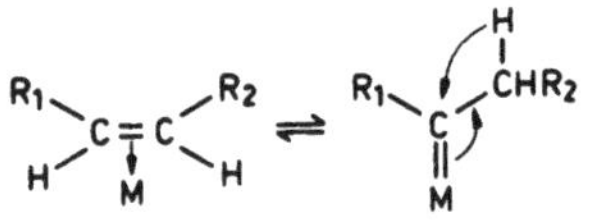

Scheme 37

The equilibrium in scheme 37 is shifted to the left as the π-adsorbed olefin is more substituted. Nevertheless, the concentration of metallacarbenes by this reaction could be made high enough at elevated temperatures to permit the formation of dicarbene or dicarbyne and the occurrence of dehydrocyclization. However, Rooney in his review [90] mentionned that this route to carbenes can hardly extend to norbornene and other similarly constrained, highly reactive cycloalkanes, since the transition state in Figure 3 requires considerable distortion from the planarity of the olefinic group. He proposed a possible variation avoiding this difficulty which occurs with a 1,2 hydride transfer *via* metal hydride (scheme 38)

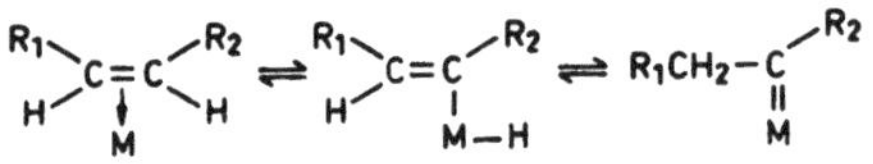

Scheme 38

3. *Hydrogenolysis*

The subject of hydrogenolysis of alkanes is a broad one and has been reviewed by Sinfelt [95, 96]. The purpose of this section is to discuss the various mechanisms involved in the carbon-carbon bond rupture of alkanes with metals having isomerization activity.

a) Platinum catalysts

On platinum catalysts the activation energy for hydrogenolysis is very much larger for ethane than for larger molecules [97]. As ethane reacts in considerably higher temperature range [98] we can reasonably conclude that a mechanism involving a 1–2 adsorbed intermediate will not be of much importance in hydrogenolysis reactions over platinum for hydrocarbons other than ethane. On the other hand a mechanism involving a 1–3 adsorbed intermediate can operate with other hydrocarbons in addition to neopentane.

Avery and Anderson [47] have pointed out that alkane isomerization is accompanied by hydrocracking and the activation energies for the two processes are identical [31]. Thus both reactions would, most probably, proceed *via* a common intermediate and they suggested the bridged intermediate (scheme 39). Thus hydrocracking may be represented by the failure to form the bond between C_1 and C_3, probably as a result of attack by a surface hydrogen atom at C_1 (scheme 39). The slow step is the formation of a bridged intermediate B from the 1–3 adsorbed precursor A. Further hydrogen attack then results in the desorption of the species C (scheme 39).

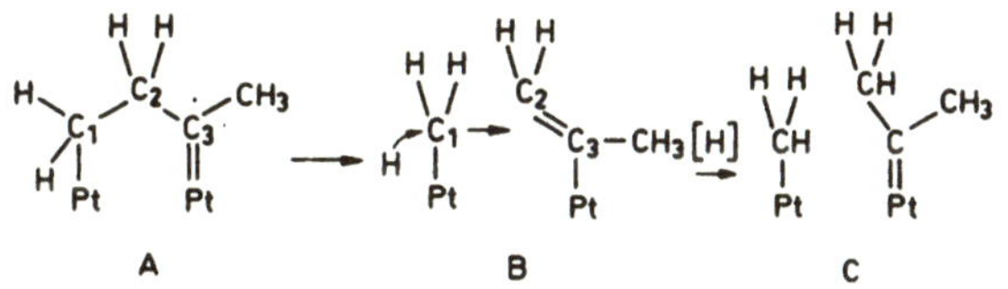

Scheme 39

We may note that this process requires two nearest-neighbour surface platinum atoms.

Anderson and Shimoyama [97] suggested another mechanism to account: i) for the formation of isopentane from 3-methylpentane, ii) for the increase of the specific hydrogenolysis rate when the metal particle size decreases. They used the essentials of Shephard and Rooney's original proposals for ring closure and applied them for C—C bond rupture on platinum (schemes 40),

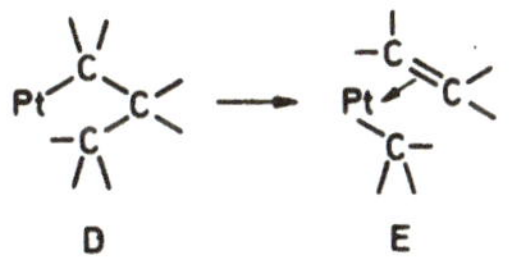

Scheme 40-1

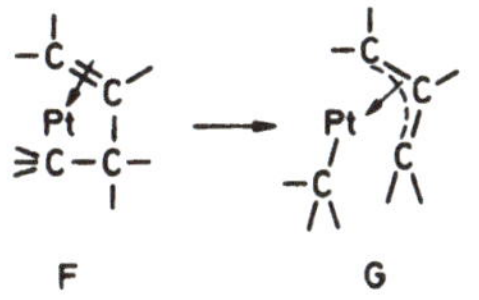

Scheme 40-2

These mechanisms 40-1 and 40-2 involving a catalytically active site consisting of a single platinum atom, were referred to as π-olefin allyl hydrogenolysis. The species in E and G are subsequently hydrogenated off the surface. This mechanism could explain that the overall hydrogenolysis rates decrease with increasing crystallite size and it is reasonable to infer that the rate of the π-olefin/allyl pathway decreases in the same way.

Leclercq, Leclercq and Maurel [99] studied the hydrogenolysis of saturated hydrocarbons and have shown that the mechanism suggested by Anderson cannot account for their results. They found that bonds in the *β-position to a tertiary carbon atom* are extensively broken, but, on the contrary, with a quaternary carbon atom, hydrogenolysis in the α position to this atom is more rapid than in the β position. They supposed that the first stage of adsorption will occur more easily on the most substituted carbon atom except for quaternary carbon. The intervention of a 1,3 intermediate will explain very well the great influence on the reactivity of carbon atoms in the α position, provided that those bonds do not include a quaternary carbon atom.

To explain why bonds in the β position to tertiary carbon atoms are more easily broken than are α bonds they proposed that the primary carbon atom which is adsorbed after the tertiary one can lose two or three atoms of hydrogen and the bond in the α position to this carbon atom doubly or triply linked with the metal is weakened and is split more easily (scheme 41).

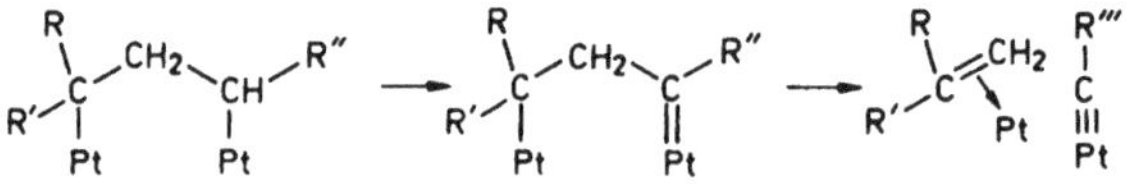

Scheme 41

From these two groups of workers we can underline that hydrogenolysis of carbon-carbon bonds proceeds through 1,1,3 adsorbed species. However, it must be pointed out that the bond contiguous to the carbon atom doubly bonded to platinum is broken in the case of Maurel's proposal and not in Anderson's proposal.

Garin and Gault proposed, from kinetic data for skeletal rearrangement and hydrocracking of alkanes, a mechanism involving the reactions of platinacyclobutanes which can explain simultaneously the internal fissions (deethylation) and the methyl shift reactions (scheme 24). This proposed mechanism could be supported by the work of Puddephatt and Tipper [100] who studied the skeletal isomerization of metallacyclobutane complexes. They observed that reactions of platinum (IV) cyclobutanes, in solution, normally involve rearrangements of the C_3 moiety or its ejection as a whole from the molecule [101]. However, under certain conditions, in particular with solvents of relatively high dielectric constant in the presence of phosphines, high yields of hydrocarbons (*e.g.* C_2H_4) from fission of the C_3 moiety are obtained [102, 103],

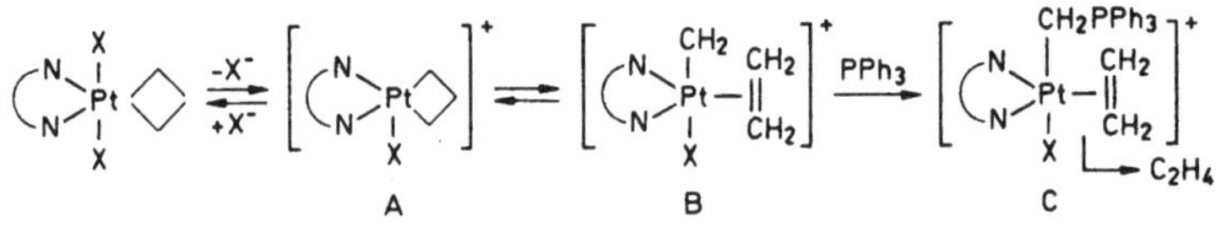

Scheme 42

The species A is coordinatively unsaturated and could rearrange to the carbene-alkene complex B followed by trapping of the carbene by phosphine to give the ylid species C, which would then undergo the loss of the ethylene [102a, 103] (a similar mechanism has been proposed by Green *et al.* [104]): Thus the metallacyclobutane rearrangement in heterogeneous catalysis has an analogue in homogeneous catalysis and can nicely explain either the skeletal rearrangement or the hydrocracking of alkanes which may be considered as a degenerate bond shift type isomerization. Finally, on highly dispersed catalysts (0.2% Pt/Al_2O_3) the kinetic data have shown that the isopentane deethylation reaction has a higher apparent activation energy compared to the *n*-pentane reaction (Table 2). On the other hand, isopentane and *n*-pentane demethylation reactions have the same apparent activation energy as the internal fission of the *n*-pentane.

Garin *et al.* [105] put forward the existence of an $\alpha\alpha\gamma$ intermediate tri-adsorbed on *one metal atom* in the hydrocracking reactions of pentanes.

They suggested that in such species the carbon atom singly bonded to the platinum will be secondary or tertiary and that the carbon is partially positively charged [106, 107]. This species leads to hydrogenolysis by the rupture of the bond contiguous to the carbon atom doubly bonded to platinum (scheme 43),

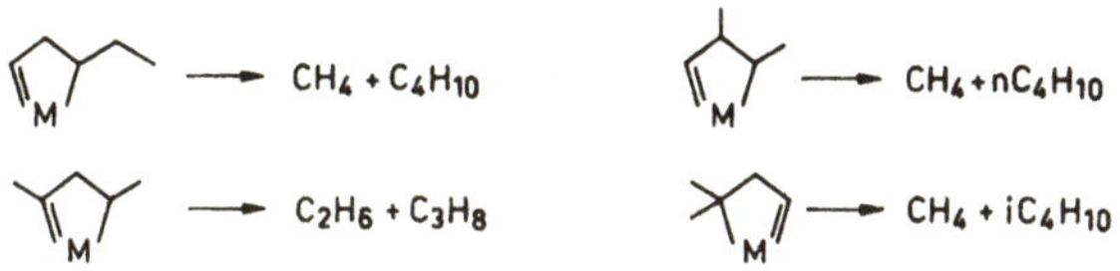

Scheme 43

This mechanism accounts for the higher apparent activation energy for the internal fission in isopentane and can also explain the selective demethylation of methyl and ethyl cyclopentanes on cluster-derived catalysts [12, 108] as we shall see later in section 4..

b) Palladium catalysts

Gault and Muller [64] have suggested an allylic precursor such as A in scheme 44, where three adjacent carbon atoms are attached to two different metal atoms by a π and a σ-bond, and an attack by hydrogen would lead to cracking,

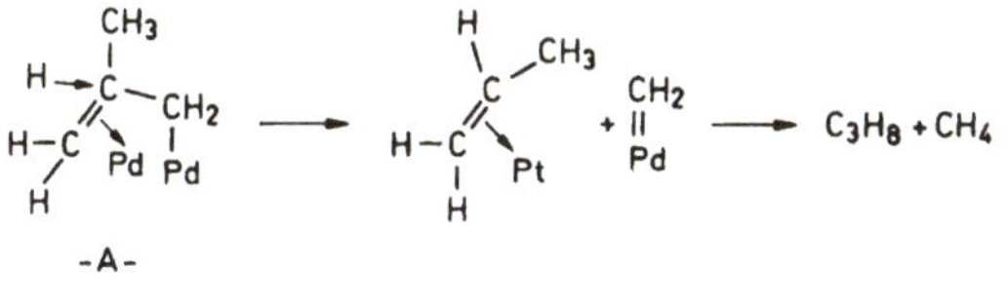

Scheme 44

c) Iridium catalysts

As the rupture of bonds between di-secondary or primary-secondary carbon atoms is highly favoured, Weisang and Gault [77] suggested that dismutation of 1–2 dicarbene species to form two metallacarbyne entities is responsible for this selectivity in hydrocracking and in methylcyclopentane ring opening,

Scheme 45

Skeletal reactions over iridium are dominated by carbon-carbon bond rupture and, Foger and Anderson [109] put forward that there are two basic hydrogenolysis mechanisms which are characterized by different activation energies and which occur at different structural units in the hydrocarbon reactant molecule [110]. These are: C_2-unit mode reaction and iso-unit mode reaction, with an activation energy of about 175 kJ mol^{-1} and 235 kJ mol^{-1} respectively.

Both of these basic hydrogenolysis modes are insensitive to catalyst structure in the sense that they are essentially uninfluenced by iridium particle size in the range < 1–7 nm.

The C_2 unit mode is the reaction for bond types C_I—C_I, C_I—C_{II}, C_{II}—C_{II}, and in this case both carbon atoms carry not less than two hydrogen atoms in the reactant molecule. For the iso-unit mode the bond types are C_I—C_{III}, C_I—C_{IV}, C_{II}—C_{IV} and C_{II}—C_{III}.

Anderson observed besides hydrocracking some isomerization activity with both neopentane and isobutane on iridium catalysts of very low dispersion [109] and on iridium-gold [111]. Both of these molecules react *via* the iso-unit mode. He concluded that *iso-unit hydrogenolysis* and bond shift isomerization both proceed from a common adsorbed intermediate and suggested the existence of an iridiacyclobutane (scheme 46),

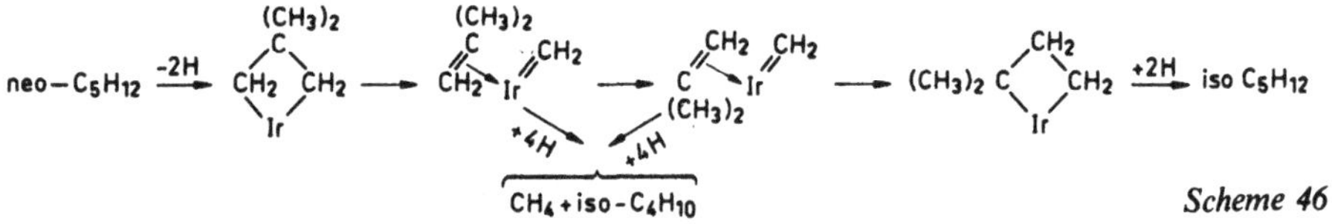

Scheme 46

This mechanism is similar to the one suggested by Gault *et al.* [44] on platinum.

With the *C_2 unit mode* reaction, Anderson *et al.* [111] suggested a parallelism with reactions involving a C_5 carbocyclic intermediate if interconversion between 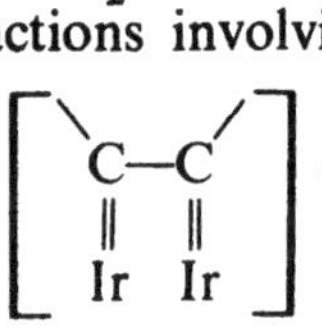 and 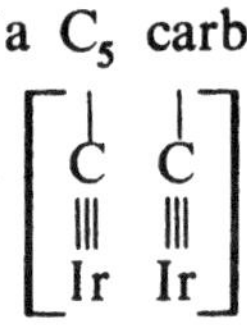 is reversible and if in a carbon chain such as —C_xH_2—C_yH_2—C_zH_2— the adsorption mode 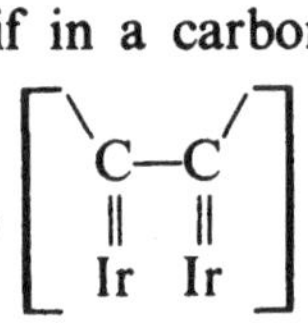 can be propagated from the pair C_x—C_y to C_y—C_z.

The relation between C_2 unit hydrogenolysis and this C_5 carbocyclic isomerization is shown thus in scheme 47,

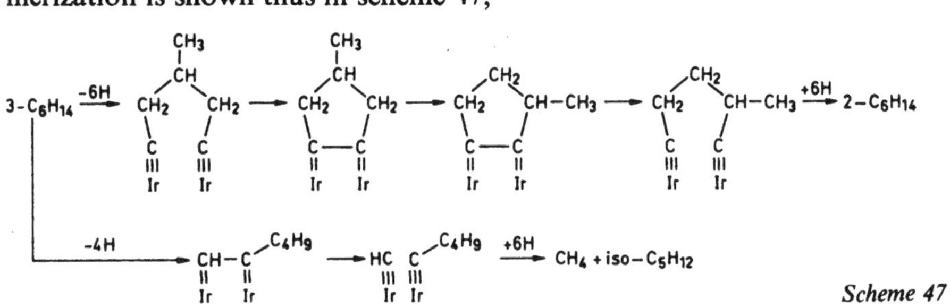

Scheme 47

The isomerization mechanism represented in scheme 47 predicts that neither of the methylpentanes can yield *n*-hexane [77, 112], and the complete lack of any isomerization component with *n*-hexane reactant [77, 112, 113]. This is in reasonably good agreement with experiments.

4. Structure Sensitivity

A. Metal Particle Size and Reaction Mechanism

Since the development of the theory of "active ensembles" by Kobozev [114], a considerable amount of work has been devoted to determining specific rates for various reactions. After the pioneer work of Boreskov and coworkers [115], the improvement of chemisorption techniques to determine the surface area of a metal component of a composite catalyst [116, 117, 118], and the preparation by adsorption of highly dispersed supported catalysts [119] have allowed accurate measurements of the specific rates for various reactions on series of catalysts with a very wide metal dispersion range. The metal dispersion is here defined as the ratio between superficial and total metal atoms.

The specific rates for a number of reactions have been shown to remain remarkably constant, *i.e.* not to depend on the metal dispersion: for instance, dehydrogenation of cyclohexane [120], hydrogenolysis of cyclopentane [121], hydrogenation of cyclohexene and of allyl alcohol [122, 123], hydrogenation of cyclopropane [124], and dehydrogenation of 1,1,3-trimethylcyclohexane [125].

The first reported case of a reaction with a large dependence of specific rate on catalyst structure was for neopentane on platinum [126]. This result led Boudart to classify reactions on metals as: a) "facile" or "structure-insensitive" reactions, for which the specific rate does not depend upon the size of the metal particle, and b) "demanding" or "structure-sensitive" reactions, for which the specific rate is highly dependent on the metal dispersion [127]. Boudart suggested that specific sites with special geometrical requirements are involved in structure-sensitive reactions, while all metal atoms of the surface are available for structure-insensitive reactions.

Since then, a few more structure-sensitive reactions have been discovered, *e.g.* exchange of benzene on iridium and nickel [128], and dehydrocyclization of *n*-heptane on platinum catalysts [129]. It is worth noting that in these examples, as in neopentane reactions, the change in metal particle size, determining the change in specific rate, is obtained by sintering the catalysts [129, 130] or modifying the nature of the carrier and the method of preparation.

All reactions discovered so far which are structure-sensitive for the rate may also be considered structure-sensitive for selectivity, since the product distributions vary widely with the metal particle size, *e.g.* cracking versus isomerization in skeletal rearrangements [129, 130], relative exchangeability of various types of hydrogen or deuterium distribution patterns in the case of exchange [128, 131]. Indeed the first example of a reaction which is structure-sensitive *for selectivity* was the hydrogenolysis of methylcyclopentane on supported platinum catalysts [45, 76]. On catalysts with large metal particles (10% Pt/Al_2O_3), the di-secondary $CH_2{-}CH_2$ bonds are selectively broken, so that 2-methylpentane and 3-methylpentane are the only reaction products, while on highly dispersed catalysts (0.2% Pt/Al_2O_3) the five cyclic bonds are randomly ruptured and the ratio of 3-methylpentane to *n*-hexane is very close to 1/2. On catalysts of intermediate dispersion both types of hydrogenolysis, selective and non-selective, contribute in various proportions to the overall reaction.

Obviously the constancy of the specific rate in catalytic reactions, *e.g.* hydrogenolysis of methylcyclopentane, cannot be explained as in the theory of "structure-sensitive" and "structure-insensitive" reactions, by the simple presence of specific or non-specific active sites on the metal surface. In order to clarify the problem, simultaneous kinetic and mechanistic studies are required, dealing with other reactions, which are structure-sensitive for the selectivity.

The isomerization of hexanes are typically structure-sensitive for the selectivity [53, 132] in the sense that the cyclic mechanism is largely predominant on highly dispersed catalysts [53], while the bond shift mechanism is the major one on supported metals with large particles [132].

1. Existence of Different Types of Sites

As will be discussed in more detail in a subsequent section (4.B.2), there is good evidence that isomerization by bond shift and by a cyclic mechanism involves different types of sites on a platinum surface. For the present, we only need to note that the percentage of cyclic mechanism in the isomerization of 2-methylpentane to 3-methylpentane remains constant as long as the mean platinum particle diameter, $\bar{d}$, is larger than about 1–2 nm and increases steeply for more highly dispersed catalysts. The observed increase in the percentage of cyclic mechanism with increasing platinum dispersion at $\bar{d} < 2$ nm clearly indicates the existence of two types of sites associated with the cyclic and bond shift mechanisms. Furthermore, for the most highly dispersed catalysts, the proportion of bond shift sites becomes quite small if not actually zero. On the other hand, the constancy of the proportion of cyclic mechanism for $\bar{d}$ between about 2 and 15 nm suggests that the sites responsible for bond shift and cyclic mechanisms are topographically similar, both involving either face or edge atoms. Since even the most highly dispersed catalysts are active for isomerization, it is likely that the two types of isomerization sites although different in nature, both include edge atoms.

We may note that, although in principle the same sort of argument may be applied to data concerning the percentage of cyclic mechanism in the isomerization of *n*-hexene to methylpentanes or of 2-methylpentane to *n*-hexane, in fact in these cases the nature of the cyclic mechanism changes with changing platinum dispersion (the ring opening of the cyclic intermediate changes from non-selective at high dispersion to selective at low and medium dispersions) and, as a consequence, the constancy of the percentage of cyclic mechanism in the platinum particle size range 2–15 nm cannot be observed.

If, as in the model outlined above, there are two types of sites on platinum, both active for isomerization and both involving edge atoms in same way, it is necessary to consider the way in which the specific reaction rates may vary with particle size. In practice the specific isomerization rate is constant with changing particle size: in principle, a model involving the supply of adsorbed species *via* a weakly adsorbed precursor allows for a variety of behaviour, and this is outlined in more detail below.

2. Weakly and Strongly Adsorbed Species

Let us assume that all superficial atoms are available for weak adsorption and may be used for reactions, such as exchange, in which only C—H bond rupture and formation take place. Suppose, on the contrary, that the sites required for reactions such as isomerization or cyclopentane hydrogenolysis, which involve strong adsorption and C—C bond rupture or formation, are located on some very specific parts of the metal crystallites-for example the edge. If one further assumes that the weakly adsorbed species are the precursors of the strongly adsorbed species, the following sequence of elementary reactions may then be proposed (scheme 48):

1) indiscriminate adsorption of a gaseous molecule on a face site of the metal crystallite to form a weakly adsorbed molecule;

2) superficial migration from the adsorption site S_A to a reactive site S_R located on some specific region of the metal particle;

3) formation on the reactive site of a highly dehydrogenated species;

4) skeletal rearrangement of this strongly adsorbed species;

5) rehydrogenation of the strongly adsorbed moiety, superficial migration, and desorption of the weakly adsorbed isomerized molecule (reverse of steps 3, 2 and 1), to yield the reaction products.

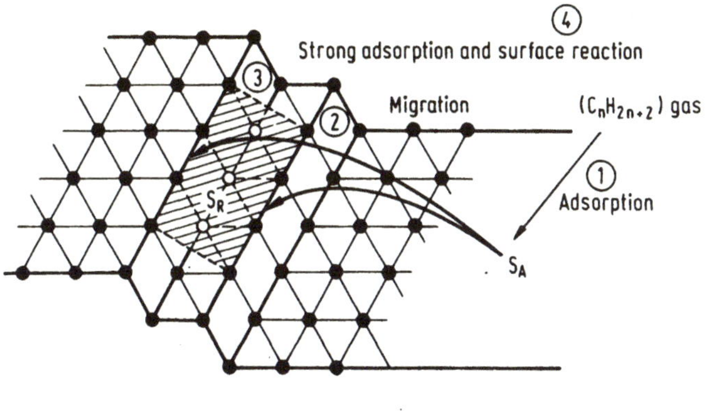

Scheme 48

On account of the different reactivities of the weakly and strongly adsorbed species, weak indiscriminate adsorption (step 1) and superficial migration (step 2) should be much faster than strong adsorption (step 3) and the surface reaction (step 4). Superficial migration indeed, which can be pictured either as a series of interconversions between mono- and diadsorbed species or as a very fast interchange between chemisorbed alkyl radicals and physically adsorbed molecules, involves only weakly adsorbed species which react much more rapidly than the strongly adsorbed and highly dehydrogenated species involved in the skeletal rearrangement itself. Therefore one expects the rate-determining step in isomerization to be one of the two reactions, (3) and (4), which occur on the reactive sites S_R. Kinetic data for the isomerization of pentanes [44] and of hexanes [133] strongly suggest that the rate-determining step is the surface reaction (step 4), while the successive dehydrogenation reactions (step 3) required to form the reactive strongly adsorbed species are equilibrated. Even if the rate-determining step were strong adsorption rather than surface reaction, as seems to be the case in the isomerization of heptane [52], the reasoning which follows would still be valid.

3. Relationship between Selectivity and Specific Rates

The model proposed above (scheme 48) can explain the various situations which may arise concerning the dependence of specific rates and selectivity on the size of the metal particles.

Let us suppose first that on one crystallite the total number N_R of reactive sites S_R is much larger than the number N_A^* of weakly adsorbed hydrocarbon molecules. This situation could be the most common one, since it is known that only a small fraction of the surface is covered by chemisorbed alkanes. Then, only some of the available reactive sites are occupied, the number of which, N_R^*, should be proportional to the number N_A^* of weakly adsorbed molecules. Since N_A^* itself is proportional to the number of superficial metal atoms, N_S, the number of reacted molecules is proportional to N_S. One therefore expects the turnover number to be independent of the size of the metal crystallite. On the other hand the distribution of the various reactive sites. N_{R_1}, N_{R_2}, ... (with $\sum_i N_{R_i} = N_R$), associated with various parallel reaction pathways, generally depends upon the size of the metal particle. This explains the present situation, where the selectivity — or the contribution of various mechanisms — widely varies with metal dispersion, while the specific rate remains constant.

A second possibility is that the number N_R of reactive sites of any type is much smaller than the number N_A^* of weakly adsorbed hydrocarbons. All the reactive sites are then occupied ($N_R^* = N_R$); consequently the number of reacted molecules is no longer proportional to the number of weakly adsorbed hydrocarbon N_A^* and to the number of superficial metal atoms. In this case, not only the selectivity, but also the specific rate depend upon the metal particle size. This situation, which is that of the so-called structure-sensitive reaction rate [127], arises especially when the change in particle size has been obtained by sintering a supported metal catalyst [126, 129]. The effect of sintering is to smooth the metal surface, thus reducing considerably the number of those irregularities (steps, kinks, etc.) with which the reactive sites may be associated.

A third possibility is that for some reason the rate of superficial migration becomes much smaller than the rate of weak adsorption or desorption. In this case only the molecules adsorbed in the vicinity of a reactive site may be strongly adsorbed and react. The reaction rate is proportional to the number of reactive sites and not to the total number of surface atoms, so that the reaction is again structure-sensitive for the rate.

B. Crystallography and Structure of Catalyst Particles

One approach to the structure sensitivity of skeletal reaction mechanisms, on which emphasis has been laid during the past few years, is the attempt to correlate them with the size of the metal particles and more generally with the structure of the metal surface [17, 53, 126, 132, 134, 135, 136]. If interpreted from a strictly geometrical point of view, particle size effects (or the sensitivity to the structure of the metal surface) could provide information about the number (or the nature) of the metal atoms involved in a particular reaction. For instance, one could expect that selective hydrogenolysis, favoured on large metal particles, requires a larger number of metal atoms than non-selective hydrogenolysis, which takes place on extremely dispersed catalysts [45]. Similarly, isomerization of 2-methylpentane to 3-methylpentane takes place predominantly according to a bond shift mechanism on catalysts of low dispersion [132] and according to a cyclic mechanism on catalysts with very small particles [53]: this again could imply a larger number of metal sites for the former than for the latter reaction.

1. Interpretation by Crystallographic Models

The interpretation of these and other particle size effects is, however, difficult. On the one hand, in the case of films or supported metals, the structure of the surface and even of the bulk is only roughly characterized. The metal dispersion defined as the ratio between surface and total metal atoms, may be obtained by hydrogen chemisorption measurements [115a, 116], while electron microscopy or small-angle X-ray scattering (SAXS) provides information on the particle size distribution [137, 138]. On the other hand, well-defined catalysts such as single crystals exposing low Miller index faces or stepped surfaces [139] represent a limited class of structures, all belonging to the regular *fcc* system.

Several alternative models have been proposed to date providing reasonable explanations of particle size effects in metal catalysis. According to the first and most simple model, different reactivities are associated with surface atoms of different coordination numbers. Following the *mitohedrical* theory developed by Poltorak [119], face atoms and edge atoms in metal crystallites may act as specific active centers. In the case of hexane isomerization, for instance, these sites would be associated with bond shift and cyclic mechanisms [136]. As Poltorak pointed out, the relative numbers of edge and face atoms vary widely when the size of the metal particles varies between 1 and 5 nm; below 1 nm the crystallites have only edge atoms or atoms of even lower coordination number. Beyond 5 nm more than 90% of the surface atoms are regular face atoms. One therefore expects the largest changes in selectivity or reaction mechanisms to appear in the 1–5 nm range.

According to a second model, groups, or ensembles, of atoms with a specific geometrical arrangement are required for some catalytic reactions. Such sites arise when incomplete layers are added over (111) and (100) faces of *fcc* crystals (Figure 5),

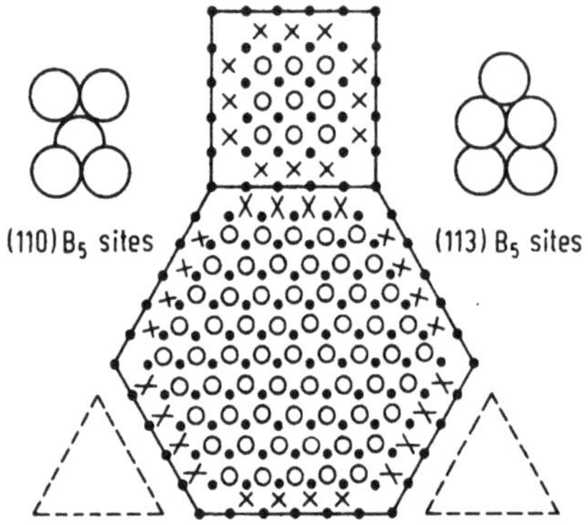

Figure 5. Schematic drawing of (111) and (100) planes in incomplete cubooctahedron. (●) Atoms of the filled layers, (○) atoms added in the incomplete layer. Crosses are B_5 sites. (Reprinted with permission from Surface Science, copyright by North Holland Publ. Co., reproduced by courtesy of Prof. R. Van Hardeveld. Ref. [140])

Of particular interest are the B_5 sites, including edge and face atoms in (110) or (311) configuration on the *fcc* cubo-octahedron. Van Hardeveld and Hartog calculated the maximum number of these B_5 sites as a function of particle size [140].

The ratio between the maximum number of B_5 sites and the number of edge and corner atoms remains practically constant for particles larger than 2 nm, and then decreases sharply to zero for the smallest particles. If one assumes that the bond shift and cyclic mechanisms involve B_5 sites and isolated edge atoms, respectively, the percentage of bond-shift should vary as does this ratio, first remaining constant and then decreasing below a critical particle size.

A third possible explanation of particle-size effects, proposed recently by Burton [141], relates the occurrence of some catalytic properties to the existence of non-lattice symmetries (D_{5h} or icosahedral) observed in the case of very small metal aggregates (Figure 6),
These pseudo crystals were assumed to arise from multiple twinning of *fcc* tetrahedra [142, 143] or from continuous growth on nuclei with D_{5h} or icosahedral symmetry (pentagonal bypyramid or icosahedron with 7 or 13 atoms, respectively) [144, 145]. Although most observations were made in the case of gold and with particles not smaller than 2 nm, one could imagine that very small platinum particles with abnormal C_5 symmetry are present in highly dispersed catalysts and that they have some special catalytic properties, favouring, in the case of hexane isomerization for instance, the non-selective cyclic mechanism.

A last explanation for particle size effects stems from Hoare and Pal's calculations on the stability of very small clusters (2–60 atoms). These authors [146] using both the Lennard-Jones and the Morse potential, showed that polytetrahedral aggregates obtained by packing 4-atom tetrahedra to-

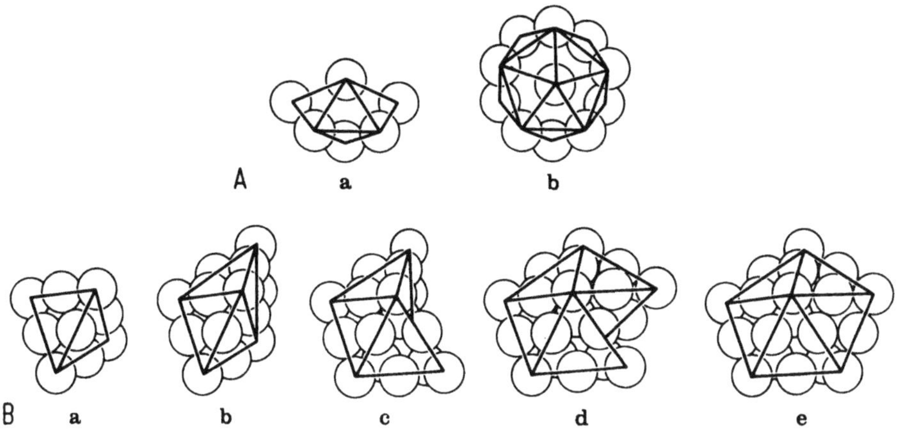

Figure 6. Pseudo-crystals with D_{5h} or I_h symmetries. **A-(a)** pentagonal bi-pyramid ($N = 7$) and (b) icosahedron ($N = 13$) **B-(a–e)** Multiple twinning sequence for generation of poly-tetrahedral clusters. (Reprinted with permission from J. Cryst. Growth, copyright by North Holland Publ. Co., reproduced by courtesy of Prof. M. Hoare. Ref. [146])

gether face-to-face were more stable than any regular lattice micro-crystallites of the *fcc* or *hcp* types. When these stable aggregates are being constructed atom by atom, polytetrahedral packing may be carried out according to three different growth sequences whose first two complete shells have tetrahedral, pentagonal D_{5h}, and icosahedral symmetry, respectively (Figure 7). When a third shell is added, however, it is no longer possible

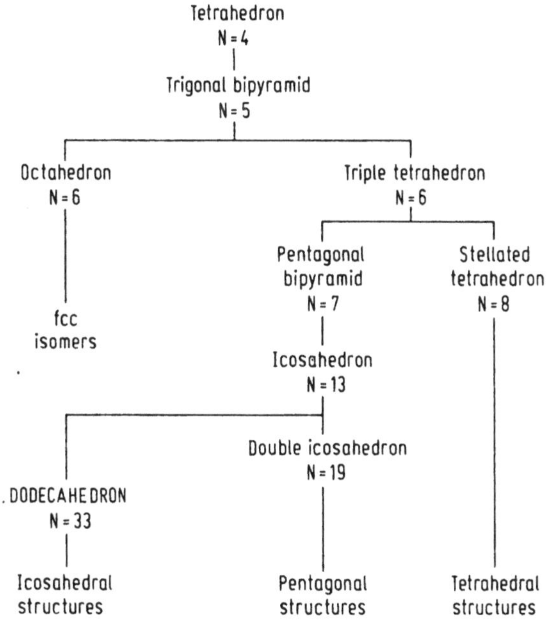

Figure 7. Generation of poly-tetrahedral clusters according to three different growth sequences leading to icosahedral, pentagonal and tetrahedral aggregates (N = Number of atoms). (Reprinted with permission from Physical Clusters, copyright by North Holland Publ. Co., reproduced by courtesy of Prof. M. Hoare. Ref. [146])

for the symmetry of the central seed structure to propagate outwards through the cluster. The polytetrahedra with an equivalent radius of 0.5–1 nm (40–60 atoms) would consist of a central highly symmetric core of twenty to thirty atoms on which the surface atoms show a distinctly liquid-phase mobility [146]. The existence of these highly mobile quasi-isolated atoms (Figure 8) supported on metal provides a new possibility for catalytic reactions on very small crystallites with the participation of mononuclear surface complexes, proposed several times to explain reaction mechanisms [17].

Figure 8. Icosahedral growth. 45-atom cluster with broken icosahedral symmetry

For clusters of more than 70 atoms, beyond a size of roughly 2 nm, the most stable aggregates are expected to recover a regular *fcc* lattice symmetry and normal catalytic behaviour.

2. Isomerization of Hexanes and Hydrogenolysis of Methylcyclopentane

In order to test the various models for the effect of metal particle size on skeletal reaction selectivity, the isomerization of labeled hexanes and the hydrogenolysis of methylcyclopentane have been investigated by Gault and coworkers over a series of platinum/alumina catalysts with a mean particle size ranging from 1 to 20 nm [147]. In these catalysts, the continuous change of metal dispersion from 0.04 to 1.0 was effected by changing the metal content of the catalyst, from 10 to 0.2% and the dehydroxylation state of the alumina. These authors have shown that the percentage of cyclic mechanism in the isomerization of 2-methylpentane to 3-methylpentane as a function of metal dispersion remains roughly constant (*ca.* 20%) over a large dispersion range (0–50%), and increases sharply above 50% dispersion without, however, reaching 100% cyclic mechanism even for the most dispersed catalysts. Careful determination by electron spectroscopy and SAXS, of metal particle size distributions shows that there are no crystallites smaller than 1 nm in the catalysts of low dispersion while they are present in increasing amounts with increasing dispersion in those catalysts for which an enhancement of the cyclic mechanism is observed [133]. The above results cast some doubt upon the "mitohedrical" theory. Indeed, if the cyclic mechanism were taking place on edge or corner atoms and the bond shift mechanism on face atoms [136], one should observe a continuous decrease of the percentage of cyclic mechanism with metal dispersion, especially in the critical range 1–5 nm of particle size. However, the percentage of cyclic mechanism is roughly constant on particles larger than about 2 nm, suggesting that the sites responsible for the bond shift and the cyclic mechanism are topographically similar, *i.e.*, both involve or do not involve, edge atoms. Since extremely dispersed catalysts are very active for isomerization, Gault *et al.* [133] proposed that both types of isomerization sites include edge atoms. Moreover, the limiting particle size

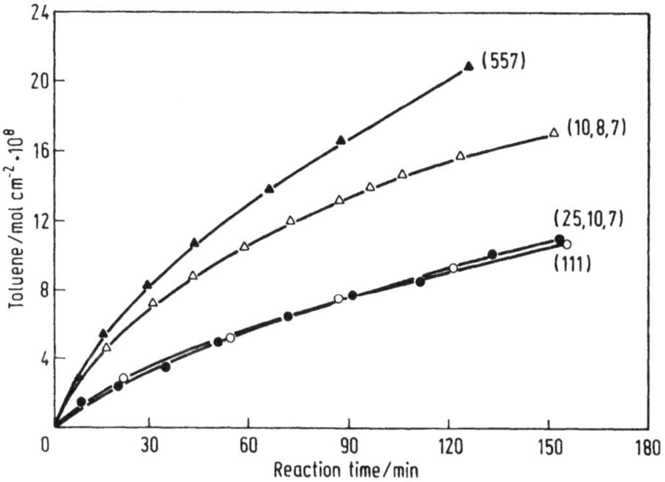

Figure 9. Toluene formation from *n*-heptane over four single crystal surfaces. Reaction conditions: *n*-heptane 2 kPa, hydrogen 64 kPa, 573 K. (Reprinted with permission from Journal of Catalysis, copyright by Academic Press, reproduced by courtesy of Prof. G. A. Somorjai. Ref. [153])

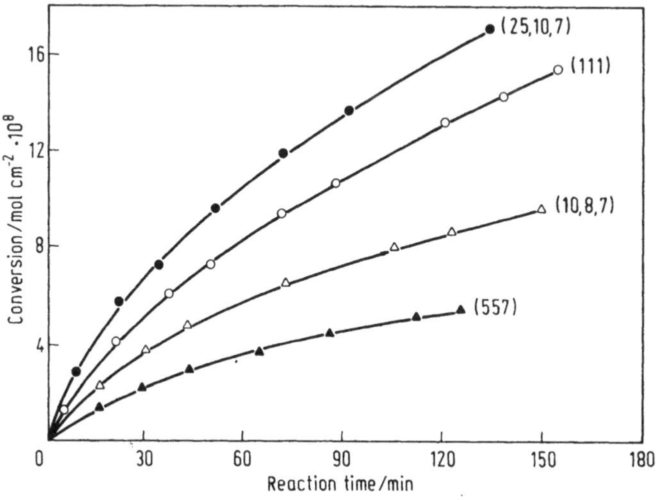

Figure 10. Hydrogenolysis of *n*-heptane over four single crystal surfaces. Reaction conditions: *n*-heptane 2 kPa, hydrogen 64 kPa, 573 K. (Reprinted with permission from Journal of Catalysis, copyright Academic Press, reproduced by courtesy of Prof. G. A. Somorjai. Ref. [153])

below which an enhancement of the cyclic mechanism, or a disappearance of bond-shift, is observed seems to favour (over the B_5-sites model) any model involving polytetrahedra, such as that of Hoare and Pal, or pseudocrystals with D_{5h} or icosahedral symmetry. On the other hand, an *upper* limit of metal particle size around 2.5 nm was defined below which selective hydrogenolysis is no longer possible [133].

The particle size effects observed for the isomerization of 2-methylpentane to 3-methylpentane, especially the sharp increase of the percentage of the cyclic mechanism on the catalysts with particles smaller than 1–2 nm are confirmed in the case of the isomerization of 2-methylpentane and 3-methylpentane to *n*-hexane. Obviously, a complete change of reaction mechanism takes place with the change from "normal" crystallites, (larger than 1–2 nm), to extremely small metal particles. This result is confirmed by experiments [148] made with single crystals exposing various stepped surfaces [6(111) × (100)] and [5(100) × (111)], according to Somorjaï's nomenclature [149].

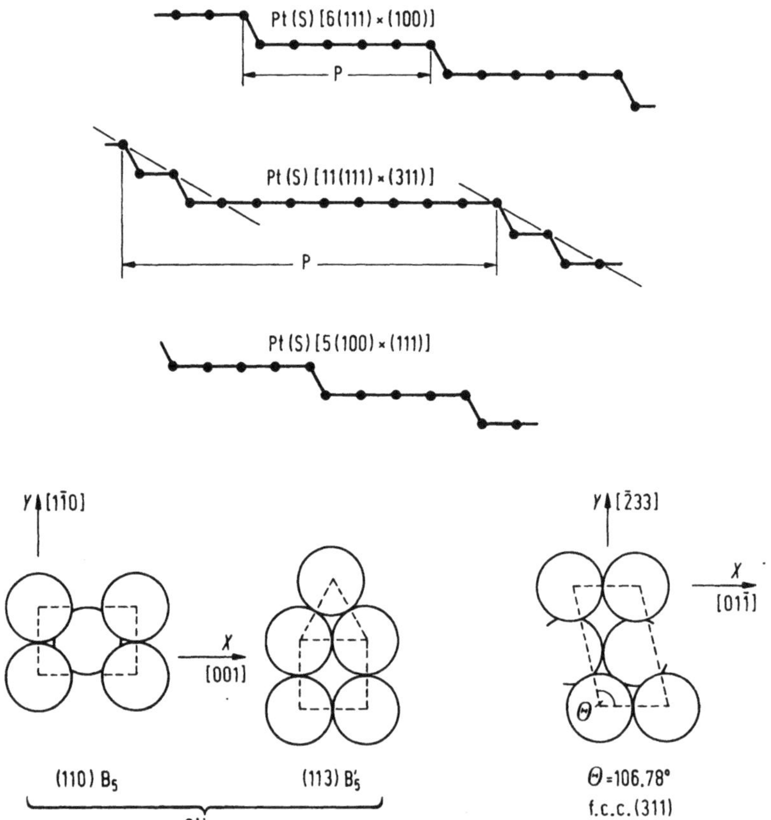

Figure 11. Schematic representation of upper: the initial stepped array Pt(S)[6(111) × (100)]; the final stepped array after exposure to hydrogen Pt(S)[11(111) × (311)]; the stepped array Pt(S)[5(100) × (111)]. Lower: the (110) B_5 and (113) B_5' sites according to Van Hardeveld and Hartog [140]; the unit mesh for a (113) plane grown on a (111) plane.

These surfaces, especially that with (111) terraces and (100) steps, simulate extremely well supported Pt/Al_2O_3 catalysts of low dispersion. Furthermore, work with stepped platinum surfaces [148] has clearly shown that a stepped surface with either (111) or (100) terraces has an enhanced activity for both bond shift and for cyclic isomerization mechanisms compared with a planar (111) surface.

In general terms, the results discussed above are at variance with the sextet-doublet model [159] for hydrogenolysis and dehydrogenation which assumes that the (111) face of platinum is of critical importance, and that reaction occurs over all of that face.

The question remains concerning that nature of the reactive sites on stepped platinum surfaces. As pointed out above, a surface with vicinal (111) planes (*i.e.* a stepped surface with (111) terraces) gives an enhanced rate of bond shift relative to a vicinal (100) plane which is itself more active than a (111) plane.

As shown by LEED [152], hydrogen induces step coalescence and terrace broadening on a stepped surface of platinum [m(111) × (100)] in the temperature range 470–770 K. Thus, under reaction conditions, the actual nature of a stepped surface has some degree of uncertainty. Nevertheless, in the cyclization of *n*-pentane (*c.f.* Figures 9 and 10) on the (557) face of platinum, it was concluded [153] that the surface structure favouring the reaction consisted of a step-terrace structure, with steps separated by six- or seven-atom-wide terraces. Following these ideas, it has been suggested [152] that the activity of stepped platinum surfaces in skeletal isomerization results from surface reconstruction in the presence of gaseous hydrogen with the generation of (113) edge sites or (113) B_5' sites (*c.f.* Figure 11).

C. Electronic Factors

In the above discussion, emphasis has been placed on geometrical and structural considerations. It should be emphasized, however, that a change in particle size or symmetry, or both, could result in a significant change in electronic properties. This is suggested by some recent calculations which showed that the local density of states and also the asphericity of the charge density around a surface metal atom vary with the particle size and are different for *fcc* and icosahedral clusters [154, 155]. The importance of this electronic factor is suggested when considering the predominant isomerization mechanism on various supported metal or alloy catalysts with large particle sizes ($\bar{d} \geqq 10$ nm). Besides platinum two metals, palladium and iridium catalyze the isomerization of 2-methylpentane to 3-methylpentane. With platinum the bond shift mechanism is largely predominant for this reaction, non-selective and selective cyclic mechanisms are favoured on palladium [156] and iridium [77], respectively. On the other hand, alloying platinum with gold or copper [157, 158] increases the contribution on the non-selective cyclic mechanism, while on platinum-nickel alloy catalysts, methylpentane isomerization takes place mostly according to a selective cyclic mechanism [159]

The reason for the drastic changes in reaction mechanisms with the nature of the metal is not understood any more than are, in coordination chemistry, the changes in stability with the metal and with the ligands of the various organometallic complexes. On account of the complexity of the carbon-metal bonding in the various intermediates of skeletal rearrangement reactions of hydrocarbons, besides the steric requirements according to which some reactions involve several metal atoms and some only one [62], the local electronic density around a surface metal atom should influence the relative stabilities and reactivities of the various precursor and intermediate species.

Taking into account the experimental and theoretical work on stepped surfaces of platinum [160–163], Garin and Maire [108, 148] have discussed the enhanced activity of the Pt (557) surface for bond shift isomerization relative to Pt (119), Pt (111) and Pt (100) in terms of a combination of geometric and electronic effects, together with the electron density in the various alkanes [106, 107]. These authors postulated that the change in the geometry of the ledges added to the presence of hydrogen along the edges can greatly modify the local density of states on, for instance, the edges A and B in Figure 12a relative to the edges A′ and B′ in Figure 12b. The authors believe that more than one single metal atom must operate in the bond shift isomerizations.

Figures 13 and 14 show models of precursor species with the different atoms involved either in the reaction of chain lenghtening or methyl shift *via* a bond shift [148]. In such models, it was assumed that multiadsorbed species

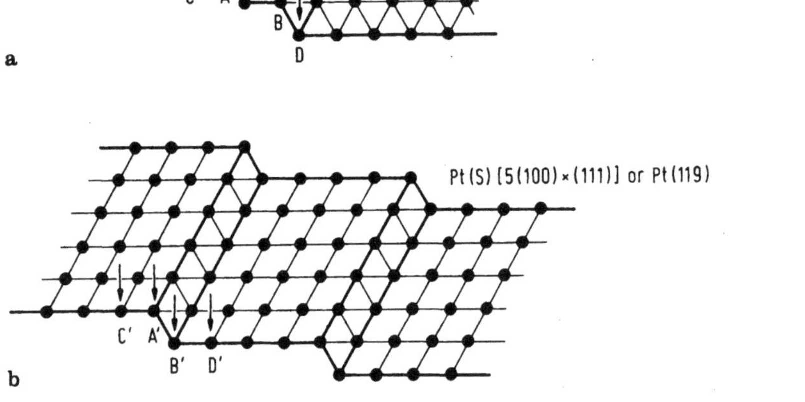

Figure 12. Representation of the different edges for the Pt(557) and Pt(119) stepped surfaces. *Hypothesis* **a**: $n(E)_A \neq n(E)_C \neq n(E)_D$. *Hypothesis* **b**: $n(E)_{A'} \neq n(E)_{B'} \neq n(E)_{D'}$ (*c.f.* text)

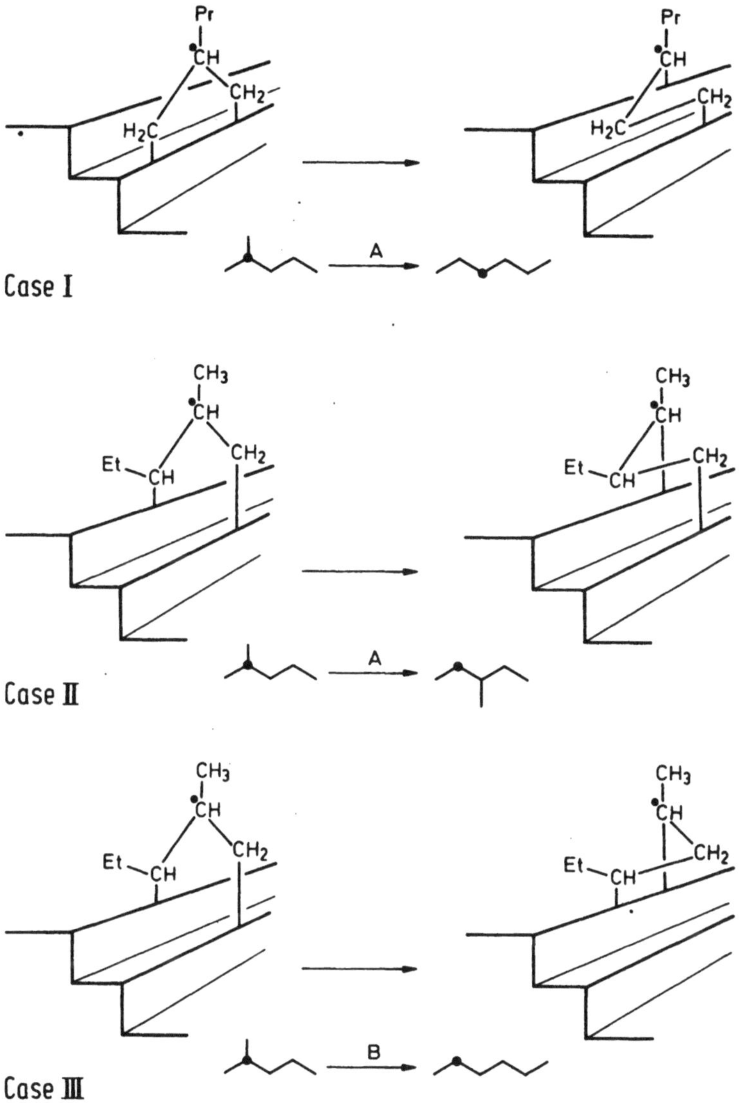

Figure 13. Models proposed for isomerization of 2-methylpentane to 3-methylpentane or *n*-hexane *via* Bond shift A or B on the reconstructed Pt(557) surface

are formed as precursors on at least two atoms, in agreement with both mechanisms of Gault and Anderson [1, 9] (Figure 15).

In general, an electronic factor is directly related to a geometrical factor. Taken together, these two factors must explain the differences between the selectivities of different catalysts for methylcyclopentane hydrogenolysis and cyclic-type isomerization, and would, for instance, regulate the formation from the metallacarbene of the reactive species (dicarbyne or dicarbene) responsible for selective and non-selective dehydrocyclization respectively.

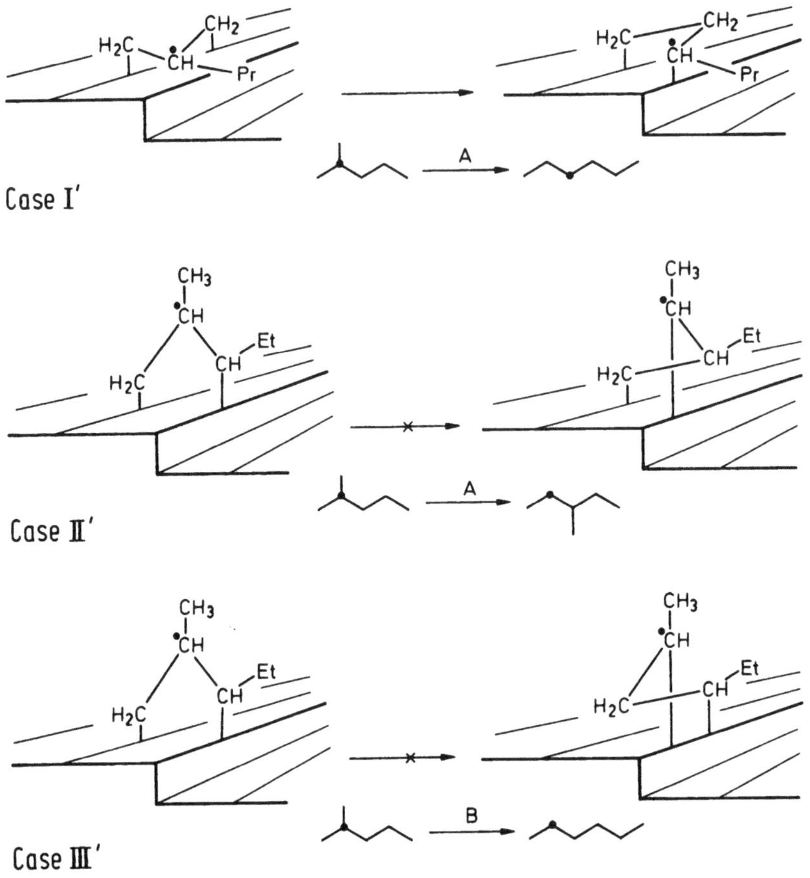

Figure 14. Models proposed for isomerization of 2-methylpentane to 3-methylpentane or *n*-hexane via Bond shift A or B on the Pt(119) surface

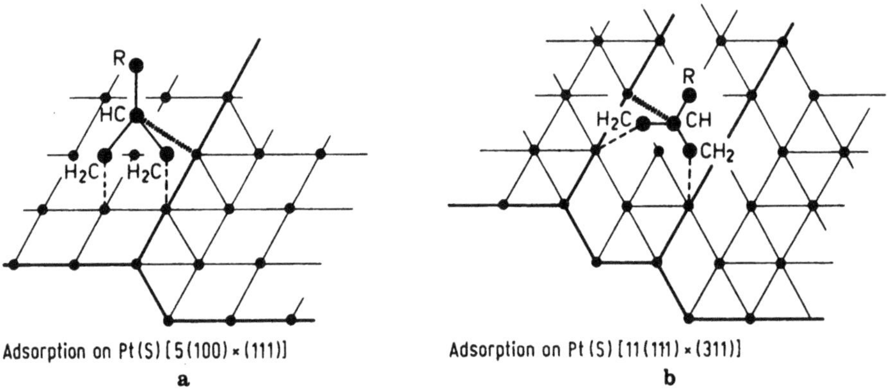

Figure 15. Representation of the two modes of adsorption proposed on Pt(S)[11(111) × (311)] and Pt(S)[5(100) × (111)] taking in account the partially positive charge of the tertiary carbon atom and the high local density of states

The peculiar properties of the highly dispersed 0.2% Pt catalyst with metal particles smaller than 1 nm are never simulted by *f.c.c.* single crystals. They may be explained by the presence of pseudo-crystals with uncommon symmetry, or by electronic properties for the very small clusters that differ from the ones in inifinite crystals.

5. Other Studies

As seen above, the proposed mechanisms with a limited number of adsorbed species for the reaction of skeletal rearrangement of hydrocarbons account for many of the experimental facts. All the proposed adsorbed species require one or several surface metal atoms with low coordination number, such as the ones located on the edges or corners of the crystallites [164]. The involvement in skeletal rearrangements of such metal atoms with several free valencies available have been demonstrated by catalytic studies on single crystals [153, 165], and by investigation of particle size effects in supported catalysts [133]. As already discussed, the relative contributions of the various reaction mechanisms in the isomerization of hexanes on platinum films and supported platinum catalysts [30, 134, 135] may be drastically changed by modifying the size of the metal particles [133].

The electronic and geometrical factors are clearly demonstrated when considering the predominant isomerization mechanism on various supported metal or alloy catalysts [157, 158, 159, 165]. More recently, supported mono- or bimetallic catalysts prepared by decomposition of inorganic complexes have made it possible to prepare highly dispersed particles of transition metals [12, 166, 167, 168]. Almost all the changes in selectivity observed with "classical" catalysts have been correlated to changes in particle size distribution, but drastic changes in selectivity occur when using new catalyst systems prepared by decomposition of inorganic complexes.

Important classes of metal catalysts comprising soluble, mono- or polynuclear metal complexes [169] are used commercially in olefin hydroformylation [170] and methanol carbonylation [171], but due to problems with stability supported metal complex catalysts have not yet found great industrial

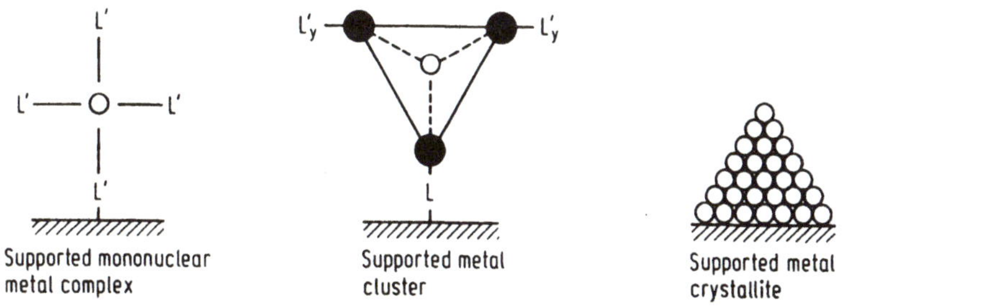

Figure 16. Classification of supported catalysts, ranging from the simplest to the most complex and most commonly applied. (Reproduced with permission from ref. [172])

applications. Figure 16 illustrates the hierarchy of catalyst systems which exists between a mononuclear complex and a metal crystal [172], depending on the degree of aggregation.

A. Reactions on Catalysts Derived from Monometallic Complexes

An important approach to the problem of size effects in the skeletal rearrangement reactions of alkanes may involve the study of the catalytic properties of "surface species" containing a known and controlled number of metal atoms. Yermakov [167] described recently a change of catalytic properties of anchored complexes containing different numbers of palladium atoms from 1 to 5. As the number of palladium atoms N increases from 1 to 5. the binding energy for the Pd $3d^{5/2}$ level measured by photoemission spectroscopy approaches gradually to that of supported metallic palladium.

At $N = 5$ the catalytic properties for olefin hydrogenation and cyclohexene disproportionation are close to that of suported metallic palladium. A Pd_3 fragment is the smallest structural unit necessary for the cyclohexene disproportionation to proceed. On the other hand [173], Takasu *et al.* observed that the UPS. spectra of palladium particles deposited on silica are drastically altered when the size of the crystallites is smaller than 2–3 nm; the narrow *d*-level emission and the non-observation of a well-defined Fermi level clearly indicate that the small metal clusters no longer exhibit metallic properties, but should rather be considered as molecular type particle. Légaré and Rhodin [174] demonstrated too that polynuclear transition metal molecular clusters, *i.e.* $Os_3(CO)_{12}$, $Os_6(CO)_{18}$, $Ru_3(CO)_{12}$, $Ir_4(CO)_{12}$, are of particular interest for the preparation of prototype heterogeneous catalysts.

Ichikawa [175] observed certain differences for the specific reactions of dehydrocyclization and isomerization of *n*-hexane over supported platinum

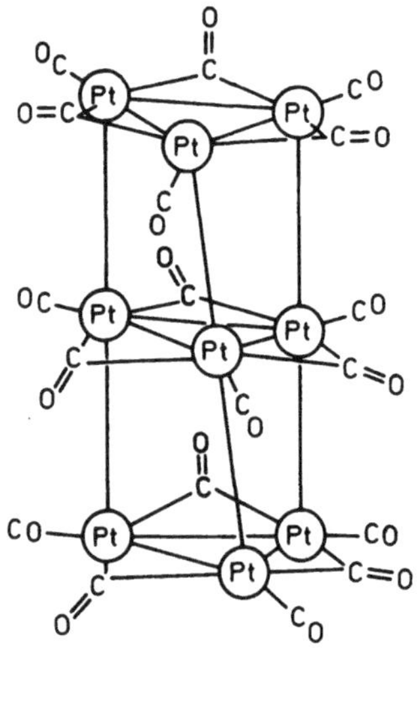

Figure 17. Chini cluster $[Pt_3(CO)_3\ (\mu_2\text{-}CO)_3]_n^{2-}$ with $n = 3$

Table 4. Isomerization of ^{13}C-labeled hexanes and hydrogenolysis of methylcyclopentane

Catalyst % Pt/Al_2O_3	Reaction temp./K	Dispersion/ %	d_{Pt}/nm a	d_{Pt}/nm b	Activity (α_T)/ mol %	Selectivity/ $\frac{\Sigma C_6 \text{ Isomers}}{\alpha_T} 100$	 % C.M. c	 % C.M. c	M.C.P. Hydrogenolysis 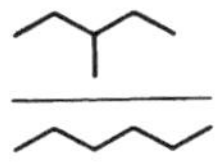 d
7.1 /B′[e]	527	35	2.4	36	—	—	16	90	0.7
4.1 /B′[e]	527	55	1.5	17	18	56	30	86	0.4
2.25/A[f]	548	4	21.2	100	—	—	44	92	0.7
5.5 /B′ (C.D. n = 4)	548	—	—	20	11	55	44	92	0.57
0.56/B′ (D.C. n = 5)	548	43	2.0	23	3.3	57	57	90	0.62
2.5 /B′	527	70	1.2	15	15	50	62	97	0.4
1.1 /B′ (C.D. n = 2)	548	48	1.8	17	3.3	41.5	67	—	0.73
0.4 /B′ (C.D. n = 3)	548	54	1.6	15	7.5	41	80	82	0.7
0.2 /A′	527	100	0.85	20	—	—	83	100	0.42

[a] From hydrogen adsorption
[b] From electron miçroscopy
[c] Percent cyclic mechanism
[d] Hydrogenolysis of methylcyclopentane
[e] B′ is an alumina A′ which has been calcined at 873 K for 210 h [133]
[f] Catalyst previously reduced at 473 K and sintered in H_2 at 673 K
[g] C.D. — cluster derived catalysts obtained from Chini clusters [177], $[Pt_3(CO)_3(\mu\text{-}CO)_3]_n^{2-}$ with n = 2, 3, 4, 5

crystallites prepared from platinum carbonyl cluster anion salts $[\{Pt_3(CO)_6\}_n]^{2-}$ — $[NEt_4]_2^+$ (n = 1–5) following Chini's [176] method (*c.f.* Figure 17). Methylcyclopentane and cyclohexane were preferentially formed over the Pt aggregates prepared from Pt_{15}, Pt_{12}, Pt_9 and Pt_6 salts. Negligible skeletal isomerization of *n*-hexane occurred in the range 523–613 K. Methylpentanes were mainly formed from the simultaneous hydrogenolysis of the methylcyclopentane produced. Skeletal isomerization of neopentane occurred over the Pt_{15} and Pt_{12} derived crystallites on γ-Al_2O_3, and hydrocracking of neopentane proceeded preferentially over Pt crystallites prepared from smaller Pt clusters such as Pt_6 and Pt_3-carbonyls. Ichikawa concluded that the smaller Pt aggregates derived from the platinum type Chini clusters consist of lower coordinated corner Pt atoms which favour the formation of the cyclic mechanism for hexanes.

Zahraa *et al.* [108], comparing the catalytic properties of "classical" and Chini cluster-derived catalysts [177] made the following observations (Table 4 and Figure 18). i) In the reaction 2-methylpentane to 3-methylpentane, the percentage of cyclic mechanism increased with the percentage of particle sizes in the range 0.5–1 nm, and this is so whatever the catalyst used

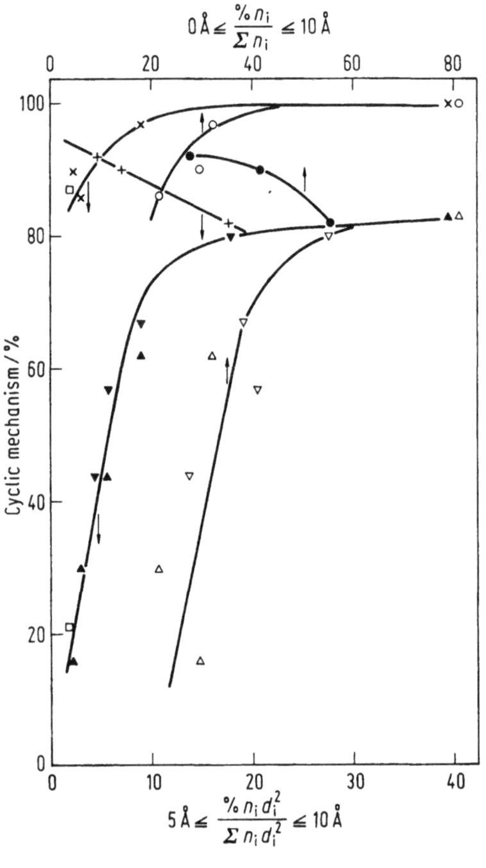

Figure 18. Percentage of cyclic mechanisms *versus* the percentages of surface and length particle sizes. 2-MP-2-^{13}C → 3-MP-3-^{13}C. % length particle sizes; △, (classical catalysts); ▽, (CD* catalysts). % surface particle sizes; ▲, (classical catalysts); ▼, (CD catalysts). 2-MP-4-^{13}C → n-H-3-^{13}C. % length particle sizes; ○, (classical catalysts); ●, (CD catalysts). % surface particle sizes; ×, (classical catalysts); +, (CD catalysts); □, Euro-Pt catalyst. *CD catalysts: "cluster derived" catalysts

(*i.e.* classical or inorganic complex-derived catalysts). ii) In the reaction 2-methylpentane to *n*-hexane, two distinct phenomena were shown: on classical Pt/Al_2O_3 catalysts the cyclic contribution increased with decreasing particle size; on the other hand, the use of inorganic complex-derived catalysts led to a decrease of the percentage of cyclic mechanism. In view of these results the authors conclude that there is no influence of the support, and only metal aggregates in the range 0.5–1 nm are responsible for the cyclic mechanism. The selectivity for the isomerization of 2-methylpentane to 3-methylpentane *via* a cyclic mechanism is the same for both classical and complex-derived catalysts. On the other hand, for isomerization of 2-methylpentane to *n*-hexane the rupture of a secondary-tertiary C—C bond is highly dependent on the catalyst preparation inducing different active sites responsible for changes in electronic or geometrical factors. Simpson *et al.* [178] also made a comparison between the activities of silica-supported ruthenium, rhodium and platinum catalysts prepared from Chini complexes and from classical analogues, towards the activation of saturated alkanes. These authors observed especially the selective cleavage of the alkyl group in ethylbenzene giving toluene and methane (together with various xylenes isomers).

These results obtained with Chini complex-derived catalysts [12, 13, 175, 178] clearly demonstrate that inorganic complex-derived catalysts can display significantly different properties from their conventionally prepared counterparts.

Various elements incorporated during the preparation of the catalysts can act as promoters, leading to specific properties: P, from the phosphine ligands [12], Na or K, counter-ions from Chini complexes [12, 108] for example.

The results obtained on catalysts prepared from Chatt complexes by Maire *et al.* [12, 13, 108, 179, 180] showed a great selectivity for the demethylation reaction, for the bond shift isomerization, and for ring opening. Until now, no classical platinum catalyst had led to such a large amount of selective demethylation [179] (around 60 %). These trends were confirmed by the hydrogenolysis of polymethylcycloalkanes and toluene. EXAFS and XPS measurements clearly show that the deposition of the Pt complex on alumina and the subsequent reduction by hydrogen modify the structure of the complex but, depending on the severity of reduction, does not necessarily give platinum crystallites. Incomplete reduction can give a product in which no Pt—Pt bonds are present as first neighbours [180] and Pt—P bonds are still present. The structure of such a platinum catalyst is very far from that of a classical catalyst. Mechanisms involving an intermediary metallacyclobutane have been proposed [108, 179, 180] to account for the skeletal reactions on these catalysts,

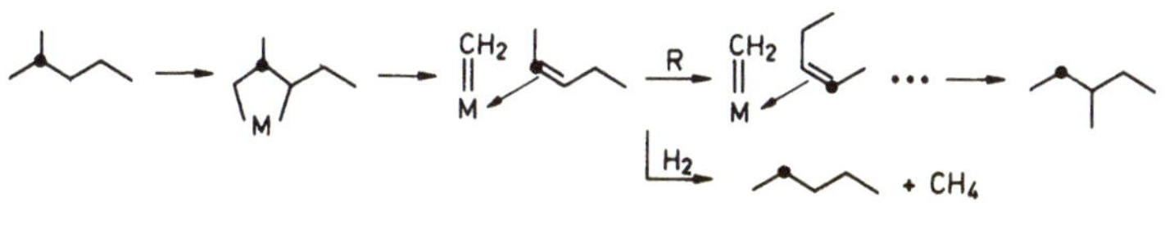

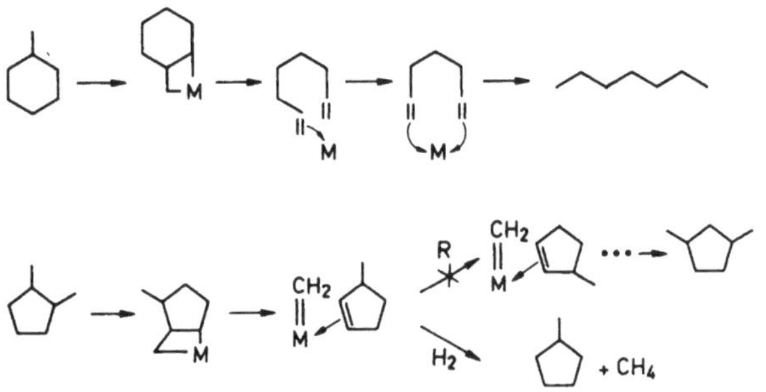

Scheme 49

Here, the metallacyclobutane precursor leads either to isomerization or to selective demethylation involving the participation of a single platinum atom [179].

B. Reactions on Supported Bimetallic Catalysts Obtained by Decomposition of Heteropolymetallic Complexes

Alloying Pt with a less active metal such as Sn has the same effect as decreasing the metal particle size [166]. However, when the Sn content is increased to the extent that the atomic ratio Pt/Sn is near unity, the active sites are blocked by the Sn (*c.f.* Table 8). Gault *et al.* [166] compared these results (obtained with classical Pt—Sn catalysts) to those obtained on Pt—Sn catalysts prepared from inorganic complexes $(C_8H_{12})_3Pt_3(SnCl_3)_2$ [182] and $[Pt(SnCl_3)_2Cl_2]$ $(NMe_4)_2$ [181]. As presented in Table 5 the inorganic complex-derived catalysts simulated very well the classical Pt—Sn catalysts.

Zahraa [184] showed that the catalysts derived from linear Pt_x—Co_y clusters at low metal loadings (*e.g.* 0.3% Pt–0.2% Co/Al_2O_3) have the same catalytic properties as highly dispersed Pt catalysts (0.2% Pt/Al_2O_3). For more heavily loaded catalysts (*e.g.* 10% Pt) derived from Braunstein's complexes [185], *trans*-$Pt[Co(CO)_4]_2$ $[C_6H_{11}NC]_2$, $Pt_2Co_2(CO)_5$ $(\mu_2\text{-}CO)_3$ $(PPh_3)_2$ and Pt_3Co_2 $(\mu_2\text{-}CO)_5$ $(CO)_4(PEt_3)_3$, the presence of Co results in an increased selectivity for demethylation ($\varrho = C_1 + C_5/C_2 + C_4$; Table 6). Although this trend can logically be attributed to the presence of cobalt, it should be noted that selective demethylation is not characteristic of cobalt alone as seen from the results given by the 6.6% Co/Al_2O_3 catalyst: thus, a synergistic effect between Pt and Co seems probable. The use of carbon 13 shows [108] that the percentage of bond shift mechanism on these bimetallic catalysts is comparable to the one obtained on classical "large size" Pt/Al_2O_3 catalysts.

6. Conclusion

In conclusion, although only indirect proofs have been provided for them, the proposed mechanisms for skeletal rearrangement of hydrocarbons account for many experimental facts and therefore seem reasonable The

Table 5. Isomerization of ^{13}C-labeled hexanes on Pt—Sn catalysts

Catalyst[e]	Pt/Sn mol ratio	Hydrogen adsorption H/Pt ratio	$\bar{d}$/nm	Reaction temp./K	% C.M. 	% C.M. 	Reaction temp./K	Ratio $\frac{3\text{-MP}}{n\text{-H}}$
			a	b			c	d
Wilkinson Pt—Sn 0.5–0.6[f]	0.5	0.05	80	658	16	22	633	0.7
Parshall Pt—Sn 8–3.3[g]	1.4	—	12	633	9	—	—	—
Pt—Sn 0.8–0.7[h]	0.7	—	55	658	15	32	658	0.55
Pt—Sn 7.5–4.5[h]	1	0.05	45	678	8	2	658	0.65
Pt 8.5%	—	0.05	90	527	16	72	493	2.4
Pt 0.2%	—	1	20	527	83	100	493	0.4

[a] From electron microscopy
[b] For isomerization of 2-methylpentane
[c] For hydrogenolysis of methylcyclopentane
[d] For hydrogenolysis of methylcyclopentane
[e] On alumina A′
[f] From Pt—Sn complex, Wilkinson [181]; 0.5–0.6 ≡ wt.% Pt and Sn on alumina
[g] From Pt—Sn complex, Parshall [182]; 8–3.3 ≡ wt.% Pt and Sn on alumina
[h] wt.% Pt and Sn on alumina

Table 6. Reaction of 2-methylpentane on inorganic complexes-derived catalysts

Catalyst[a]	Reaction temp./K	Activity (α_T)/ mol %	Selectivity/% b	$\varrho = \frac{C_1 + C_5}{C_2 + C_4}$	3-MP/ mol %	n-H/ mol %	MCP/ mol %	$\bar{d}$/nm c
Chini, $n = 3$[d]								
0.4% Pt/Al_2O_3 B′	548	7.5	41	0.84	16.3	19	6.6	1.5
Chini, $n = 3$[d]								
0.12% Pt/Al_2O_3 B′	548	8.4	37·	0.81	13.8	17	5.9	1.7
Chini, $n = 3$[d]								
0.03% Pt/Al_2O_3 B′	548	3.5	37	1.30	11.8	11.6	14.2	2.3
Chatt[e]								
2% Pt/Al_2O_3 B′	573	20.2	4.5	3.20	1.5	1.9	0.7	2.2
0.2% Pt/Al_2O_3 A′[f]	527	12	60	1.48	20	34	6.5	2.0
10% Pt/Al_2O_3 A′[f]	527	10.9	65	1.49	40.7	20.1	4.1	8.0
$PtCo_2/Al_2O_3$ A′[g] (10% Pt — 6.6% Co)	527	7.7	14	5	8.3	3.5	2.2	—
Pt_2Co_2/Al_2O_3 A′[g] (10% Pt — 3% Co)	573	7	18	2.36	5.3	4.8	8	—
Pt_3Co_2/Al_2O_3 A′[g] (10% Pt — 2% Co)	573	1.4	28	2.80	10	5.9	11.8	—
6.6% Co/Al_2O_3 A′[f]	527	20.2	ε	1.56	ε	ε	ε	—

[a] Alumina support. B′ is an alumina A′ which has bean calcined at 873 K for 210 h [133]
[b] To isomers in mol %
[c] From electron microscopy
[d] Cluster derived catalyst from Chini cluster with $n = 3$, $[Pt_3(CO)_3(\mu\text{-}Co)_3]_n^{2-}$
[e] Cluster derived catalyst from Chatt cluster [183], $Pt_3(CO)_3L_4$ with $L = PEt_3$
[f] Catalyst prepared normal impregnation methods
[g] Catalyst prepared from platinum-cobalt bimetallic complex [185]

discussion focuses attention on metallacarbenes and metallacarbynes as intermediates, unlike the low-temperature reactions of olefins, for which much less dehydrogenated species such as σ-alkyl, π-olefinic and π-allylic species, are required. On account of the complexity of the carbon—metal bonding in these intermediates, besides the steric requirements according to which some reactions involve several metal atoms and some only one, the local electronic density around a surface metal atom should influence the relative stabilities and reactivities of the various precursor and intermediate species. A better understanding of these geometric and electronic factors is therefore required to provide a consistent explanation of the very high specificity of metals in skeletal isomerization.

Isomerization of hexanes on monometallic or heteropolymetallic complex-derived catalysts exhibits various selectivities in isomerization or demethylation depending on the size of the particles on the ratio of the two metals for bimetallic catalysts, and on changes in the geometric and electronic factors. A study of the reactivities of ligands in stable inorganic complexes can provide insight into the reaction pathways and chemical intermediates in reactions on surfaces. In this respect the approach of inorganic complex chemistry from the view point of reactivity and transformations provides a new source of information in understanding mechanisms of reactions on surfaces and may provide a tool to fill the gap between current ideas in homogeneous and heterogeneous catalysis.

From these results on polymetallic catalysts, it is obvious that the use of heteropolymetallic complexes of transition metals as precursors to obtain dispersed metal particles is an attractive direction in this field. It is especially interesting and important to develop methods for controlled synthesis of polymetallic active centers.

Acknowledgments

The authors are grateful to Drs. L. Hilaire and F. Luck, Université Louis Pasteur de Strasbourg, for helpful comments and for a critical reading of the manuscript. For the detailed work involved in the preparation of the data, our sincere thanks go to Mr. S. Wechsler, who drew the figures, and to Mrs. M. Wechsler, who typed the manuscript. They have all carried out tedious tasks with meticulous care. Finally, we would like to acknowledge all authors and publishers who gave permission to reproduce published material.

References

1. Anderson, J. R.: Advan. Catal. **23**, 1 (1973)
2. Dowden, D. A.: The reactions of hydrocarbons on multimetallic catalysts. In: Catalysis, Ed. The Chemical Society, Burlington House, London, **2** (1978)
3. Basset, J. M., Ugo, R.: Aspects of Homogeneous Catalysis. Reidel, Dordrecht, III (1976)

4. a. Naito, S., Sakurai, Y., Shimizu, H., Onishi, T., Tamaru, K.: Bull. Chem. Soc. Jap. **43**, 2274 (1970)
b. Ueno, A., Onishi, T., Tamaru, K.: Trans Faraday Soc. **66**, 756 (1970)
c. Naito, S., Shimiza, H., Hagiwara, E., Onishi, T., Tamaru, K.: Trans. Faraday Soc. *67*, 1529 (1971)
5. Fischer, E. O.: Advan. Organomet. Chem. **14**, 1 (1976)
6. Casey, C. P.: Org. Chem. **33**, 189 (1976)
7. Herisson, J. L., Chauvin, Y.: Makromol. Chem. **141**, 161 (1970)
8. Clark, D. N., Schrock, R. R.: J. Amer. Chem. Soc. **100**, 6774 (1978)
9. Gault, F. G.: Advan. Catal. **30**, 1 (1981)
10. O'Donohoe, C., Clarke, J. K. A., Rooney, J. J.: J. Chem. Soc. Faraday I, **76**, 345 (1980)
11. Whyman, R.: Transition metal clusters. In: Johnson (ed.), John Wiley, New York, chapter 8 (1980)
12. Zahraa, O., Garin, F., Maire, G.: Proc. Faraday Discussion No 72 Paper No 3, Roy. Soc. Chem. (1982)
13. Garin, F., Zahraa, O., Crouzet, C., Schmitt, J. L., Maire, G.: Surface Sci. **106**, 466 (1981)
14. Emmett, P. H. (ed.): Catalysis, Reinhold, New York (1958)
15. a. Weisz, P. B., Prater, C. D.: Advan. Catal. **6**, 143 (1954)
b. Ciapetta, F. G.: Ind. Eng. Chem. **45**, 162 (1953)
c. Mills, G. A., Heinemann, H., Milliken, T. H., Oblad, A. G.: Ind. Eng. Chem. **45**, 134 (1953)
d. Weisz, P. B., Swegler, E. W.: Science, **126**, 31 (1957)
16. Anderson, J. R., Baker, B. G.: Nature **187**, 937 (1960)
17. Barron, Y., Cornet, D., Maire, G., Gault, F. G.: J. Catal. **2**, 152 (1963)
18. Condon, F. E.: Catalysis, Reinhold, New York, **6**, 141 (1958)
19. Newham, J.: Chem. Rev. **63**, 123 (1963)
20. Oblad, A. G., Hindin, S. G., Mills, G. A.: J. Amer. Chem. Soc. **77**, 535 (1955)
21. Stevenson, D. P., Wagner, C. D., Beeck, O., Otwos, J. W.: J. Amer. Chem. Soc. **74**, 3269 (1952)
22. Roebuch, A. K., Evening, B. L.: J. Amer. Soc. **75**, 1631 (1953)
23. Evening, B. L., Waugh, R. C.: Ind. Eng. Chem. **43**, 1820 (1951)
24. Topchieva, K. V., Battalova, Sh., Panchenkov, G. M.: Dokl. Akad. Nauk. SSSR, 78 (1951); Chem. Abs. **45**, 7860 (1951)
25. Voge, H. H., Good, G. M., Greensfilder, B. S.: Proc. 3rd World Petr. Congr. The Hague, **IV**, 124 (1951)
26. Bloch, H. S., Pines, H., Schmerling, L.: J. Amer. Chem. Soc. **68**, 153 (1946)
27. Liberman, A. L., Bragin, O. V., Kazanskii, B. A.: Dokl. Akad. Nauk. SSSR **156**, 1114 (1964)
28. Zelinski, N. D., Kazanskii, B. A., Plate, A. F.: Ber. Dtsch. Chem. Ges. **66**, 1415 (1933)
29. Kazanskii, B. A., Bulanova, T. F.: Dokl. Akad. Nauk. SSSR **62**, 83 (1948)
30. Anderson, J. R., Baker, B. G.: Proc. Roy. Soc. Ser. A **271**, 402 (1963)
31. Anderson, J. R., Avery, N. R.: J. Catal. **5**, 446 (1966)
32. Boudart, M., Ptak, L. D.: J. Catal. **16**, 90 (1970)
33. Karpinski, Z.: Nouv. J. Chim. **4**, 561 (1980)
34. Barron, Y., Maire, G., Muller, J. M., Gault, F. G.: J. Catal. **5**, 428 (1966)
35. Brunelle, J. P., Sugier, A., Montarnal, F.: Summer school on catalysis by metals, Lyon (1975). Ed. C.N.R.S. and unpublished results
36. Muller, J. M., Gault, F. G.: Proc. Int. Congr. Catal., 4th, Moscow, paper No 15 (1968)
37. Levy, R. B., Boudart, M.: Science, **181**, 547 (1973)
38. Horiuti, J., Polanyi, M.: Trans. Faraday Soc. **30**, 1164 (1934)
39. Kemball, C.: Advan. Catal. **11**, 223 (1959)
40. Anderson, J. R., Avery, N. R.: J. Catal. **2**, 542 (1963)
41. Gault, F. G., Rooney, J. J., Kemball, C.: J. Catal. **1**, 255 (1962)
42. Anderson, J. R., Avery, N. R.: J. Catal. **7**, 315 (1967)
43. Zimmerman, H. E., Zweig, A. J.: J. Amer. Chem. Soc. **83**, 1196 (1961)
44. Garin, F., Gault, F. G.: J. Amer. Chem. Soc. **97**, 4466 (1975)
45. Maire, G., Plouidy, G., Prudhomme, J. C., Gault, F. G.: J. Catal. **4**, 556 (1965)

46. Rooney, J. J., Gault, F. G., Kemball, C.: Proc. Chem. Soc. 407 (1960)
47. Shephard, F. E., Rooney, J. J.: J. Catal. **3**, 129 (1964)
48. Chambellan, A., Dartigues, J. M., Corolleur, C., Gault, F. G.: Nouv. J. Chim. **1**, 41 (1977)
49. Parayre, P., Amir-Ebrahimi, V., Gault, F. G., Frennet, A.: J. Chem. Soc. Faraday I, **76**, 1704 (1980)
50. Amir-Ebrahimi, V., Gault, F. G.: J. Chem. Soc. Faraday I, **76**, 1735 (1980)
51. Wei, J., Prater, C. F.: Advan. Catal. **13**, 203 (1962)
52. Parayre, P., Amir-Ebrahimi, V., Gault, F. G.: J. Chem. Soc. Faraday I, **76**, 1723 (1980)
53. Corolleur, C., Corolleur, S., Gault, F. G.: J. Catal. **24**, 385 (1972)
54. Corolleur, C., Corolleur, S., Gault, F. G.: Bull. Soc. Chim. Fr. 842 (1969)
55. Cimino, A., Boudart, M., Taylor, H. S.: J. Phys. Chem. **58**, 796 (1954)
56. Sinfelt, J. H., Taylor, W. F : Trans. Faraday Soc. **64**, 3086 (1968)
57. Cokelbergs, R., Delaunois, Y., Frennet, A., Lienard, G.: J. Chim. Phys. **61**, 1167 (1964)
58. Frennet, A., Lienard, G., Crucq, A., Degols, L.: J. Catal. **53**, 150 (1978)
59. Muller, J. M., Gault, F. G.: In Catalysis: Proc. 5th Inter. Congr. Catal. Miami. ed. Hightower, J. W.: **1**, 743 (1972)
60. a. McKervey, M. A., Rooney, J. J., Samman, N. G.: J. Chem. Soc. Chem. Commun. 1185 (1972)
b. Grant, D., McKervey, M. A., Rooney, J. J., Samman, N. G., Step, G.: J. Chem. Soc. Chem. Commun. 1186 (1972)
61. Quinn, H. A., Graham, J. H., McKervey, M. A., Rooney, J. J.: J. Catal. **26**, 333 (1972)
62. McKervey, M. A., Rooney, J. J., Samman, N. G.: J. Catal. **30**, 330 (1973)
63. Clarke, J. K. A., Rooney, J. J.: Advan. Catal. **25**, 125 (1976)
64. Muller, J. M., Gault, F. G.: J. Catal. **24**, 361 (1972)
65. Ephritikhine, M., Green, M. L. H., McKenzie, R. E.: J. Chem. Soc. Chem. Commun. 619 (1976)
66. Byrne, J. W., Blaser, H. V., Osborn, J. A.: J. Amer. Chem. Soc. **97**, 3871 (1975)
67. Tipper, C. F. H.: J. Chem. Soc. 2045 (1955)
68. a. Binns, S. E., Cragg, R. H., Gillard, R. D., Hearton, B. T., Pilbraw, M. F.: J. Chem. Soc. A, 1227 (1969)
b. Gillard, R. D., Keeton, M., Mason, R., Pilbraw, M. F., Russel, R. D.: J. Organometal. Chem. **33**, 247 (1971)
c. McGinnety, J. A.: J. Organometal. Chem. **59**, 429 (1973)
d. McQuillin, F. J., Powell, K. G.: J. Chem. Soc. Dalton 2123 (1972)
e. Rajaram, J., Ibers, J. A.: J. Amer. Chem. Soc. **100**, 829 (1978)
69. Adams, D. M., Chatt, J., Guy, R. G., Sheppard, N.: J. Chem. Soc. 738 (1961)
70. Puddephatt, R. J., Quyser, M. A., Tipper, C. F. H.: J. Chem. Soc. Chem. Commun. 626 (1976)
71. Casey, C. P., Scheck, D. M., Shusterman, A. J.: J. Amer. Chem. Soc. **101**, 4233 (1979)
72. Paquette, L. A., Leichter, L. M.: J. Amer. Chem. Soc. **94**, 3653 (1972)
73. a. Luck, F., Garin, F., Maire, G.: unpublished results
b. Luck, F., Thèse d'Etat, Strasbourg (1983)
74. Zimmerman, H. E.: Accounts Chem. Res. **5**, 393 (1972)
75. Gault, F. G.: Thèse d'Etat, Paris (1958)
76. Gault, F. G.: C.R. Acad. Sci. Paris, **245**, 1620 (1957)
77. Weisang, F., Gault, F. G.: J. Chem. Soc. Chem. Commun. 519 (1979)
78. Kazanskii, B. A., Rumyantseva, Z. A., Batuev, M. I.: Izv. Akad. Nauk. SSSR, Ser. Khim. 483 (1947)
79. Kazanskii, B. A.: Uspekhi Khim, **17**, 655 (1948)
80. Raik, S. E.: Vestnik. Mosk. Univ. **6**, Ser. Fiz. Mat. i Est. Nauk. **1**, 67 (1951), C.A. **46**, 8617 (1952)
81. Toledano, C. A., Levisalles, J., Rudler, M., Rudler, H., Daran, J. C., Jeanin, Y.: J. Organomet. Chem. 227, C7 (1982)
82. Herrington, E. F. G., Rideal, E. K.: Proc. Roy. Soc. Ser. A, **184**, 434, 447 (1947)
83. Pitkethly, R. C., Steiner, H.: Trans. Faraday Soc. **35**, 979 (1939)
84. Twigg, G. H.: Trans. Faraday Soc. **35**, 1006 (1939)

85. Amir-Ebrahimi, V., Garin, F., Weisang, F., Gault, F. G.: Nouv. J. Chim. 3, 529 (1979)
86. Muetterties, E. L., Busch, M. A.: J. Chem. Soc. Chem. Commun. 754 (1974)
87. Muetterties, E. L.: Inorg. Chem. **14**, 951 (1975)
88. Cooper, N. J., Green, M. L. H.: J. Chem. Soc. Chem. Commun. 761 (1974)
89. Calderon, N., Lawrence, J. P., Ofstead, E. A.: Advan. Inorganomet. Chem. **17**, 449 (1979)
90. Rooney, J. J., Stewart, A.: Chem. Soc. Spec. Period. Rep. Catal. **1**, 277 (1977)
91. Laverty, D. T., McKervey, M. A., Rooney, J. J., Stewart, A.: J. Chem. Soc. Chem. Commun. 193 (1976)
92. Kemball, C.: Proc. Roy. Soc. Ser A, **217**, 376 (1953)
93. a. Diaz Guerrero, G.: Thèse d'Etat Strasbourg 1982
b. Diaz Guerrero, G., Garin, F., Maire, G.: J. Catal. **82**, 13 (1983)
94. Fischer, E. O., Held, W.: J. Organomet. Chem. **112**, C59 (1976)
95. Sinfelt, J. H.: Catal. Rev. **3**, 175 (1969)
96. Sinfelt, J. H.: Advan. Catal. **23**, 91 (1973)
97. Anderson, J. R., Shimoyama, Y.: Proc. Int. Congr. Catal. 5th, Miami, ed. Hightower **1**, 695 (1973)
98. Anderson, J. R., Baker, B. G.: in Chemisorption and Reactions on metallic films Ed. Anderson, J. R., Academic Press, London, **2**, 63 (1971)
99. Leclercq, G., Leclercq, L., Maurel, R.: J. Catal. **50**, 87 (1977)
100. Al-Essa, R. J., Puddephatt, R. J., Quyser, M. A., Tipper, C. F. H.: J. Amer. Chem. Soc. **101**, 364 (1979)
101. Puddephatt, R. J.: Coord. Chem. Rev. **33**, 149 (1980)
102. a. Perkins, D. C. L., Puddephatt, R. J., Rendle, M. C., Tipper, C. F. H.: J. Organometal. Chem. **195**, 105 (1980)
b. Al-Essa, R. J., Puddephatt, R. J., Perkins, D. C. L., Rendle, M. C., Tipper, C. F. H.: J. Chem. Soc. Dalton, Trans. 1738 (1981)
103. Rendle, M. C., Tipper, C. F. H.: J. Organometal. Chem. **224**, 321 (1982)
104. Ephritikhine, M., Green, M. L. H.: J. Chem. Soc. Chem. Commun. 926 (1976)
105. a. Garin, F., Gault, F. G., Maire, G.: Nouv. J. Chim. **5**, 553 (1981)
b. Garin, F., Maire, G., Gault, F. G.: Nouv. J. Chim. **5**, 563 (1981)
106. a. Pople, J. A., Gordon, M.: J. Amer. Chem. Soc. **89**, 4253 (1967)
b. Hall, L. H., Kier, L. B.: Tetrahedron **33**, 1953 (1977)
107. Hoffmann, R.: J. Chem. Phys. **39**, 1399 (1963)
108. Maire, G., Zahraa, O., Garin, F., Crouzet, C., Aeiyach, S., Legare, P., Braunstein, P.: J. Chim. Phys. **78**, 951 (1981)
109. Foger, K., Anderson, J. R.: J. Catal. **59**, 325 (1979)
110. Anderson, J. R.: Am. Chem. Soc. Div. Pet. Chem. Prepr. **26**, 361 (1981)
111. Foger, K., Anderson, J. R.: J. Catal. **64**, 448 (1980)
112. Karpinski, Z., Clarke, J. K. A.: J. Chem. Soc. Faraday Trans. I, **71**, 2310 (1975)
113. Plunkett, T. J., Clarke, J. K. A.: J. Catal. **35**, 330 (1974)
114. a. Kobozev, N. I.: Zh. Fiz. Khim. **27**, 761 (1953)
b. Kobozev, N. I.: Ousp. Khim. **25**, 545 (1956)
115. a. Boreskov, G. K., Chesalova, V. S.: Zh. Fiz. Khim. **30**, 2560 (1956)
b. Boreskov, G. K., Slinko, M. G., Chesalova, V. S.: Zh. Fiz. Khim. **30**, 2787 (1956)
116. Spenadel, L., Boudart, M.: J. Phys. Chem. **64**, 204 (1960)
117. Gruber, H. L.: J. Phys. Chem. **66**, 48 (1962)
118. Benson, J. E., Boudart, M.: J. Catal. **4**, 704 (1965)
119. Poltorak, O. M., Boronin, V. S.: Zh. Fiz. Khim. **39**, 1476 (1965); **40**, 2671 (1966)
120. Mitrofanova, A. N., Boronin, V. S., Poltorak, O. M.: Vestn. Mosk. Univ. Ser. II, Khim. **5**, 40 (1966)
121. Mitrofanova, A. N., Boronin, V. S., Poltorak, O. M.: Vestn. Mosk. Univ. Ser. II, Khim. **1**, 95 (1967)
122. Dmitrienko, N. M., Boronin, V. S., Poltorak, O. M.: Vestn. Mosk. Univ. Ser. II, Khim. **6**, 106 (1966)
123. Poltorak, O. M., Boronin, V. S.: Zh. Fiz. Khim. **40**, 2671 (1966)

124. Boudart, M., Aldag, A., Benson, J. E., Dougharty, A. N., Harkins Girvin, C.: J. Catal. **6**, 92 (1966)
125. Maurel, R., Leclercq, G., Leclercq, L.: Bull. Soc. Chim. Fr. 491 (1972)
126. Boudart, M., Aldag, A., Ptak, L. D., Benson, J. E.: J. Catal. **11**, 35 (1968)
127. Boudart, M.: Advan. Catal. **20**, 153 (1969)
128. Van Hardeveld, R., Van Montfoort, A.: a) Proc. 4th Intern. Congr. Catal. Moscow (1968), Preprint 70, Akad. Kiado, Budapest, **11**, 295 (1971); b) Advan. Catal. **22**, 75 (1972)
129. Maat, H. J., Moscou, L.: Proc. 3rd Inter. Congr. Catal. Amsterdam (1964); North Holland Pub. Co. Amsterdam (1965)
130. Dorling, T. A., Moss, R. L.: J. Catal. **5**, 111 (1966)
131. Crawford, E., Roberts, M. W., Kemball, C.: Trans. Faraday Soc. **58**, 1761 (1962)
132. Corolleur, C., Tomanova, D., Gault, F. G.: J. Catal. **24**, 401 (1972)
133. Dartigues, J. M., Chambellan, A., Corolleur, S., Gault, F. G., Renouprez, A., Moraweck, B., Bosch-Giral, P., Dalmai-Imelik, G.: Nouv. J. Chim. **3**, 591 (1979)
134. Maire, G., Corolleur, C., Juttard, D., Gault, F. G.: J. Catal. **21**, 250 (1971)
135. Corolleur, C., Gault, F. G., Juttard, D., Maire, G., Muller, J. M.: J. Catal. **27**, 466 (1972)
136. Anderson, J. R., McDonald, R. J., Shimoyama, Y.: J. Catal. **20**, 147 (1971)
137. a. Adams, C. R., Benesi, H. A., Curtis, R. M., Meisenheimer, R. G.: J. Catal. **1**, 336 (1962)
b. Flynn, P. C., Wanke, S. E., Turner, P. S.: J. Catal. **33**, 233 (1974)
138. Renouprez, A., Hoang-Van, C., Compagnon, P. A.: J. Catal. **34**, 411 (1974)
139. Somorjaï, G. A.: Advan. Catal. **26**, 2 (1977)
140. a. Van Hardeveld, R., Van Montfoort, A.: Surface Sci. **4**, 396 (1966)
b. Van Hardeveld, R., Hartog, F.: Surface Sci. **15**, 189 (1969)
141. Burton, J. J.: Catal. Rev. Sci. Eng. **9**, 209 (1974)
142. Ino, S.: J. Phys. Soc. Jap. **21**, 346 (1966)
143. Allpress, J. G., Sanders, J. V.: Surface Sci **7**, 1 (1967)
144. a. Kimoto, K., Nishida, I.: J. Phys. Soc. Jap. **22**, 940 (1967)
b. Komoda, T.: Jap. J. Appl. Phys. **7**, 27 (1968)
145. Gillet, E., Gillet, M.: a) J. Cryst. Growth **13/14**, 212 (1972); b) Thin Solid Films **15**, 249 (1973)
146. Hoare, M. R., Pal, P.: a) Nature Phys. Sci. **230**, 5 (1971); b) Phys. **20**, 161 (1971); c) Nature Phys. Sci. **236**, 35 (1972); d) J. Cryst. Growth **17**, 77 (1972)
147. Dartigues, J. M., Chambellan, A., Gault, F. G.: J. Amer. Chem. Soc. **98**, 856 (1976)
148. Garin, F., Aeiyach, S., Legare, P., Maire, G.: J. Catal. **77**, 323 (1982)
149. Lang, B., Joyner, R. W., Somorjaï, G. A.: Surface Sci. **30**, 440 (1972)
150. Bragin, O. V., Olfer'Eva, T. G., Preobrazhenskii, A. V., Liberman, A. L., Kazanskii, B. A.: Dokl. Akad. Nauk. SSSR, **202**, 339 (1972)
151. Bragin, O. V., Olfer'Eva, T. G., Kazanskaya-Koperina, A. V., Liberman, A. L.: Izv. Akad. Nauk. SSSR, Ser. Khim. 704 (1971); 1340 (1977)
152. Maire, G., Bernhardt, P., Legare, P., Lindauer, G.: Proc. 3rd Inter. Conf. Solid. Surfaces, Vienne, 861 (1977)
153. Gillespie, W. D., Herz, R. K., Petersen, E. E., Somorjaï, G. A.: J. Catal. **70**, 147 (1981)
154. Gordon, M. B., Cyrot-Lackmann, F., Desjonqueres, M. C.: 1st Europ. Conf. Surface Sci., E.C.O.S.S. 1, Abst. 294 (1978)
155. Gordon, M. B., Cyrot-Lackmann, F., Desjonqueres, M. C.: Surface Sci. **68**, 359 (1977)
156. Hajek, M., Maire, G., O'Cinneide, A., Corolleur, C., Gault, F. G.: J. Chim. Phys. **71**, 1329 (1974)
157. O'Cinneide, A., Gault, F. G.: J. Catal. **37**, 311 (1975)
158. De Jongste, H. C., Ponec, V., Gault, F. G.: J. Catal. **63**, 395 (1980)
159. Aeiyach, S.: Thèse d'Etat, Strasbourg (1982)
160. Besocke, K., Krahl-Urban, B., Wagner, H.: Surface Sci. **68**, 39 (1977)
161. Tsang, J. W., Falicov, L. M.: a) Solid State Comm. **16**, 1201 (1975); b) J. Molecular Catal. **3**, 351 (1977)
162. Collins, D. M., Spicer, W. E.: Surface Sci. **69**, 85 (1977)
163. Baro, A. M., Ibach, H.: Surface Sci. **92**, 237 (1980)

164. Gault, F. G., Garin, F., Maire, G.: Growth and properties of metal clusters, J. Bourdon (ed.), Elsevier, Netherlands, 451 (1980)
165. Aeiyach, S., Almon, J. J., Légaré, P., Garin, F., Hilaire, L., Maire, G.: Unpublished results
166. Gault, F. G., Zahraa, O., Dartigues, J. M., Maire, G., Peyrot, M., Weisang, E., Engelhard, P. A.: Proc. 7th Inter. Congr. Catalysis, Ed. Seiyama, T. Tanabe, K.: Tokyo **1**, 199 (1981)
167. Yermakov, Yu. I., Kuznetsov, B. N.: J. Mol. Catalysis, **9**, 13 (1980)
168. Yermakov, Yu. I.: Proc. 7th Inter. Congr. Catalysis, Ed. Seiyama, T., Tanabe, K.: Tokyo, **1**, 57 (1981)
169. Parshall, G. W.: J. Mol. Catalysis, **4**, 243 (1978)
170. Pruett, R. L.: Advan. Organomet. Chem., **17**, 1 (1979)
171. Forster, D.: Advan. Organomet. Chem., **17**, 255 (1979)
172. Gates, B. C., Lieto, J.: CHEMTECH., **10**, 195 (1980)
173. Takasu, Y., Unwin, R., Tesche, B., Bradshaw, A. M.: Surface Sci. **77**, 219 (1978)
174. Légaré, P., Sakisaka, Y., Brucker, C. F., Rhodin, T. N.: Surface Sci. 138 (1984)
175. Ichikawa, M.: J. Chem. Soc. Commun. 11 (1976)
176. Calabrese, J. C., Dahl, L. F., Chini, P., Longoni, G., Martinengo, S.: J. Amer. Chem. Soc., **96**, 2614 (1974)
177. Chini, P., Longoni, G., Albone, V. G.: Advan. Organometal. Chem. **14**, 285 (1976)
178. Simpson, A. F., Whyman, R.: J. Organometal. Chem., **213**, 157 (1981)
179. Maire, G., Hilaire, L., Zahraa, O.: Metal-support and Metal-additive effects in Catalysis, Imelik, B. et al. (editors), Elsevier Sc. Publish. Compagny 285 (1982)
180. Maire, G., Hilaire, L., Zahraa, O., Ravet, M. F.: Proc. Inter. Conf. on EXAFS and Near Edge Structures, Frascati (Italy), september (1982) Springer Verlag Publish. Compagny (1982)
181. Young, J. F., Dillard, P. D., Wilkinson, G.: J. Chem. Soc. 5176 (1964)
182. Lindsey, R. V., Parshall, G. W., Stolberg, V. G.: Ing. Chem. **5**, 109 (1966)
183. Booth, G., Chatt, J.: J. Chem. Soc. A, 2131 (1969)
184. Zahraa, O.: Thèse d'Etat, Strasbourg (1980)
185. a. Barbier, J. P., Braunstein, P.: J. Chem. Res., S. 412 (1978) — M, 5029 (1978)
b. Barbier, J. P., Braunstein, P., Fischer, J., Ricard, L.: Inorg. Chem. Acta., **31**, L. 361 (1978)

Chapter 4

Dispersed Metal Catalysts

K. Foger

CSIRO Div. of Materials Science
Catalysis and Surface Science Lab.
University of Melbourne
Parkville, Victoria 3052, Australia

Contents

1. Introduction

Transition metals catalyse a large variety of important industrial processes in the chemical industry, in petroleum refining, in exhaust emission control and gas purification, and in alternative fuel production. Bulk metal catalysts, *e.g.* powders, reactor coatings or honeycomb structures, pose severe drawbacks in terms of catalyst efficiency (activity per weight of active component) and are therefore costly. In addition such catalysts show low thermal stability and further surface area losses occur by sintering processes. To overcome those disadvantages the active metal component is very often supported on high surface area carriers, *e.g.* refractory inorganic oxides. This not only produces highly efficient catalysts with maximum specific metal surface areas, and improves the thermal stability and therefore the life of the catalyst significantly (due to (i) separation and dilution of metal particles, and (ii) interaction of such particles with the support), but also may result in more selective catalysts depending whether or not the reaction is structure sensitive (change of reaction pattern with structure of the catalyst particle and crystallographic faces exposed). Further improvements of activity, selectivity and regenerability of supported metal catalysts may be achieved by a careful choice of carriers or by incorporation of additional catalyst functions.

Over the past decade marked progress has been made in the understanding of preparation procedures in terms of the metal precursor distribution, as a function of impregnation conditions, the migration of precursor species during drying and calcination, and the reduction process itself. The application of highly sensitive techniques like high resolution electron microscopy, Mössbauer spectroscopy, EXAFS and XPS to highly dispersed metal catalysts lead to a better understanding of structural and electronic properties of supported small particles and allowed a determination of the bulk and surface composition of bimetallic microclusters. Progress in the field of organometallic chemistry provided the catalytic chemist not only with useful precursor compounds for catalyst preparations, but high nuclearity metal clusters may provide defined models for extremely small metal particles. Bi- and multimetallic catalysts have experienced a revival and some systems have gained significant industrial importance. New concepts regarding the influences of supports on metal particles are emerging after intense research into the subject over the past few years.

The literature on supported metal catalysts is voluminous and reviews summarising the subject or particular aspects have appeared [1–6]. This chapter is an attempt to present a logical framework within which the problems of supported mono- and multimetallic catalyst may be explored. However, it is not intended as an exhaustive account of the literature on supported metal catalysts. The emphasis is directed towards catalyst design rather than *ad hoc* preparation methods, the importance of thorough characterisation is pointed out, concepts emerging from the comparison with cluster chemistry are addressed, and a rationalisation of support effects is attempted.

2. Support Materials

A. General

The selection of carriers is the first important step in designing supported metal catalysts and is determined by parameters like surface area, characteristic porosity, thermal stability, and if catalytic functions of the supports are utilized, maximum activity and selectivity. Support materials can be classified into

(a) Inert Supports; like SiO_2 supplying high surface area for dispersion of the active component.
(b) Catalytically Active Supports; like aluminas, silica-aluminas, zeolites, etc. Those materials make up about 80% of all industrial supports (see U.S. Sales 1977 — Chem. Tech. (April) 1979).
(c) Supports Influencing Active Component by Strong Interaction; *e.g.* partially reducible oxides, TiO_2, Nb_2O_5, V_2O_5, etc.
(d) Structural Supports; which are of increased importance for exhaust gas purification (monoliths).

Support materials in a variety of physical forms can be obtained from support manufacturers, the most common ones are spheres, granules, extrudates, cylinders and powders. Whereas cold pressing of a powder and regrinding to obtain the desired grain size may be sufficient for laboratory experiments, such supports would not satisfy the mechanical properties required for industrial processes. Thus sintering or fusion processes with or without binders are generally employed to produce supports of particular physical forms, and details of those processes are covered by extensive patent literature. In the following chapter, supports of commercial importance or of special scientific interest are dealt with in more detail in terms of preparation, structure and surface properties.

B. Preparation and Structure of Supports

1. Silica

The preparation and physical and chemical properties of silicas have been reviewed in detail recently by R. K. Iler in "Chemistry of Silica", John Wiley and Sons, 1979 [7].

From the *natural silicas* only *Kieselguhr* is of commercial importance. Although it offers only relatively low surface areas (below 20 $m^2\ g^{-1}$), its open structure containing mainly macropores may be advantageous if access of large molecules to the pore structure is important for the process.

Synthetic silicas are generally preferred in commercial applications since porosity and surface areas can be controlled by the preparation conditions, *e.g.* *pH* of precipitation, ageing of precipitate, speed of precipitation, and calcination conditions. Commercial grade silicas are generally classified according to their preparation into:

(i) *Silica gels*, which consist of three dimensional networks of contiguous particles of colloidal silica. They are prepared either by polymerization of silicic acid or aggregation of colloidal silica. If the polymerization of silicic acid is carried out at low *pH* values, high surface area hydrogels with lower pore volume are obtained. If however, the precipitation is carried out at higher *pH* values, hydrogels of lower surface area and larger pore volume are formed [8]. Upon dehydration of the hydrogel, "Xerogel" forms, often accompanied by structural changes. To minimise pore shrinkage and loss of surface area the water is often replaced with a liquid of lower surface tension before the drying process, or alternatively, the water is removed at a temperature above its critical point, the so formed product is known under the name "Aerogel".

Xerogels and Aerogels are commonly available with surface areas from 100 $m^2\ g^{-1}$ to 1000 $m^2\ g^{-1}$ and average pore diameters in the range 2–10 nm.

(ii) *Pyrogenic silicas* are made by condensing silica from the vapour phase at elevated temperatures. The silica vapour can be produced by volatilisation of silica in an arc or plasma, reduction of silicon dioxide to silicon monoxide and reoxidation, oxidation of silicon compounds (esters, halides), or vapour phase hydrolysis of silicon halides. The best known products "Aerosils" and "Carbosils" are obtained by burning silicon tetrachloride vapour within a hydrogen-oxygen flame. Those products not only offer high surface areas (200–400 $m^2\ g^{-1}$) but in addition the advantage of extreme high purity.

High surface area silicas prepared as gels or *via* the pyrogenic route are X-ray amorphous and reasonably thermally stable up to temperatures of 800 K, however, significant surface area losses due to collapse of the internal pore structure are observed on heating above such temperatures.

2. Alumina

Aluminas represent the most important commercial support materials and their preparation and properties have been extensively researched and dealt with in numerous publications. Informative reviews on the preparation, properties and structure have been published by Lippens and Steggerda [9] and Wefers and Bell [10]. Aluminas of high surface areas (100–600 $m^2\ g^{-1}$) are generally prepared by thermal decomposition of well crystallized hydroxides or precipitation of colloidal gels. The latter process is often favoured since it yields aluminas of controlled surface areas and porosities depending on the choice of gelation conditions [11].

Upon dehydration of the precipitates a variety of crystalline modifications form, depending on the starting hydroxides or oxyhydroxides and the dehydration conditions (air or vacuum and temperature). Lippens [9] summarized the sequence as follows:

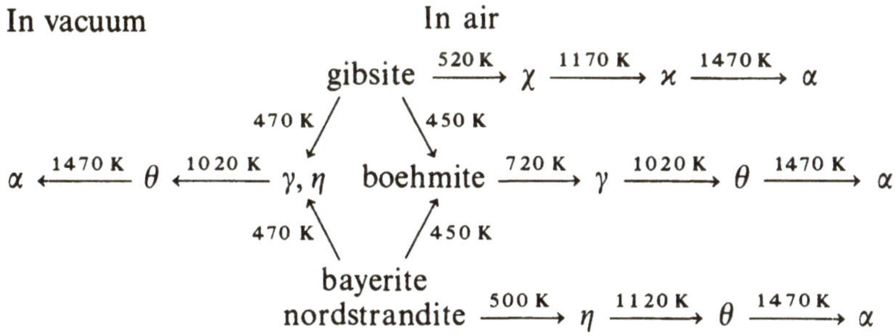

All of those structures are based on more or less close packed oxygen lattices with aluminum ions in the octahedral or tetrahedral interstices and have been classified by Kirschner [12] according to the arrangement of close packed oxygen layers into three series:

(i) α-series with hexagonal close packed stacking (ABAB) represented by α-Al_2O_3 (corundum).
(ii) β-series with alternating close packed stacking (ABAC-ABAC).
(iii) γ-series with cubic close packed stacking (ABC-ABC).

The most commonly used alumina supports are the low temperature representatives of the γ series, γ- and η-alumina. Both aluminas possess defect spinel structures which are slightly tetragonal distorted. The unit cell is built of 32 close packed oxygen atoms and $21^1/_3$ aluminum ions are distributed over the 24 cation positions, leaving $2^2/_3$ sites vacant. γ-Al_2O_3 is a more ordered structure compared to η-Al_2O_3 where stacking faults are very common. This is reflected in values for the average Al—O bond length calculated from radical electron distributions, of 1.818–1.820 Å for γ-Al_2O_3 and 1.825–1.838 Å for η-Al_2O_3 [13]. The cation distribution is slightly different for both modifications, with a somewhat higher fraction of Al^{3+} ions present in tetrahedral positions in the case of γ-alumina.

In practice aluminas are available in a wide range of surface areas (100–600 $m^2\ g^{-1}$) and pore sizes and are characterized by excellent thermal stability, however, heating above 850 K resulted in the closure of micropores and reduction of surface area [14].

3. Silica-alumina

Amorphous silica aluminas. Various methods are employed in preparing silica-aluminas, *viz.* (i) chemisorption of alumina hydroxide on a silica surface, (ii) precipitation of an alumina hydrogel in the presence of a freshly prepared silica hydrogel, (iii) co-gelling of aqueous solutions of alkali silicate and alumina salts, (iv) hydrolysis of esters. Whereas methods (i) and (iv) are mainly used for laboratory preparations, the methods (ii) and (iii) are common preparation procedures for industrial catalysts. The products obtained after washing free of electrolyte, drying and calcining at temperatures around 700 K are X-ray amorphous, offer surface areas of 200–600 $m^2 \times g^{-1}$ and a pore structure consisting of pores in the size range 2.5 to 12 nm. More detailed preparation procedures together with effects of precipitation conditions on surface areas, pore structures and acidity are given in a review by Ryland *et al.* [15].

The co-polymerization of alumina hydroxides and silicic acid can be represented by the following scheme:

Accordingly, three main structure defects are present in silica-aluminas [16] to accommodate the different valence requirements of aluminum and silicon:
(i) aluminum ions substituting silicon cations in tetrahedral coordination.
(ii) aluminum cations in perturbed tetrahedral arrangement, and
(iii) silicon cations in a perturbed tetrahedral arrangement.

Defects of type (i) occur over the entire composition range and are responsible for Brønsted acidity, whilst defects of type (ii), probable in the composition range 30 to 60% alumina, and type (iii) above 50% alumina, are related to Lewis acidity. In the composition range up to 30% alumina a homogeneous distribution of aluminum ions can be expected, while separate "free alumina" phases have been observed at higher alumina contents. In this region the maximum acid strength has been reported [17, 18] and industrial silica-aluminas generally fall within that composition range.

"*Semicrystalline*" *Silica-aluminas: Clays.* Natural clays, which have been used as catalyst supports are kaolinite and montmorillonite. Kaolinite has a chemical composition $Al_2O_3:SiO_2$ or 1:2 whereas in montmorillonites significant variations in Si/Al ratios and coordination environment are observed. Both mineral groups are composed of stacks of layers consisting of sheets of oxide and hydroxide ions coordinated to Si^{4+} and Al^{3+} cations, and hydrogen bonds between the oxide sheet and the hydroxide sheet hold the layers together. Initially, the acid strength of natural clays is weak but proper acid treatment enhances the acidic properties by removing aluminum ions from octahedral positions resulting in a negative lattice charge balanced by a H^+ ion [19]. In recent years, synthetic alumino-silicates with clay like layer structures have attracted renewed interest, the best known material of this kind being Synthetic Mica-Montmorillonite (SMM). The preparation and properties of those materials have been the subject of a recent review by Swift [20].

Crystalline alumino-silicates: Zeolites. The three-dimensional framework of zeolites is based upon the continuous linkages of all four vertices of $[SiO_4]$ and $[AlO_4]$ tetrahedra. The anionic framework caused by the isomorphous substitution of aluminum for silicon atoms is counterbalanced by a variety of cations. Numerous possibilities for linking $(SiAl)O_4$ tetrahedra exist, and a variety of secondary building units like chains, rings and polyhedra can be identified within the zeolite structure. The polyhedra, hollow themselves, are joined by rings and chains, which creates further cavities and channels with openings of 0.25–1.0 nm, causing the molecular sieve effect.

Some zeolites occur naturally, but a large number of new structures have been synthesized, following Barrer's discovery of their specific adsorption properties [21]. It is not intended in this report to discuss the enormous literature on zeolites in detail, and an excellent review has been presented by Breck [22].

Most important catalytically are the zeolites from the faujasite group (X and Y); the chabasite zeolites-offretite, erionite, Linde L and zeolite omega; the mordenites; and more recently the pentasil zeolites with their prominent members ZSM-5 and ZSM-11. The silicon to aluminum ratio varies from the lower limit of 1 for *e.g.* zeolite A up to values of >50 for the ZSM series, and since each aluminum ion induces a negative framework charge, the cation content changes accordingly.

4. Titania

Titania exists in three crystalline modifications, anatase, rutile and brookite with the former two being commonly used as adsorbants, pigments and catalysts. Synthetic titanias are prepared by [23–25]:

(i) *hydrolytic methods*: Hydrogels of titanic acid are originally obtained if the hydrolysis process is carried out in aqueous media at room temperature, but on calcination the hydrogels transform into crystalline modifications, generally anatase, accompanied by significant surface area losses. If titania is precipitated from boiling titanium (IV) salt solutions, anatase is formed from sulfate and rutile from chloride solutions.

(ii) *pyrogenic methods*: Titanium tetrachloride or titaniumtetraisopropoxide is hydrolysed in an H_2—O_2 flame or directly oxidized at 1000–1200 K. Both processes primarily produce anatase, but rutile can be obtained in the direct oxidation process in the presence of silicon tetrachloride or aluminum chloride.

Most commonly available anatase and rutile powders offer surface areas in the range 5–100 $m^2\ g^{-1}$, but samples with areas up to 200 $m^2\ g^{-1}$ may be obtained. Anatase and rutile form a tetragonal lattice, whereas brookite is orthorhombic. The structural unit in all three forms is a TiO_6 octahedron and the different crystal structures are caused by different stacking of those octahedra [26]. Crystallites in rutile powders are believed to be terminated by (110), (100) and (101) planes of which the (110) plane represents 60% of the surface. In the case of anatase the (100) plane is atomically most densely packed and should therefore predominantly terminate anatase crystals.

Titanium also exists in the valence states 3+ and 2+ and treatment of titania under reducing conditions may cause the formation of non-stoichiometric oxide phases [27, 28].

5. Chromia

Chromia xerogels are usually prepared by slow addition of ammonia to aqueous solutions of CrIII salts, or by simmering of a solution of urea containing a chromium salt. The process causes condensation of $Cr(H_2O)_5(OH)^{2+}$, similar to the condensation of silicic acid [29], to form polymers of the type:

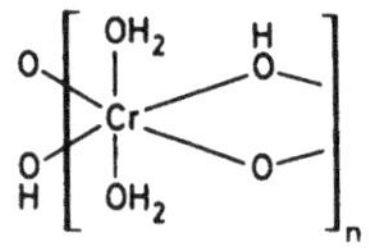

After filtering the green precipitate, drying at 400 K, followed by heating in an inert atmosphere at temperatures up to 700 K, a black powder is obtained which is X-ray amorphous. Surface areas of 300 $m^2\ g^{-1}$ are common and the majority of pores are micropores of diameters around <2 nm [30]. If however, the heat treatment is carried out in hydrogen, microcrystalline α-Cr_2O_3 will form around 670 K [31], and in air the hydrogel undergoes a violent conversion to the green coloured α-Cr_2O_3 [32]. Recrystallization to α-Cr_2O_3 results in significant reductions in surface area, due to the destruction of the micropore structure, mainly because the fast recrystallization as has been observed during calcining in air leads to large temperature rises. However, α-Cr_2O_3 of surface areas around 80 $m^2\ g^{-1}$ are obtained if the transformation occurs more slowly, as during heating in hydrogen or inert gases [33].

The crystal structure of α-Cr_2O_3 consists of a hexagonal close packed oxygen lattice with 2/3 of the octahedral sites occupied by Cr^{3+} ions and the (001) face is assumed to predominate the external surface [34].

Mixed oxides containing chromia in combination with silica or alumina [35] have been prepared by:

(i) deposition impregnation of a chromium salt onto the oxide surfaces followed by decomposition to chromia,
(ii) precipitation of chromium hydroxide onto the alumina or silica surface,
(iii) coprecipitation of the two oxides.

The chromia contents usually vary from 5–20 wt-% and surface areas range from 50 to 300 $m^2\ g^{-1}$. Catalysts prepared by impregnation mainly consist of "chromia coated" alumina particles, whereas coprecipitated samples are generally heterogeneous mixtures of alumina, chromia and various chromia alumina solid solutions.

6. Magnesia

The formation of MgO by dehydration of the oxide has been studied in detail by Anderson *et al.* [36–38]. Depending on dehydration conditions — temperature, atmosphere — powders of various surface areas and crystallite sizes are obtained. Detailed information on the formation of magnesia from magnesite and nesquehonite are reported by De Vleeschauwer [39]. Those authors reported that complete conversions of magnesite was achieved at 900 K, whereas only 800 K was required for the decomposition of nesquehonite. Powders of average surface areas in the range 30 $m^2\ g^{-1}$ were obtained and macropores (formed by interstices between crystals) and mesopores (within magnesia crystals) were measured.

Magnesium oxide forms a cubic lattice of the NaCl type and it has been assumed that (100) faces are predominantly exposed.

7. Carbon

Carbon exists in many diverse morphological forms of which charcoal, carbon blacks, graphite and molecular sieve carbons are most commonly used as support materials. Charcoal is the oldest substance known with adsorptive properties and is generally prepared by pyrolysis of natural or synthetic organic polymers and subsequent activation by partial oxidation. Charcoals are generally highly porous and offer surface areas around 1000 $m^2\ g^{-1}$. Carbon blacks prepared usually by controlled burning of gaseous or liquid hydrocarbons form aggregates of very small dense spherical particles and surface areas and porosity vary greatly, depending on thermal treatment conditions. Graphites occur naturally but very high purity graphites are produced by high temperature (~3000 K) treatment of carbons. In contrast to carbon blacks they are essentially non-porous.

Because of their importance as adsorbents, all three types of carbons have been studied in detail and have been reviewed extensively [40–43].

More recently glassy carbons have been introduced [44, 45], which can be produced with high purity from certain polymers and which contain molecular size pores, therefore the name molecular sieve carbons. The carborization of polymerised furfuryl alcohol has been successfully used to prepare molecular sieve carbons, with pores in the range 0.4–0.6 nm [44]. The formation of carbons with a pore structure consisting of larger pores in addition to molecular size micropores has been reported by Vannice [46]. They used a combination of a carbon-yielding monomer, an organic liquid pore former, a dispersion agent and a polymerization catalyst according to the method reported by Hucke [45]. After polymerization, the samples were pyrolyzed under controlled conditions at temperatures up to 1000 K in an inert atmosphere and the resulting glassy carbons showed a variety of surface areas, and porosities, depending on the choice of appropriate starting materials and preparation conditions.

8. Monolithic Supports

Monoliths are continuous unitary structures composed of many parallel channels of varying shapes (circular, hexagonal, square, triangular, etc.).

They have gained immense importance for exhaust gas purification (*e.g.* in emmission control), where conventional catalyst beds are difficult to use because of their role as flow restrictors. Originally monoliths were manufactured by moulding processes, but currently extrusion and stacked sheet methods are the two major processes yielding monolithic structures. To obtain optimal physical properties — shock resistance, thermal conductivity, crush resistance and flow characteristic — monoliths of a large variety of materials like oxides, nitrides, carbides and metals, and different geometric shapes have been produced. The surface area of monoliths is usually in the range 0.1–1.0 $m^2\,g^{-1}$, but coating of the monolithic structure with high surface area materials (usually oxides) can result in an increase in surface area to values up to 40 $m^2\,g^{-1}$. Three methods have been used mainly to produce high surface area washcoats on a monolithic structure. One involves dipping the monolith into a slurry of oxide powder, the second one consists of impregnation of the monolith with a salt of the desired component followed by decomposition, and in the third process the high area coating is precipitated onto the support as *e.g.* hydroxide, and heated to give the desired coating.

The most common application of monoliths is still in exhaust gas purification but more recently monoliths have been tested successfully for reactor applications (*e.g.* catathermal combustion, trickle phase reactors, and even methanation). Some of the extensive literature now available on monolithic supports has been summarized in a recent review by DeLuca and Campbell [47].

C. Surface Properties

1. Surface Charge in Solution

When oxide particles are suspended in aqueous solutions a surface polarization results in net electrical surface charges, which have been associated with two processes [48], (i) dissociation of surface hydroxyl groups, or (ii) readsorption of hydroxo complexes formed by partial dissolution of the oxide particle. Both processes involve H^+ and OH^- ions and thus are controlled by the *pH* value of the solution. In *acid medium* the surface is most likely *positively charged* (δ-$OH_2^{\oplus}$) and will preferentially adsorb *anions* while in *alkaline solutions* the particles carry a *negative* surface charge (δ-O^-) and adsorb *cations*. At some intermediate *pH* value a zero net charge will result. That value is called zero point of charge (ZPC) or isoelectric point (IEP) and is generally determined by electrophoretic velocity measurements. Colloid chemists have acquired detailed knowledge of the oxide-solution interface and a very comprehensive collection of isoelectric points of oxides and hydroxides was published by Parks [49]. Isoelectric points on different samples of the same oxide may vary markedly. This has been attributed to factors like impurity levels, surface crystallinity, dehydration and ageing. A recent example which demonstrates how surface modifications by ion adsorption changes the surface charge of Al_2O_3 has been presented by

Jiratova [50], who found isoelectric points in the range 4.5 and 9.5 depending on the type of ion present on the surface.

Isoelectric points of support materials are of significance in the preparation of catalysts regarding the choice of metal compound (cationic or anionic complex) and deposition conditions (*pH* of impregnating solutions and additions of competing ions). A careful consideration of the surface charge of support particles in the impregnating solution may avoid unsuccessful preparations due to attempts of adsorbing cations in acid media and vice versa and illuminates problems arising in the preparation of multimetallic catalysts by co-impregnation of cationic and anionic metal compounds.

According to their surface charge in solution, support materials can be classified within the *pH* range 1–14 as:

(i) cation adsorbers: silicas, silica-aluminas, zeolites.
(ii) anion adsorbers: magnesia.
(iii) amphoteric supports: alumina, chromia, titania and zirconia.

2. Surface Functionalities and Catalytic Properties

The surfaces of inorganic oxides will expose oxygen ions, hydroxyl groups and incompletely coordinated cations in various configurations in close relation to the chemical and thermal treatment of the oxides, while on carbon surfaces the main functionalities originate from carboxyl groups, phenolic hydroxyls and carbonyl groups. Those functionalities are of twofold interest in supported metal catalysis:

a) to supply anchoring sites for metal precursor compounds during catalyst preparations — ligand displacement reactions between functional groups on the carrier surface and metal compounds (metal carbonyls, organometallic compounds, alkoxides, halides, etc.).

b) to supply active sites in multifunctional catalysis. Catalytically most support materials exhibit acid-base properties [51–56] and/or oxidation-reduction functions [30, 57–60].

The importance of those materials as catalysts in their own right, as supports for active phases (*e.g.* metals, oxides, halides, carbonyl compounds) as adsorbants, etc., ensured continuous interest into their surface properties and comprehensive reviews on the characterisation, identification, estimation and reactivity of surface functional groups [61], the determination of surface acidity [55], and the catalytic properties [30, 54, 56, 59, 62, 63] have been published.

3. Preparation of Catalysts

Catalyst manufacture was regarded more as an art (knowhow), where successful recipes were acquired by trial and error rather than scientific understanding of the processes involved. However, in the past ten years scientists became more interested in the "design" of catalysts with well defined properties and the controlled preparation is a necessary prerequisite.

The various processes involved in the preparation of supported metal catalysts link inorganic chemistry, colloid and surface chemistry, cluster chemistry, etc. and the increasing research efforts have justified the organization of three international symposia on catalyst preparation [64–66]. The "supreme" aim of a successful preparation method is to distribute the active phase (metal) in the most efficient way (*e.g.* in a highly dispersed form to obtain large specific surface areas and thus maximum activity per weight of active compound) over the support surface.

Although it is possible to deposit a metal directly from the vapour phase onto the support and the method has been used to prepare model catalysts in the laboratory especially for electron microscopy studies [67, 68], the common industrially important preparation methods are multistep processes consisting of:

(i) Distribution of a precursor compound (for example a metal salt, or a cluster compound) over the support surface either by impregnation and ion exchange, precipitation and coprecipitation, reaction of organometallic cluster compound with functional groups on the support surface, or vapour phase deposition of a precursor compound.

(ii) Drying and calcining of the catalyst.

(iii) Transformation of the precursor compound into the active metallic phase by reduction in the gas or liquid phase.

A. Introduction of the Precursor Compound

1. Impregnation — Adsorption

a) General

To manufacture catalysts in an efficient and reproducible process, it is essential to gain control over parameters like metal loading, metal dispersion and metal location in the final product. This may be achieved for a given support by the right choice of metal precursor compound and impregnation conditions [69].

Two types of impregnation techniques may be deployed:

(i) Incipient Wetness Impregnation: In this case the support is wetted with a quantity of solution containing the metal compound corresponding to the pore volume of the particular support and the obtained product is subsequently dried. This process has the advantages of technical simplicity, low cost, reproducible metal loadings. Its disadvantage that metal loadings are limited by the solubility of the metal compound can be overcome by multiple impregnation steps.

(ii) Dipping Impregnation (wet soaking impregnation) where the support is immersed in a solution of the metal compound. The slurry is stirred for a predetermined time, filtered, and the product subsequently dried. The mother liquid may be readjusted and recycled. The technique is mainly applied if the precursor interacts with the support. Thus the metal loading is governed by the concentration of adsorption sites on the support surface. Although industrially the process is more capital intensive due to lower productivity and complex mother liquid recycle systems, it has the advantage of producing catalysts with "designed" metal concentration profiles within a catalyst pellet. This is achieved by chromatographic adsorption techniques (competing ions are added to the impregnation solution).

b) Interaction of metal compound with substrate surface

The common substrates used to support metal particles are inorganic oxides, and *ion exchange processes* (cation, anion) with protons or surface hydroxyl groups are predominantly used to fix the metal compound to the support surface. In addition *ligand exchange reactions* between surface hydroxyls and ligands bound to the metal ion may occur, resulting in even stronger precursor-support interaction.

In principle the interactions are controlled by [70]:

(i) the type of support and state of the surface:
number of functional groups, their acidity and/or basidity;

(ii) the impregnating solution:
pH, type and concentration of metal compound, and presence of competing ions.

In a previous section we discussed the surface polarization of oxide particles immersed in solution. The surface of such particles carries a positive or negative charge depending on their environment, and thus interacts with ether anions or cations. It was pointed out that a useful parameter for describing the ion exchange behaviour of an oxide is its *pH* of net zero charge (ZPC) and that such values for a large list of oxides have been summarised by Parks [49]. Of the most common support materials zeolites, silica-aluminas and silicas adsorb cations; aluminas, titanias and chromia are amphoteric (they adsorb anions in acid and cations in alkaline solutions) and magnesia adsorbs anions. In addition, the number of adsorption sites depends on the thermal treatment of a support and in general decreases after very high temperature calcinations.

The adsorption equilibria can be represented by the following equations:

$$\text{S—OH} + \text{C}^+ \rightleftharpoons \text{S—O}^-\text{C}^+ + \text{H}^+ \quad \text{Cation adsorption}$$

$$\text{S—OH} + \text{A}^- + \text{H}^+ \rightleftharpoons \text{S—OH}_2^+\text{A}^- \quad \text{Anion adsorption}$$

For a given support, equilibria and strength of interaction in both processes are determined by the *pH* of the impregnating solution (cation uptake increases with increasing *pH*, anion uptake decreases with increasing *pH*) and by the affinity of the functional group for the particular ion.

For cation exchange this affinity is a function of charge and radius of the cation and therefore follows the general rule:

$$\text{M}^{4+} > \text{M}^{3+} > \text{M}^{2+} > \text{M}^{1+} \approx \text{H}^+$$

The affinity for anions increases with the polarisability of the anions and the anionic charge, e.g. halides:

$$I^- > Br^- > Cl^- > F^-$$

$$SO_4^{2-} > Cl^-$$

Predictions about ionic adsorption parameters (amounts adsorbed, strength of adsorption) are further complicated if complex ions are used as metal precursor compounds or if such complexes are formed in the impregnating solution. In this case equilibria are dependent on complexing agent and stability of the complex ion in addition to parameters like *pH* and ion affinity. Kennedy [71] presents in a recent publication tables of the stability of metal ions at various *pH* values and the adsorption strength of metal cations and anions in various environments, which may prove very useful in selecting optimal impregnation solutions and conditions. Although each individual adsorption process has to be considered separately the general rules on anionic adsorption outlined above have been demonstrated in a study by D'Aniello [72] on the adsorption of transition metal oxalate and cyanide complexes carrying different charges on a γ-alumina surface. In the absence of ligand displacement reactions the results have been interpreted with an adsorption mechanism of the type

$$-Al-OH + H^+ \rightleftharpoons -Al-OH_2^+$$

$$y(-Al-OH_2^+) + ML_a^{-n} \leftrightarrow (Al-OH_2^+)_y\ ML_a^{-n}$$

involving simultaneous coadsorption of protons and anions. In the absence of chemical reactions the amounts of anion adsorbed could be easily controlled by the amount of acid present in the impregnating solution; and the strength of adsorption depended on the ionic charge (*c.f.* Figures 1 and 2).

However, not all ion exchange processes follow the simple scheme outlined above but involve a series of sequential reactions. For example, the

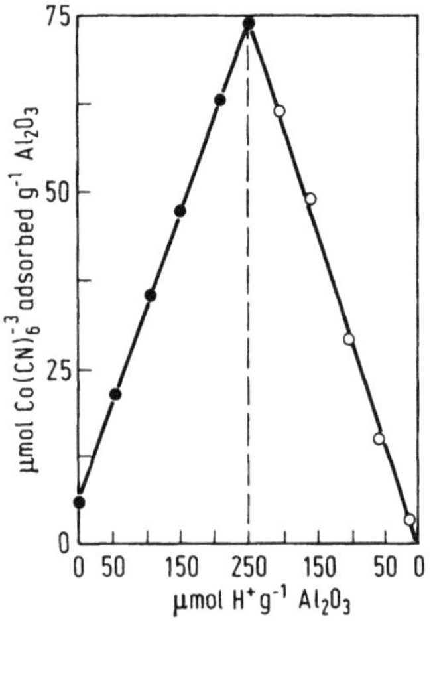

Figure 1. Adsorption of $Co(CN)_6^{-3}$ as function of $HClO_4$ addition and desorption on addition of KOH. (Reproduced with permission from ref. [72])

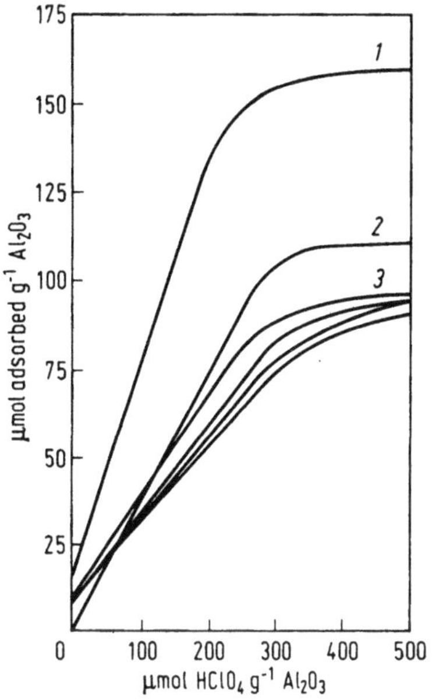

Figure 2. Adsorption of differently charged ions on alumina as function of acid addition. *1*, Co (EDTA)$^-$; *2*, $Pt(CN)_4^{-2}$; *3*, in order of decreasing slope $RhOx_3^{-3}$, $Co(CN)_6^{-3}$, $CoOx_3^{-3}$ and $CrOx_3^{-3}$. Initial complex concentration: 10^{-2} M. (Reproduced with permission from ref. [72])

adsorption of H_2PtCl_6 on γ-Al_2O_3 has been associated [73, 74] with a hydrolytic dissolution of alumina and formation of $Al(OH)^{2+}$ ions followed by a complex formation between those ions and $PtCl_6^{2-}$ and readsorption of this complex according to the scheme:

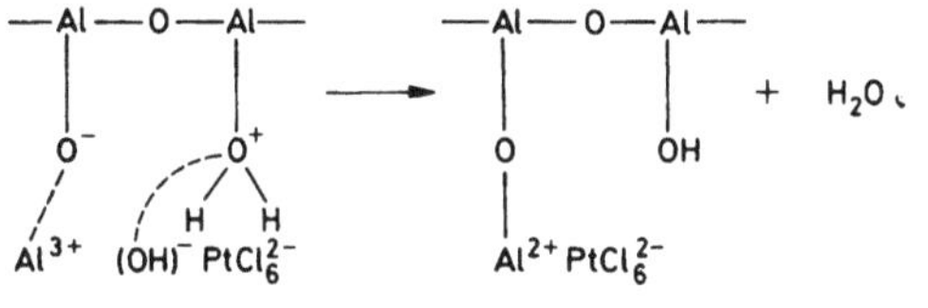

Uptake of some transition metal complexes by oxide surfaces can be associated with a *ligand exchange reaction* between a surface hydroxyl group and a ligand on the metal complex, with the result that the surface is incorporated into the complex and some of the original ligand is released into the solution:

$$S{-}OH + ML_a^{-n} \rightarrow ML_{a-1}^{-(n-1)} \cdots {}^{-}O{-}S + HL$$

The interaction of a series of noble metal halide complexes with alumina falls into that category and was studied by Summers and Ausen [75]. The reactivity depended basically on two factors, (i) the dissociation of a metal-ligand bond and (ii) the ease of ligand displacement. For a given metal both are determined by the ligand bond strength which follows for octahedral complexes the series:

$$CN^- > NO_2^- > NH_3 > H_2O > OH^- > Cl^- > Br^- > I^-$$

Thus PtI_6^{2-} was found to decompose on contact with alumina, $PtBr_6^{2-}$ was taken up rapidly from solution and $PtCl_6^{2-}$ was less reactive. Exchange of one ligand yields monodentate complexes but if two ligands are exchanged bidentate complexes result,

c) Metal distribution within catalyst pellets

Most catalyst preparations aim to produce catalysts where the active metals phase is uniformly distributed throughout a support granule. However, catalysts with non-uniform metal concentration profiles are important, since it is evident from theoretical predictions [76, 77] and experimental data [78, 79] that for some reactions such catalysts may possess superior catalytic properties compared to catalysts with uniform metal distribution.

Although the active phase may redistribute during the reduction step, in general the distribution of metal is determined by the concentration profile of precursor compound within a catalyst granule. Non uniform distributions of active phase precursors originate either (i) in the impregnation step, or (ii) from a redistribution during the drying step or both, depending on the adsorption strength of the compound on the support surface [80–82].

When a porous support is contacted with a solution containing the metal compound in a concentration C, the solution is fast drawn into the pore system due to capillary forces. The equilibria between the adsorbed phase and the solute phase is determined by the adsorption isotherm and accordingly two limiting cases have to be considered:

In case 1 a *strong interaction* between the metal compound and the support is assumed-adsorption type catalysts. Therefore most of the compound adsorbs near the pore mouth and a large concentration gradient develops within the pellet pore. To obtain a uniform distribution it is necessary:

(i) to supply enough compound to saturate every adsorption site;
(ii) to leave the support for long times in contact with the impregnating solution (only successful, if desorption is possible) [83];
(iii) to add an agent competing for the same adsorption sites to the solution (chromatographic adsorption) [84–87].

Because of the strong precursor support interaction, redistribution during the drying step is unlikely and the final distribution of active phase is mainly determined in the impregnation step. One of the best examples was recently reported by Shyr and Ernst [84] who studied the effect of different competing agents on the platinum concentration profile within alumina pellets. The results are summarized in Table 1.

Basically nine types of Pt distributions were observed depending on the co-ingredient and the impregnation time. Kulkarni *et al.* [85] developed

Table 1. Properties of co-impregnated Pt/Al_2O_3 catalysts[a, b]

Co-ingredient	Type of profile			
	1 hr contact time		22 hr contact time	
None, $AlCl_3$, HCl, NaCl, NaF, $NaNO_3$, Na benzoate	I	Outer shell sharply defined	II	Outer shell diffuse
Acetic acid	III	Outer shell diffuse to centre		Uniform
NaBr	I	Outer shell sharply defined	IX	Linearly increasing from centre
Citric acid	IV	Inner shell sharply defined	VI	Inner shell diffuse to centre
HF	V	Inner shell diffuse	V	Inner shell diffuse
Tartaric acid	VI	Inner shell diffuse to centre	VII	Core sharply defined
Na_3PO_4, Na citrate	VII	Core sharply defined	VIII	Core diffuse

[a] Ref. [84]
[b] Metal content approx. 0.5 wt.%; concentration of co-ingredient 0.01 M

a model which aims to predict the distribution obtained in a co-adsorption impregnation from adsorption data of the single components: Adding a coingredient of similar adsorption strength suppresses the adsorption of one species, adding a faster adsorbing species will result in a shift of the metal profile towards the centre of the pellet due to blocking of the external adsorption sites.

Case 2 deals with impregnation type catalysts, where only a *weak interaction* exists between the metal compound and the support surface. In this case the amounts of adsorbed compared to dissolved compound are small and a uniform concentration profile throughout a pellet pore is achieved during the impregnation step. However, since the compound is only weakly adsorbed or still in solution within a pore, redistribution during the drying step is common, and depending on the type of pore system and the speed of the drying process the active phase accumulates either in the cluster of the pellet or at the external surface [81, 82, 87–89].

If a pellet with uniform pore systems is heated, a temperature gradient between the external surface and the interior of the pellet develops. Evaporation starts at the external surface and the gas liquid-interfaces proceeds towards the interior. The precursor concentration increases at the menisci and the compound is deposited on the pore walls. On the other hand if the drying process is slow enough the compound diffuses into the remaining liquid, resulting in its deposition in the centre of the pellet. If a pellet contains interconnected macro- and micropores, vaporization begins from the macropores, and the gas-liquid interface recedes. Eventually the external surface of the pellet reaches a high enough temperature to enable evaporation from the micropores. However, as long as there is liquid left in the macropores the liquid interface does not recede in the micropores, but instead the solution lost by evaporation is drawn in from the macropores by capillary forces. As a consequence highly concentrated liquids accumulate in the micropores and since all the evaporation takes place close to the external surface the compound is deposited there. Two options have been reported to minimize segregation effects:

(i) high heating rates force the evaporation zone to move continuously towards the granule cluster [90];

(ii) an increase in the viscosity of the solution slows down the redistribution process due to diffusional limitations [91].

In the preparation of *bimetallic catalysts* one aims to bring both precursor compounds into close contact in order to produce bimetallic clusters on reduction. However, considering the preceding discussion on the deposition of precursor compounds on supports, some problems, which one may face in achieving this goal become apparent immediately. In order to avoid physical separation of the two components in the impregnation and drying step, it is essential to chose precursor compounds of similar adsorption properties and solubilities.

2. Metal Cluster Compounds as Active Precursors

Metal cluster compounds (carbonyls, organometallic compounds) are increasingly used for catalyst preparations in laboratory studies [92–97] because they offer a variety of interesting prospects:

(i) metal catalysts prepared from carbonyls or organometallic compounds exhibit generally high dispersions, since the ligands are easily removed and high temperature treatment can be avoided;

(ii) such catalysts contain no halide ions, which may mask the catalytic properties of the pure metals;

(iii) another interesting aspect of cluster derived supported metal catalysts has been demonstrated by Ichikawa who reported distinct differences in the catalytic behaviour of supported platinum prepared from Pt_6, Pt_9, Pt_{12} and Pt_{15} carbonyl clusters [98], of rhodium prepared from Rh_2, Rh_4, Rh_6 and Rh_{13} clusters [99], and of nickel catalysts prepared from Ni_1, Ni_2 and Ni_3 clusters [100]. For most catalytic important metals carbonyl clusters of different sizes and structures are known [101, 102]. Such clusters, decomposed under mild conditions, may retain their integrity [95, 103, 104] and the resulting metal aggregates would reflect the structure of the originating cluster, thus exposing active sites of distinct different properties compared to conventional metal catalysts [105, 106];

(iv) in the preparation of bimetallic catalysts problems encountered with conventional type preparation techniques like spatial separation of the components and inhomogeneous composition throughout the catalyst can be overcome by using either well defined heteronuclear organometallics as precursors [92, 94], or by preparation of such complexes on the surface of the support [94, 96]. Numerous heteronuclear metal clusters of varying $metal_1$ to $metal_2$ ratios have been reported [102]. Not all of them may prove useful as precursors for bimetallic catalysts, since some of the ligands may act as poisons for active metal sites (*e.g.* phosphines, arsines). It is further possible to prepare heteronuclear complexes on the support surface by anchoring an organometallic complex of one metal on a low valent ion of the second element. Surface bound low valent metal ions can be prepared either by interaction of metal complexes in low oxidation states with surface hydroxyl groups [94, 107], or by anchoring a complex containing the metal in a higher oxidation state followed by treatment in hydrogen at elevated temperatures [94, 108].

Two methods are commonly used to anchor carbonyls, organometallic compounds or alkoxides on a support surface [94]:

(i) Direct interaction with surface hydroxyl groups

$$(S{-}OH)_m + MX_m \rightarrow [S{-}O]_n MX_{m-n} + nXH$$

The metal concentration, which can be introduced by this method is controlled by (a) the concentration of hydroxyl groups on the support surface and this changes drastically with thermal pretreatments; (b) the stoichiometry of interaction, and (c) the number of metal atoms within the cluster.

(ii) Introduction of functional anchoring groups and subsequent reaction with metal compounds. A large number of functions can be utilized and details about the selection of such groups and their further reactions can be found in Yermakov's extensive review [94].

A selection of cluster compounds [102] useful to prepare monometallic or multimetallic catalysts is presented in Tables 2(a) and 2(b).

Table 2a. Selected compounds for the preparation of monometallic catalysts

Organometallic complexes	
Nickel	$Ni(C_3H_5)_2$, $Ni(C_5H_7O_2)_2$, $Ni(C_5H_5)_2$
Palladium	$Pd(C_3H_5)_2$, $Pd(C_5H_5)(C_3H_5)$, $Pd(OCOCH_3)_2$, $Pd(C_5H_7O_2)_2$
Platinum	$Pt(C_4H_7)_2$, $Pt(C_5H_7O_2)_2$
Rhodium	$Rh(C_5H_7O_2)_3$, $Rh_2(O_2CCH_3)_4$
Ruthenium	$Ru(C_5H_7O_2)_3$, $Ru(C_5H_5)_2$
Mononuclear and multinuclear carbonyl clusters of catalytic important metals	
Cobalt	$Co(CO)_3^{3-}$, $CoH(CO)_4$, $Co_2(CO)_8$, $Co_3(CO)_{10}^-$, $Co_4(CO)_{12}$, $Co_6(CO)_{16}$, $Co_8(CO)_{18}C$
Iridium	$Ir_4(CO)_{12}$, $Ir_6(CO)_{16}$, $Ir_8(CO)_{22}^{2-}$
Nickel	$Ni(CO)_4$, $[Ni(CO)_3]_2H$, $Ni_2(CO)_6^{2-}$, $Ni_3(CO)_8^{2-}$, $Ni_6(CO)_{12}^{2-}$, $Ni_6(CO)_{11}^{2-}$, $Ni_9(CO)_{18}^{2-}$, $Ni_{12}(CO)_{21}^{4-}$, $Ni_{16}(CO)_{12}^{2-}$
Palladium	$Pd(CO)_4$, $[Pd(OAc)\,CO]_4$
Platinum	$Pt_3(CO)_3(PEt_3)_4$, $Pt_6(CO)_{12}$, $Pt_9(CO)_{18}^{2-}$, $Pt_{12}(CO)_{24}^{2-}$, $Pt_{15}(CO)_{30}^{2-}$, $Pt_{15}(CO)_{30}^-$, $Pt_{19}(CO)_{22}^{4-}$, $Pt_{26}(CO)_{32}^{2-}$, $Pt_{38}(CO)_{44}H_x^{2-}$
Rhodium	$Rh(CO)_3^{2-}$, $Rh_4(CO)_{12}$, $Rh_5(CO)_{15}^{2-}$, $Rh_5(CO)_{15}^-$, $Rh_6(CO)_{16}$, $Rh_8(CO)_{19}C$, $Rh_{12}(CO)_{25}C_2$, $Rh_{13}(CO)_{24}H_x^{5-x}$, $Rh_{14}(CO)_{22}^{2-}$, $Rh_{14}(CO)_{26}^{2-}$, $Rh_{15}(CO)_{27}^{2-}$, $Rh_{22}(CO)_{37}^{4-}$
Ruthenium	$Ru(CO)_5$, $Ru_2(CO)_9$, $Ru_3(CO)_{12}$, $Ru_4(CO)_{12}H_4$, $Ru_4(CO)_{13}H_2$, $Ru_5(CO)_{15}C$, $Ru_6(CO)_{18}H_2$

Table 2b. Preparation of bimetallic catalysts

(i) Heteronuclear clusters [102]

Platinum-Iron	$FePt(CO)_3NO(CNBu^t)$, $Fe_2Pt(CO)_{12}$, $Fe_3PtH(CO)_{11}(PPh_3)$, $Fe_4Pt(CO)_{16}^{2-}$, $Fe_3Pt_3(CO)_{15}^{2-}$
Platinum-Iridium	$IrPt(CO)_{12}(PPh_3)_2$, $PyPtIr_6(CO)_{15}$, $Py_2PtIr_2(CO)_7$
Platinum-Manganese	$Mn_2Pt(CO)_{12}$
Platinum-Rhodium	$PtRh_4(CO)_{12}^{2-}$, $Rh_4Pt(CO_{14}^{2-}$, $Rh_5Pt(CO)_{15}$
Platinum-Ruthenium	$Pt_2Ru(CO)_{18}$, $PtRu_2(CO)_8(PPh_3)_2$
Platinum-Tin	$H_4Pt_3Sn_8Cl_{20}$, $[Pt(SnCl_3)]^{3-}$, $[PtCl_2(SnCl_3)_2]^{2-}$
Rhodium-Cobalt	$Rh_3Co(CO)_{12}$, $RhCo_3(CO)_{12}$, $Rh_2Co(CO)_{12}$, $Rh_4Co_2(CO)_{16}$
Rhodium-Iridium	$Rh_3Ir(CO)_{12}$
Rhenium-Osmium	$ReOsH(CO)_9$, $ReOs_2H(CO)_{12}$, $ReOs_3H_5(CO)_{12}$, $ReOs_3H(CO)_{15}$, $Re_2Os(CO)_{14}$, $Re_2Os_3H_2(CO)_{16}$, $Re_2Os_3H_2(CO)_{20}$
Cobalt-Ruthenium	$Ru_3Co(CO)_6Cp$, $Ru_3CoH(CO)_{12}$, $Ru_3CoH_3(CO)_{12}$, $Ru_3CoH(CO)_{13}$, $Ru_3CoH_2(CO)_{13}$, $RuCO_3H(CO)_{12}$, $RuCo_3H(CO)_{13}$
Cobalt-Iridium	$Co_2Ir(CO)_{12}$
Cobalt-Nickel	$CoNi(CO)_5Cp$, $Co_2Ni(CO)_8(PPh_3)_2$, $Co_3Ni(CO)_9Cp$, $Co_3Ni(CO)_{11}^-$, $CoNi_2(CO)_2Cp_3$, $CoNi_3(CO)_{11}$
Cobalt-Iron	$FeCo(CO)_8$, $Fe_2Co(CO)_8Cp$

(ii) Cluster preparation on support surface [109]

Bimetallic phase		Anchored Compound	Reacting Compound
Rhenium	— Platinum	$[Re(OC_2H_5)_3]_3$	$Pt(C_4H_7)_2$
Tin, Lead	— Platinum	$Sn(OOCCH_3)_2$, $Pb(OOCCH_3)_2$	$Pt(C_4H_7)_2$
	— Rhodium	$Sn(OOCCH_3)_2$, $Pb(OOCCH_3)_2$	$Rh_2(CO)_4Cl_2$
	— Nickel	$Sn(OOCCH_3)_2$, $Pb(OOCCH_3)_2$	$Ni(C_3H_5)_2$
Molybdenum, Tungsten	— Platinum	$Mo(OC_2H_5)_5$, H_2 600 °C, $W(C_4H_7)_4$	$Pt(C_4H_7)_2$
	— Rhodium	— „ —	$Rh_2(CO)_4Cl_2$
	— Palladium	— „ —	$Pd(C_3H_5)_2$
	— Nickel	— „ —	$Ni(C_3H_5)_2$

3. *Other Preparation Methods*

Dispersed nickel and cobalt catalysts on MgO [110], ZrO_2 [111] and Al_2O_3 [112–114] have been prepared by coprecipitation of metal oxides and support followed by reduction of the resulting solid solutions. In general only part of the metal oxide reduces to zerovalent metal leaving a large proportion of the metal buried in the support. Furthermore, the reducible fraction decreases significantly as the metal oxide concentration in the solid solution is lowered.

A more exotic preparation method has been reported by Klabunde *et al.* [115–118] who demonstrated that evaporation of metal atoms in the presence of complexing solvents yield metastable "solvated metal atoms". This solution was brought into contact with the support at low temperatures. The resulting slurry was warmed up to room temperature, evacuated to remove the excess solvent followed by a mild reduction treatment in hydrogen. The method proved especially useful for metals like Ni [117, 118] which when prepared by conventional methods require high temperature reduction treatments for activation.

B. Preactivation Treatment

1. Drying

Drying constitutes a mild thermal treatment in the temperature range 350 K to 500 K aimed at the removal of the solvent used in the precursor deposition step. Slow drying can be carried out in drying rooms, slow to medium fast drying in fluidised bed dryers and fast drying is achieved in a spray drying process. In choosing the most suitable drying conditions one has to consider the influence of drying step upon the dispersion and the spatial distribution of the metal in the final catalyst. In a previous section we discussed the problem in detail and pointed out that

(i) redistribution of the precursor compound during the drying step occurs if the interactions between precursor compound and support are weak;
(ii) depending on the support pore structure and the speed of the drying process, the metal compound either redistributes towards the centre or the surface of the catalyst pellet [82].

2. Calcination

This constitutes a medium high to high temperature treatment with the aim to decompose the precursor compound. In principle it is possible with some precursor compounds (*e.g.* carbonyls) to obtain the metallic state by thermal treatment alone [95, 119], but more commonly a subsequent reduction process is required to activate the catalyst. Calcination is generally carried out in oxidising atmospheres, and brings about the following transformations;

(i) decomposition of the precursor compound and formation of an oxide species;
(ii) reaction of the formed oxide with the support; and
(iii) sintering of the precursor compound or the formed oxide species.

Accordingly, calcination may have a marked effect on parameters like reducibility of the metal [120–122], dispersion and distribution of the metal in the final product [123, 124] or alloy formation in bimetallic systems [125, 126].

The effect of calcination on the reducibility of the catalyst is only of importance if the high temperature treatment leads to extremely stable solid solutions between the formed metal oxide and the support, as has been reported for Ni/alumina [128] and Co/alumina catalysts [128]. Otherwise a decrease in the ease of reduction can readily be compensated by an increase in reduction temperature and in most cases, a strengthening of the metal precursor-support interaction upon calcination is desired in order to achieve high dispersions of the metal in the final catalyst [124]. On the other hand, calcination is avoided if it leads to the formation of large oxide crystals, which results in poorly dispersed metal catalysts on reduction. Special consideration has to be given to bi- and multimetallic systems where one component forms a volatile oxide species. The easier oxidised component will form large oxide crystals well separated from the second metal and on reduction the catalyst contains large metal particles of one component and small particles of the other one rather than small bi- or multimetallic clusters. This has been clearly demonstrated for Pt—Ir [126], Pt—Ru [125] and may also apply to combinations containing Pd.

Considering the profound effect thermal prereduction treatment has on the state of a supported metal catalyst (see Figure 3), surprisingly few

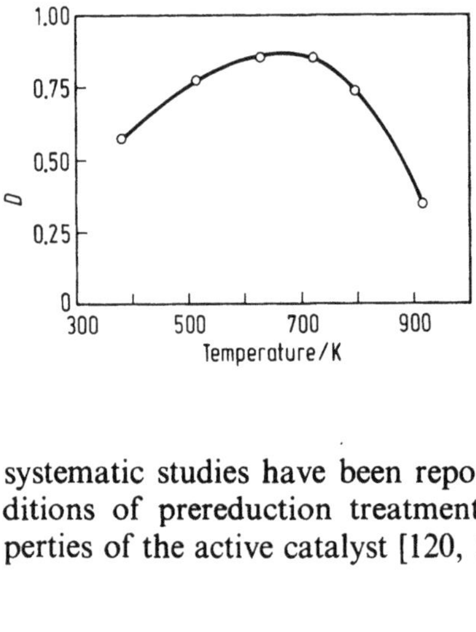

Figure 3. 0.6% Pt on alumina: Variation of dispersion as function of calcination temperature. Conditions: Dried air; GHSV: 2000; Time: 2 hrs. (Reproduced with permission from ref. [124])

systematic studies have been reported in the literature which link the conditions of prereduction treatment (temperature, atmosphere) to the properties of the active catalyst [120, 124, 127, 128].

C. Activation Process

The final activation of a supported metal catalyst consists of the transformation of the metal precursor compound or its oxide into the metallic state (metal atoms, small metal clusters). Apart from few cases, where thermal treatment alone is useful to activate the catalyst, reduction in hydrogen is commonly employed for catalyst activation. The metal compounds present on the support before reduction are either oxides, formed from a metal salt during a preceding calcination step or the salts themselves (*e.g.* halides). The reduction process of oxides and halides can be represented by the following equations

$$MO_{(s)} + H_{2(g)} \rightarrow M_{(s)} + H_2O_{(g)} \qquad (I)$$

$$2\,MX_{(s)} + H_{2(g)} \rightarrow 2\,M_{(s)} + 2\,HX_{(g)} \qquad (II)$$

Standard free energy values for the reduction of oxides are obtained from equation (I)

$$\Delta G_{\mathrm{I}} = \Delta G_{\mathrm{I}}^{\circ} + RT \log \left[\frac{p_{\mathrm{H_2O}}}{p_{\mathrm{H_2}}} \right]$$

and for halides from equation (II)

$$\Delta G_{\mathrm{II}} = \Delta G_{\mathrm{II}}^{\circ} + RT \log \left[\frac{p_{\mathrm{HX}}^2}{p_{\mathrm{H_2}}} \right]$$

Reduction is thermodynamically only possible if ΔG becomes negative, either due to a strong negative ΔG° value or a very small partial pressure of H_2O or HX, resulting in large negative $RT \log \left[\frac{p_{\mathrm{H_2O}}}{p_{\mathrm{H_2}}} \right]$ or $RT \log \left[\frac{p_{\mathrm{HX}}^2}{p_{\mathrm{H_2}}} \right]$ values. Plots of $\log \left[\frac{p_{\mathrm{H_2O}}}{p_{\mathrm{H_2}}} \right]_{\mathrm{equil}}$ and $\log \left[\frac{p_{\mathrm{HCl}}^2}{p_{\mathrm{H_2}}} \right]_{\mathrm{equil}}$ at 673 K for a series of transition metal oxides and chlorides are shown in Figures 4 and 5. As experimental conditions we assumed a $p_{\mathrm{H_2}}$ value of 101 kPa and partial pressures of H_2O and HCl of 0.13 Pa (represented by the horizontal line on the graphs). From those graphs it becomes immediately clear that reduction to the metal is thermodynamically only feasible if the equilibrium values are larger than the experimental values.

The kinetic aspects of oxide reductions by gas-solid reactions have been extensively studied and are subject of a recent book by Boldyrev *et al.* [129]. Reduction processes fall into the category of topochemical reactions, which are reactions where the initiation process proceeds at distinct sites ("potential centres") on the surface of a solid, followed by propagation of the reaction zone from such a centre through the solid, until complete conversion is achieved. Upon contact of a metal oxide with hydrogen, oxygen ions are

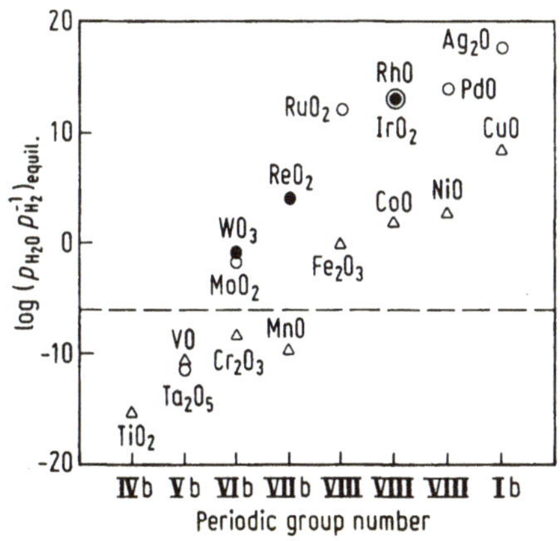

Figure 4. Thermodynamics of metal oxide reduction at 673 K. Horizontal line (– – –) corresponds to estimated experimental conditions of $p_{\mathrm{H_2}} = 101$ kPa, $p_{\mathrm{H_2O}} = 0.13$ Pa. Reduction is thermodynamically feasible for oxides with $(p_{\mathrm{H_2O}} p_{\mathrm{H_2}}^{-1})_{\mathrm{equil.}}$ values greater than experimental values. $(p_{\mathrm{H_2O}} p_{\mathrm{H_2}}^{-1})_{\mathrm{equil.}}$ values taken from ref. [1]

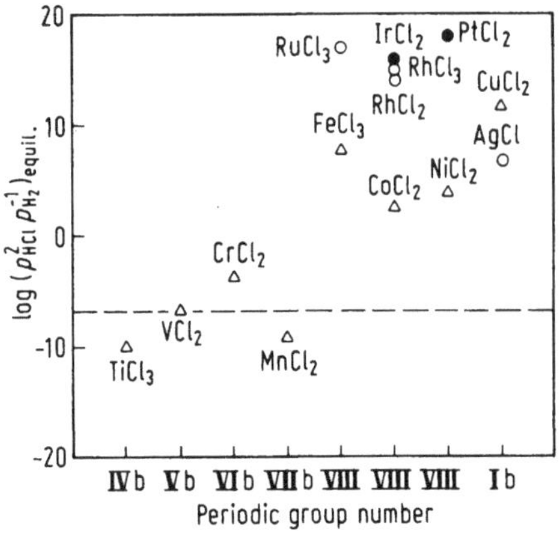

Figure 5. Thermodynamics of metal chloride reduction at 673 K. Horizontal line corresponds to estimated experimental conditions of p_{H_2} = 101 kPa and p_{HCl} = 0.13 Pa. Reduction thermodynamically feasible for chlorides above experimental line. Values for $(p^2_{HCl}p^{-1}_{H_2})_{equil.}$ taken from ref. [1]

removed from the oxide lattice and vacancies are created. As soon as the concentration of such vacancies reaches a critical value the lattice rearranges and metal nuclei are formed. The nuclei grow, new ones appear and the reaction continues until all the oxide (or halide) is consumed. Two models have been proposed for reduction reactions:

(a) In the *nucleation model* the reaction limiting step is the nucleation process. After an induction period (t_i) the reaction rate increases due to the appearance of new nuclei and the growth of existing ones. At a certain reaction stage they contact each other and form a continuous metal skin around the oxide particle and the reaction rate starts to decrease.

(b) The *contracting sphere model* assumes that the reaction proceeds evenly on the entire surface and that the interface decreases continuously as the reaction proceeds. Thus the model may be applied to the final process in the nucleation model (after the formation of a continuous film) or to a reduction reaction with instantaneous nucleation. The exact mathematical treatment of both models can be found in a variety of publications [130, 131], but characteristic features of the nucleation model and the contracting sphere model are shown in Figure 6.

It is possible to accelerate a reduction reaction either by promoting nucleation, which can be achieved by increasing the number of potential nucleation sites, or by supplying a more active reducing agent *e.g.* atomic hydrogen, which requires the presence of group VIII metals able to activate hydrogen. Accordingly many reduction reactions are autocatalytic. The literature on such effects is substantial and has been extensively reviewed [129].

For supported oxides reduction rates are generally lower compared to their unsupported bulk phases due to (i) dispersion effects (nucleation phenomena are particle size dependent, limited mobility of activated hydrogen), or (ii) interaction with the support. It is possible to accelerate the reduction of supported oxides by incorporating promoters, which act in a similar way as discussed for bulk oxide reductions. However, one has to keep in mind that a promoter will only be effective if it is brought in close

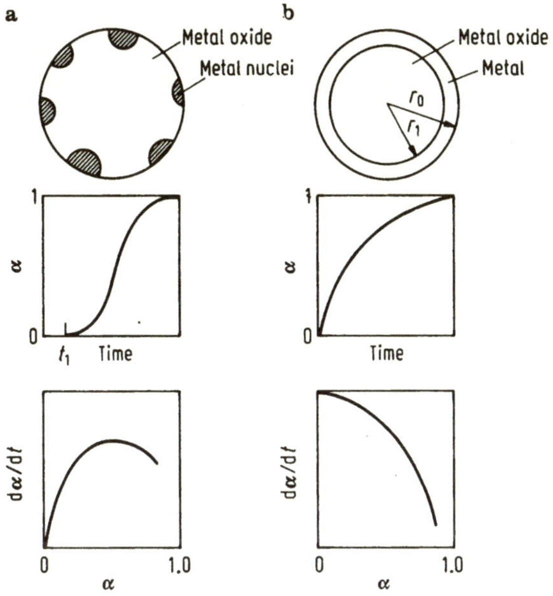

Figure 6. Mechanism of metal oxide reduction. α represents the degree of reduction and $\frac{d\alpha}{dt}$ the reduction rate. **a** nucleation model; **b** contracting sphere model. Ref. [129]

proximity to the dispersed metal oxide, a not always simple task. A systematic treatment of the reduction of a supported oxide (NiO on silica) was presented by Coenen [132], who demonstrated the influence of crystallite size and size distribution, reducing conditions (gas flow, dry and wet hydrogen) and pretreatment conditions on the reduction process.

A technique which is increasingly employed to study reduction reactions is temperature programmed reduction (TPR). Robertson *et al.* [133] first reported TPR profiles of nickel and nickel-copper catalysts and since then many catalyst systems have been investigated. Hurst *et al.* [134] recently published a comprehensive review which deals with the theoretical and experimental aspects of the technique and discusses the results obtained on various catalysts. The publication stresses the importance of a carefully designed linear programmable flow reactor and strictly controlled experimental conditions in order to avoid artefacts and to obtain meaningful results. According to flow reactor theory an increase in flow rate is expected to result in a decrease in T_{max} of the reduction peak due to a lowering of conversion and thus increase in reactant (hydrogen) concentration. On the other hand T_{max} should be unaffected by a change in the mass of catalyst. An apparent dependency of the reduction reaction on the mass of catalyst [135] is an indication that the reactor set up suffers from temperature- and concentration gradients.

Temperature programmed reduction experiments can be used:

(i) to find the most efficient reduction conditions;

(ii) to identify the supported precursor phases and their interaction with the support;

(iii) to characterize bimetallic systems — to determine the role of the second component and to establish alloy formation.

The usefulness of TPR experiments in catalyst preparation and characterization is demonstrated with the following example. TPR experiments with H_2PtCl_6 impregnated onto SiO_2 and Al_2O_3 followed by two hours thermal treatment in oxygen at 373 K, 473 K, 573 K and 773 K (Figures 7 and 8) combined with UV-diffuse reflectance spectroscopy and XRD enabled an identification of the precursor phases. Whereas the Pt^{4+} state was preserved on Al_2O_3, increasing amounts of large $PtCl_2$ crystals are formed on SiO_2 upon heating. This is clearly reflected in the TPR traces, which are practically unchanged for alumina supported platinum but for silica supported platinum the reduction profile is significantly narrower after high temperature heat treatment and its maximum had shifted from around 500 K to 320 K, a change expected if large $PtCl_2$ crystals are present on the silica.

Apart from hydrogen, carbon monoxide has been used as reducing agent in gas-solid reduction reactions [110, 136]. Metal powders have been prepared by reductions in the liquid phase with formaldehyde, hydrazine hydrochloride [138] or sodium borohydride [138]. If a support material is present in the liquid, the formed metal particles deposit on the support surface. An interesting variation of this technique has been described by De Jong

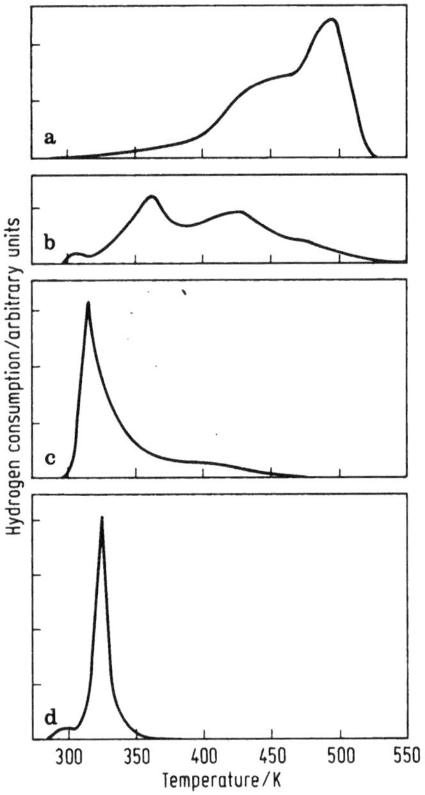

Figure 7. Temperature programmed reduction of H_2PtCl_6 on SiO_2 (Aerosil) after 2 hour treatment in 1% O_2 in He at — **a**: 373 K; **b**: 473 K; **c**: 573 K; **d**: 773 K. The TPR peak at T_{max} < 400 K originates from the reduction of $PtCl_2$

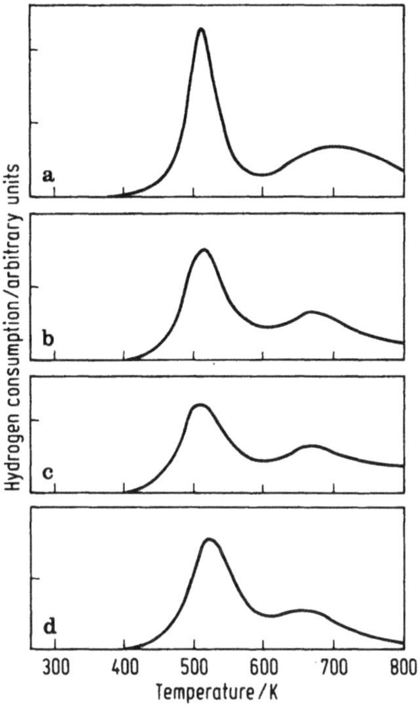

Figure 8. Temperature programmed reduction of H_2PtCl_6 on alumina after 2 hour treatment in 1% O_2 in He at — **a**: 373 K; **b**: 473 K; **c**: 573 K; **d**: 773 K

and Geus [139] who prepared platinum-silver and ruthenium-silver bimetallic catalysts by contacting a Pt/SiO_2 or Ru/SiO_2 catalyst with a solution containing $Ag(NH_3)_2^+$ and hydrazine hydrochloride. Since the particles of the first metal catalyse the reduction reaction, the second metal is exclusively deposited on those metal particles.

4. Characterization of Catalysts

Catalyst characterization is an integral step in catalyst design and is performed either as (i) quality control operation after preparation, (ii) "post mortem" autopsy of a failed catalyst, or (iii) with the aim to find a link between catalyst performance and the catalysts structural and electronic properties. Considering the three major reaction steps; transport of reactants to active sites, reaction at the active site, and the removal of the products, it seems convenient to define two sets of parameters to assess the performance of a catalyst:
The physical properties (mechanical strength, texture and macrodistribution of the metal phase) influence the lifetime of the catalyst, mass and heat transfer phenomena as well as accessibility of active sites. Whereas the active site parameters link the reaction behaviour with the number and nature of active sites.

A. Physical Properties

In the practical application of a catalyst system it is essential to estimate its mechanical stability, which critically affects catalyst lifetime and operation conditions. The two main parameters, which define the mechanical stability are the crushing strength and the attrition resistance and both can be quantified utilising well known tests from materials science [140]. A detailed knowledge of the catalyst texture (total surface area and pore structure) and of macrodistribution of the active phase is of vital importance in optimising the reaction conditions. The pore structure and the metal distribution control the accessibility of active sites for various reactants, the transport of products away from active sites as well as heat and mass transfer phenomena within catalyst pellets.

Physisorption of gases and mercury porosimetry are the most common techniques to determine total surface areas and pore structures, and both methods have been amply reviewed [141–145]. Electron probe microanalysis has proved the most useful tool to characterize the macrodistribution of the active metal phase within a catalyst pellet [146].

Table 3 summarizes the methods, their limits and relevant references.

Table 3. Physical properties of supported catalysts

Parameter	Experimental method	Evaluation	References
Mechanical stability			
crushing strength	grain by grain crushing	crushing force in kg, kg mm^{-1}, or kg cm^{-2}	140
attrition resistance	bed crushing	pressure to produce 0.5% fines	
	I.F.P. test	% fines produced after 5 mins.	
	Socony Test	catalyst wear versus time	
	fluid system attrition	% fines versus time	
Surface area and porosity			
nonporous solid	Physisorption of gases	BET method	147
macroporous solid	Physisorption of gases	BET method	147
(pores > 50 nm)	mercury porosimetry for pores up to 75 nm	mercury penetration as function of externally applied pressure	141, 145
mesoporous solids (pores < 50 nm > 2.0 nm)	mercury porosimetry for pores > 10 nm		141, 145
	Physisorption of gases	BdB analysis	148
		Kelvin equation	141
		Brunauer method	149
microporous solids	Physisorption of gases	*n*-method	150
(pores < 2.0 nm)		*t*-method	151
		α_s-method	152
	n-nonane preadsorption		153

B. Characterization of Active Phase

Although most important industrial catalysts exhibit multifunctionality utilising the acid-base properties of the support in addition to the function of the dispersed metals, only the characterization of active metal sites are discussed in this article. For the characterization of the former the reader is referred to an excellent review by Benesi and Winquist [55] dealing with the types of acid centres on oxide supports and the methods to estimate their number, strength and strength distribution.

The effects of structural and electronic properties of metal particles on a wide range of reactions have been the subject of extensive research, and concepts like structure sensitivity [154], and ensemble and ligand effects [155] are generally accepted. To rationalize reaction data and to enable a comparison between catalysts, the determination of the number of active sites is not sufficient but in addition a detailed knowledge of their nature is required. In the past ten years with continuous improved physical characterization techniques (high resolution TEM, FTIR, XPS) and the development of new techniques (STEM, EXAFS) the catalytic scientist has been provided with powerful research tools, even applicable to extremely small metal particles. These techniques enabled a characterization of bi- and multimetallic catalysts, revealed metal-support interactions (both electronic and structural), and structural and electronic properties of metal particles <2 nm. Table 4 lists the methods and the information they supply. Subsequently the more important techniques are briefly discussed.

Table 4. Characterization of the metal phase

Monometallic catalysts	
Number of metal surface atoms Metal surface area	Directly: from Chemisorption Indirectly: from Particle Size Measurements
Particle size Particle size distributions	XRD, SAXS, TEM, Chemisorption
Structure of particle	TEM, XRD, EXAFS, STEM-Microdiffraction
Electronic properties Metal support interactions	Directly: XPS, EXAFS Indirectly: Adsorption of Probe Molecules (TPD, IR)
Multimetallic catalysts	
Composition of metal phase	XRD, EXAFS, STEM-Microanalysis
Surface composition	Specific chemisorption, XPS, IR- of Adsorbed Probe Molecules

1. Chemisorption Measurements

Selective chemisorption has been used extensively since the early sixties to measure the number of surface metal atoms, and to obtain metal surface areas and average metal particle sizes. A selected gas is chemisorbed under conditions which permit the formation of a monolayer on the metal without any significant contribution of the support. In the selection of a suitable adsorbate, a table (Table 5) published by Myasaki [156] summarizing the chemisorptive properties of transition metals may prove valuable.

Table 5. Chemisorption properties of transition metals [156]

Metal	Dissociative adsorption	Associative adsorption
Hf, Ta, Zr, Nb, W, Ti, V, Mn, Cr, Mo	H_2, O_2, N_2, NO, CO	
Fe, Re	H_2, O_2, N_2, NO, CO	NO, CO
Ni, Co, Tc	H_2, O_2, NO, CO	NO, CO
Os, Ir, Ru, Pt, Rh, Pd	H_2, O_2, NO	NO, CO

After an adsorbate has been chosen, the monolayer coverage has to be measured either by volumetric, gravimetric, chromatographic techniques or titrations [6, 157, 158]. Adsorption stoichiometries as far as they are unknown can be obtained by comparison of chemisorption surface areas with BET surface areas of metal powders, films or foils, or by comparison with calculated metal surface areas from particle sizes measured by methods like TEM, XRD-line broadening, SAXS. The latter approach is in general more reliable, especially since various studies showed a particle size dependency of chemisorption stoichiometries, but considerable care has to be exercised in comparing the various techniques, since XRD-line broadening is weighted towards larger particles and particle size determinations by electron microscopy may be discriminative in favour of smaller particles. If bi- or multimetallic catalysts are being studied by selective chemisorption, three cases have to be considered:

(i) the chemisorption properties of both metals are similar and thus chemisorption determines the surface area of both metals. This situation is most common with combinations of noble metals, *e.g.* Pt—Ir [159], Pt—Rh, Pt—Pd [160], etc.

(ii) only one of the two metals chemisorbs either hydrogen, CO, etc. In this case the total metal surface area is available only by calculation from metal particle sizes obtained by other methods (XRD, TEM, SAXS). Chemisorption then supplies information about the surface area of the chemisorption-active metal. Group VIII—Group Ib combinations commonly fall into this category, *e.g.* Pt—Au [161, 162], Ru—Cu [163], Os—Cu [125], Ir—Au [164], etc.

(iii) both metals exhibit different chemisorption properties towards the adsorbing gases, and a suitable choice of adsorbates, and adsorption conditions may supply surface compositions of the alloy particles [165].

Interactions of adsorbates with the metal can be studied by temperature programmed desorption [166, 167]. The adsorbate is desorbed into an inert gas stream (He, Ar, N_2) and analysed by thermal conductivity cells or by a mass spectrometer. Adsorbate-metal interactions are intimately connected to the geometric and electronic surrounding of the adsorption site, therefore TPD of suitable adsorbates supplies information about the structure and electronic nature of active sites [168, 169]. However, a quantitative analysis of the desorption process is generally difficult to achieve due to the porous nature of a supported catalyst [170, 171].

Table 6 lists chemisorption stoichiometries and atomic surface areas for catalytic important metals [6, 157, 158, 172].

2. X-Ray Methods

X-ray diffraction line broadening is widely used for measuring metal particle sizes and particle size distributions. Detailed theoretical and experimental

Table 6. Chemisorption of H_2, CO, O_2 on supported metal catalysts

Catalyst	chemisorption stoichiometries			Surface atom concentration [atoms m^{-2}]
	H/M	CO/M	O/M	
Pt				1.25×10^{19}
SiO_2, Al_2O_3	1.0	1.0	1.0	
SiO_2	1.75 [173]	—	—	
Y-Zeolite	2.0 [174]	—	—	
Al_2O_3	1.25 [175]	0.6–1.0 [6]	0.5–1.0 [6]	
Pd				1.27×10^{19}
SiO_2, Al_2O_3	1.0 [6]	0.6 [6]	—	
Rh				1.33×10^{19}
Al_2O_3, SiO_2				
$\bar{d}_{Rh}$ >2.0 nm	1.0	1.0	1.0	
$\bar{d}_{Rh}$ <2.0 nm	1–2 [176, 177]	2.0 [178, 179]		
Ru				1.63×10^{19}
SiO_2, Al_2O_3	1.0 [180]	1.0 $\bar{d}_{Ru}$ > 2.0 nm >1.0 $\bar{d}_{Ru}$ < 2.0 nm [183]	1.0 $\bar{d}_{Ru}$ > 2.0 m, >1.0 $\bar{d}_{Ru}$ < 2.0 nm [181, 182]	
Ir				1.30×10^{19}
SiO_2, Al_2O_3				
$\bar{d}_{Ir}$ > 1.5 nm	1.0 [184]	1.0 [185]	1.0 [184]	
$\bar{d}_{Ir}$ < 1.5 nm	up to 2.0 [186, 187]	1.0 [187]		
Ni				1.54×10^{19}
SiO_2, Al_2O_3	1.0 [188, 189]	problematic due to carbonyl formation [188]	oxidation? [189]	
Fe				1.63×10^{19}
Carbon	1.0 [190, 191]	0.5 [191, 192]		
Ag				1.15×10^{19}
Al_2O_3, SiO_2	—	—	T_{ads} 420 K 0.8 [193]	
Au				1.15×10^{19}
SiO_2, Al_2O_3	—	—	T_{ads} 470 K 0.5 [194] T_{ads} 570 K 1.0 [194]	

information can be found in Klug and Alexander [195]. The simplest method to evaluate XRD profiles is the application of the Scherrer equation to obtain average crystallite sizes. Crystallite size distributions are obtained by profile shape analysis [196, 197]. The limits of the technique are generally stated to lie between 3 nm and 50 nm, however, with careful experimentation it is possible to study metal particles <2 nm [198]. The researchers however have to remember that data obtained by XRD line broadening suffer from various uncertainties. Firstly line broadening may be caused by other factors like strain and crystal faults, which are often difficult to separate [199]. An attempt to obtain crystallite size distributions free of factors of strain and crystal faulting was reported by Ganesan [200] for NiO and Al_2O_3. Secondly, particles >10 nm generally consist of several crystals and XRD-line broadening measuring crystallite sizes can be expected to underestimate particle sizes.

Small-angle-X-ray scattering in contrast determines true particle sizes [201–204] and furthermore is applicable to highly dispersed metal catalysts, as demonstrated for 1.0 nm Pt particles in Y-zeolite supercages [205]. The main difficulty, which has to be overcome, is the elimination of support scattering centres (voids). This can be achieved by filling those voids with an organic substance of similar electron density (CH_2I_2 is commonly used for Al_2O_3 and SiO_2).

X-ray diffraction can be employed to study alloy formation in bi- and multimetallic catalysts from the position and shape of diffraction peaks [206]. However, similar limitations apply as discussed in the use of XRD-line broadening for crystallite size determinations. In addition the appearance of a symmetric diffraction peak at a position expected from the alloy composition by Vegard's law does not guarantee a uniform particle composition as has been recently demonstrated by Sinfelt for Pt—Ir bimetallic catalysts [207].

Extended X-ray Absorption Fine Structure Spectroscopy and Near-Edge Spectroscopy are increasingly applied to catalyst characterization problems [208–210]. Both are element specific (the support does not interfere), equally applicable to crystalline and amorphous phases or even extremely highly dispersed materials, and can be performed in vacuum or at high pressures [211]. Thus both X-ray spectroscopies are uniquely suited for studying real industrial catalysts (low metal content, high metal dispersions). Near-edge spectroscopy looks at the fine structure in the neighbourhood (—20 to +50 eV) of the absorption edge and supplies information about electron densities of the absorbing atoms [212, 213], whereas the EXAFS region (>50 to 1500 eV) yields data about the structural environment of the absorbing centre (interatomic distances, types of neighbours, coordination numbers, and disorder parameters). The high X-ray fluxes needed to record a spectrum in reasonable scan times, which only storage rings can provide, make the experiment rather exclusive. However, some "in house" designs of EXAFS spectrometers based on more conventional high energy X-ray sources (rotating disc X-ray generators) have been reported by Knapp [214] and Cohen [215]. More recently Khalid *et al.* [216] described a high resolution EXAFS spectrometer capable of supplying spectra of comparable quality to synchrotron radiation based systems. Although significantly longer scan times are required to record good EXAFS spectra with EXAFS spectrometers based on rotating disc anode generators, such systems have the advantage that they can be built in every laboratory for a reasonable cost. The method, although still in its infancy state, has been successfully applied to characterize highly dispersed platinum [212, 217–219], Ru [219], Cu [220] and Ni, Co [221] and the few studies on bimetallic catalysts like Pt—Ir [207], Ru—Cu [222], Os—Cu [223] point to the enormous potential of the technique.

3. *Electron Microscopy*

High resolution electron microscopes, combined with accessories like microdiffraction and microanalytical facilities offer a unique opportunity to observe, measure, size and analyse metal particles of virtually all sizes on supports. The instrumentation and theory of electron microscopy has been the subject of a series of books and detailed publications [224–228]. Of the numerous methods used to prepare microscope specimens, the most common procedures involve (i) grinding of a catalyst granule and deposition of the fine powder either dry or from an ultrasonically dispersed suspension onto carbon covered specimen grids [229] or (ii) embedding the catalyst in a matrix and cutting thin sections with an ultramicrotome [230]. Using transmission electron microscopy for *particle size determinations*, the researcher faces several difficulties:

(a) the visibility of very small metal particles is strongly affected by support microcrystallinity [231], orientation of particles and imaging conditions [232]. Various research groups attempted to sort out contrast effects from small metal particles experimentally and theoretically and tried to elucidate scattering mechanisms [227, 228, 234]. The general consensus of the published work seems to be that bright field techniques are increasingly unreliable for particles <2.0 nm, but other contrast methods — hollow-cone dark field method [235], other dark field methods [236] and the Z-contrast technique [237] — are useful for extremely small particles.

(b) Statistically significant data can only be obtained if several thousand particles are counted and measured — generally a very tedious task. The wide range of contrast (support, carbon film, metal particles) on a micrograph of a supported catalyst in general prohibits the use of automatic image analysers. It is however, possible to simplify the system by removing the metal particles from the support by ultrasonic treatment [228] or dissolution of the support, but such treatments destroy other valuable information like spatial distributions of metal particles and metal-support interaction.

Apart from particle size determinations electron microscopy is increasingly applied to study *metal particle morphology* (crystal shape, crystal habit and defect structure) as well as *metal-support interactions* [238].

New opportunities to obtain direct structural information of small metal particles opened up with the development of scanning transmission microscopes [239]. Microdiffraction pattern of single metal particles have been recorded in a STEM [240, 241] and Lynch *et al.* [243] checked on epitaxial relationships between platinum particles and the support. STEM's with microanalytical attachments (X-ray fluorescence for elements of higher atomic number, and electron energy loss spectroscopy for the lighter elements) fulfilled another "dream" of catalytic scientists by offering the opportunity to analyse single metal particles, an extremely valuable method to determine the composition of bi- and multimetallic particles [243].

It is expected that the technique is applied to other problems like poisoning, role of promoters, and metal support interactions.

Another interesting development for modelling sintering processes, carbon deposition and other changes occurring with catalyst treatment, was the introduction of "in situ" electron microscopy or Controlled Atmosphere Electron Microscopy — CAEM [244, 245].

4. *Electron Spectroscopy*

Of the various electron spectroscopic methods only photoelectron spectroscopy (XPS) has been widely applied to supported catalyst systems [246, 247]. Since the excitation process in XPS involves core electrons, the technique is element specific, and elements can be identified according to their binding energy. The exact binding energy values reflect the chemical environment (oxidation state, ligands) of atoms, which makes XPS a powerful tool to monitor chemical changes during various treatment procedures, *e.g.* in catalyst preparation [246].

In addition the technique is highly surface sensitive; its sampling depth is limited by the escape depth of the ejected core electron which is in general not more than 2.0 nm. Since an ESCA analysis is carried out in ultrahigh vacuum, most ESCA spectrometers are equipped with "in situ" high pressure treatment chambers, where the catalyst activation treatment can be reproduced and the sample transferred into the analysis chamber without exposure to air. Like other characterization methods XPS suffers from a series of limitations like signal overlap (*e.g.* Pt and Al), sample degradation in the beam, insensitivity to subtle changes in chemical environment (often oxidation states cannot be easily identified), and difficulties in acquisition of quantitative results. The main areas where XPS has proved extremely valuable are in catalyst preparation (identification of precursor species after impregnation and precipitation and after activation treatment), and in the characterization of multimetallic catalysts (state of additional component, surface segregation). More recently the technique has been successfully applied to monitor electronic changes due to metal-support and metal-promoter interaction [248–250].

5. *Mössbauer Spectroscopy*

The method is based on Mössbauer's discovery in the late fifties that emitting nuclei within a solid matrix could undergo recoil-free emission and resonant absorption of γ-rays. A large number of catalytically important metals are Mössbauer active and are thus open to investigation by Mössbauer spectroscopy [251, 252]. However, some isotopes are very shortlived and of low natural abundance, which makes their use costly and impractical for catalyst characterization. A method which helps to overcome those limitations is the incorporation of Mössbauer probe nuclei (*e.g.* Fe^{57}) in small amounts [253].

A Mössbauer spectrum contains a variety of information:

(i) changes in the electron density at the nucleus by the chemical environment causes shifts in the position of Mössbauer bands (isomer shift) which reflect changes in oxidation state, co-ordination number, and ligand type.

(ii) Quadrupole splitting caused by nuclei surrounded by nonspherical charge distributions, reflects the structural neighbourhood of the emitting nuclei.

(iii) Magnetic dipole splitting senses ferromagnetic ordering in Fe and its alloys — a particle size dependent parameter. Thus Mössbauer spectroscopy is particle size sensitive.

An extensive list of publications and reviews exists on the theory, experimentation and interpretation of Mössbauer spectroscopy [252, 255] and its application to catalyst problems [254, 255]. Scanning the literature on the subject, the method has been most successful in:

(a) determining alloy formation in bi- and multimetallic catalysts containing Fe in combination with Pt [256], Pd [256], Ni [257]; Pt—Sn [258] or Pt—Ir with Fe^{57} as probe nuclei [253].

(b) characterization of the state of Fe in zeolites [259].

(c) studying the influence of chemisorption on Mössbauer active elements [254].

(d) monitoring changes in catalyst nature during "in situ" reactions — Fischer Tropsch [254], and ammonia synthesis [254].

6. *Infrared Spectroscopy*

Since Eischens [260] adapted infrared spectroscopy to the study of supported metal catalysts, its value as a characterization tool was soon recognised. The frequencies and intensities of IR-bands of adsorbed probe molecules like CO and NO provide abundant evidence on the state of the supported metal (dispersion, structure, support-metal interaction, metal-metal interaction, etc.). In catalyst preparation the technique is ideally suited to monitor the interactions of the precursor compounds (*e.g.* metal carbonyls) with the support and their decomposition by various heat treatments [261].

Originally, IR-spectroscopy of adsorbed species was plagued with sensitivity problems, slow data acquisition and thus sample heating, and difficulties arising from the low transmittance of catalyst samples. The development of Fourier Transform IR-instruments not limited by the above problems, induced a revival of the technique as a characterization tool [262]. IR-spectroscopy is known for its versatility. Catalyst pellets can be mounted in "in situ" IR-cells [263] and spectra recorded in vacuum or at high pressures, at low or at high temperatures or even under reaction conditions.

The literature on IR-spectroscopy of adsorbed molecules is extensive and two books and a series of review articles survey the field [264–266], and recently Angell published a review dealing exclusively with the study of catalysts by Fourier Transform Infrared Spectroscopy [267].

7. Other Characterization Methods

A series of other techniques are employed for catalyst characterization, but most of them are not as universally applicable as the ones discussed earlier or involve elaborate instrumentation. Nickel and Ni—Cu alloys have been extensively studied by *magnetic measurements* [268] a method applicable also to iron and cobalt. *Neutron scattering and neutron diffraction* [269] — equivalent to similar X-ray methods — can be used to extract information about the textural (pore structure, metal particle size) and structural properties of the catalyst. In a similar way various *ion beam methods* [270] have been used to characterize catalysts.

5. Structure and Properties of Dispersed Metals

A. General

Since the development of characterization methods (selective chemisorption, X-ray methods, electron microscopy, electron spectroscopy, etc.) allowed catalytic scientists to determine metal surface areas, particle sizes and particle size distributions, and to calculate *intrinsic activities* (activity per metal atom), considerable effort has been made to find links between those activities and the size and morphology of metal particles. A critical examination of the voluminous literature on the subject — see recent reviews by Moss [6], Gault [271], Clarke [272] and Davis and Klabunde [273] — reveals numerous discrepancies, indicating that such studies are not without problems. In some studies crystallite sizes were varied only within a limited range and the catalysts were poorly characterized, however, even in the most carefully planned experiments, one can never be sure about the "side effects" of the procedure used to vary particle sizes; *e.g.* if the crystallite size is changed by increasing the metal loading in the catalyst, it is conceivable that small impurity amounts affect catalysts of low metal loading to a larger extent than catalysts with large metal contents, thus simulating a particle size effect. When particle size changes are brought about by high temperature treatment in oxygen and hydrogen, metal support interactions as well as impurity effects may influence the catalytic properties.

Nevertheless adsorption studies on well characterized metal surfaces [274] indicate the importance of geometric structure and electronic properties of adsorption sites. Thus one has to expect a crystallite size dependence of a series of catalytic reactions, especially if complex adsorbed intermediates are involved.

Boudart [275] first suggested a classification of catalytic reactions into two categories: *structure sensitive* or *demanding* reactions show a change of specific activities with particle size, whereas *structure insensitive* or *facile* reactions are not affected by particle size or particle morphology. Katzer and Manogue [276] suggested a subclassification of structure sensitivity into (i) primary structure sensitivity, a term used for reactions which require special arrays of metal atoms (*e.g.* B_3, B_4, B_5 sites) as active centres; (ii) secondary structure sensitivity brought about by particle size dependent self poisoning; and (iii) primary size sensitivity, which reflects differences in structure and electronic properties of very small metal particles compared to bulk metal properties (changes in chemisorption stoichiometries on various metals have been observed with catalysts containing metal particles smaller than 2.0 nm).

In alloy catalysis Sachtler [277] introduced two terms — *ensemble* effect and *ligand* effect. The first one represents a geometric effect and relates to what Katzer and Manogue called primary structure sensitivity, whereas the latter one reflects changes in the electronic surrounding of the active site.

Of the large number of reactions studied, hydrogenation and dehydrogenation reactions appeared unaffected by size and morphology variations of metal crystallites. Reactions which involve carbon-carbon bond formation and rupture seem to fit into the category of primary structure sensitive reactions, and the structure sensitivity of various oxidation reactions is often explained by preferential oxidation of the small particles and would therefore classify as secondary structure sensitive reaction. Valuable information about structure sensitivity can further be obtained from reaction studies on selectively poisoned catalysts [278, 279] and on alloy catalysts containing a combination of an active and inactive metal, *e.g.* a group VIII–Ib alloy [272, 280, 281]. Poison molecules and inactive metal atoms act in a similar fashion; both will suppress reactions requiring large atom ensembles as active sites more than reactions which occur on a single metal atom. In addition, low coordination sites like corner atoms, edge atoms, etc. may be deactivated preferentially. In conclusion it has to be pointed out that irrespective of the numerous studies carried out to understand the relationship between the nature of the catalyst and its reaction behaviour, many questions regarding the structure and nature of active sites are still unanswered, mainly because of the lack of sensitive enough direct investigation tools necessary to study a highly dispersed metal phase. But more refined characterization methods, theoretical calculations on small metal particles, and comparison with molecular metal clusters recently prepared in sizes containing up to 55 atoms, may provide a clearer insight into the reaction behavior of dispersed metal catalysts.

B. Crystal Structures of Bulk Metals

Three crystal structures are formed by transition metals: The *hexagonal close packed* structure consists of close packed layers in a ABABA stacking sequence. The *cubic close packed* or *face centred cubic* structure originates from a stacking sequence ABCABC. Both structures are characterized by a coordination number of 12 and a space filling density of 74%. The *body centred cubic* structure is a less densely packed arrangement (68% space filling) and each metal atom is bound to only 8 nearest neighbours. Table 7 lists the crystal structures and lattice spacings of catalytically important metals and demonstrates the prominent role face centred cubic metals play in catalysis.

Table 7. The crystal structure of catalytically important

Metal	Crystal structure	*a*/nm	*c*/nm	Nearest neighbour distance/nm
γ-Fe	bcc	0.365	—	0.258
α-Fe	bcc	0.287	—	0.248
β-Co	fcc	0.355	—	0.252
α-Co	hcp	0.251	0.407	0.250[a]
Ni	fcc	0.352	—	0.250
Cu	fcc	0.361	—	0.255
Re	hcp	0.276	0.446	0.274[a]
Ru	hcp	0.270	0.428	0.267[a]
Rh	fcc	0.380	—	0.268
Pd	fcc	0.389	—	0.274
Ag	fcc	0.408	—	0.288
Os	hcp	0.273	0.432	0.270[a]
Ir	fcc	0.383	—	0.271
Pt	fcc	0.392	—	0.277
Au	fcc	0.408	—	0.288

[a] For hcp metals an average value is quoted as nearest neighbour distance

C. Structure of Small Metal Particles

1. *Naked Microclusters*

Various approaches have been pursued to model minimum energy packing arrangements in microclusters [282]. The *potential energy minimization* of selected *trial structures* (polyhedra obtained by dissecting the lattice of a bulk crystal) by suitable relaxation procedures is a common method to elucidate stable cluster structures. To simplify such calculations, central forces between the cluster atoms are assumed and angle dependent contributions as well as three body terms are generally neglected. Energy minimizations utilising as interatomic pair potentials well potentials of the type of Mie, Lennard-Jones, or Morse potentials showed that the most stable microcluster configuration is the regular icosahedron [283–285], and that the corresponding fcc and hcp cuboctahedra relax spontaneously into the icosahedral structure (*c.f.* Figure 9). The higher stability of polytetrahedral packing over regular fcc or hcp configurations is associated with a maximization of nearest neighbour bond contacts (42 in a 13 atom icosahedron compared to 36 for the equivalent cuboctahedron) and minimization of the

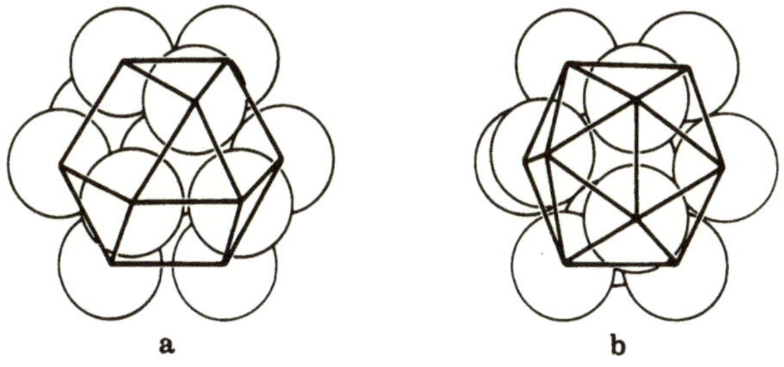

Figure 9. Cuboctahedral **a** and icosahedral **b** 13 atom particle

surface energy (the icosahedron is bounded solely by close packed surfaces, while the surface of a cuboctahedron contains 43% higher energy square arrays).

The establishment of *minimal growth sequences* is an alternative approach to estimate cluster shapes [283]. A single atom or a seed structure is used as a starting point and atom by atom added with full optimisation after each step. Such a growth sequence is shown in Figure 10. Two atoms form a dimer, three a equilateral triangle, four a tetrahedron. The energetically most favourable sites to add to a tetrahedron are the midpoints of the tetrahedron faces leading to the trigonal bipyramid for $N = 5$, the tripyramid for $N = 6$ and the pentagonal bipyramid for $N = 7$. The addition of five atoms on the

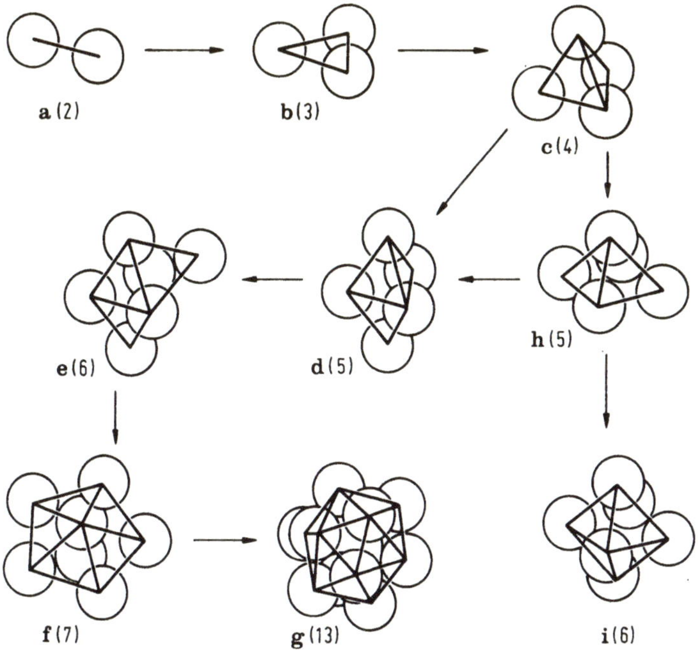

Figure 10. Growth sequence to form 13 atom cluster. The pentagonal growth sequence **c**(4)→**d**(5)→**e**(6)→**f**(7)→**g**(13) is energetically favoured over the octahedral sequence **c**(4)→**h**(5) →**i**(6). Ref. [283]

five upper faces of the pentagonal bipyramid plus one atom at the fivefold symmetry axis produces the 13 atom icosahedron. It is noteworthy that the growth sequence does not include the octahedron for $N = 6$, although its potential energy is lower than that of the tripyramid. However, except for the octahedron itself none of the same sized isomers in the octahedral growth sequence can compete with the structures of the tetrahedral sequence. To start the octahedral sequence the formation of a square pyramid for $N = 5$ is required, a structure which would spontaneously relax into a trigonal bipyramid.

Molecular orbital calculations on 13 atom clusters, Li_{13} [286] and Ni_{13} [287] also recognise the higher stability of icosahedral packing compared to cuboctahedral configurations for microclusters.

But to what extent are those predictions supported by experimental evidence? Electron micrographs of metal particles produced by gas evaporation techniques (GET), a preparation technique which satisfies the assumptions of the theoretical models best, frequently show particles with five fold symmetries like pentagonal bipyramids and icosahedra [288].

Within the catalytic important metals such structures seem to be common for Ag, Au, Cu, and Ni [289–293], but less common for Pt and Pd [291]. MO calculations [287, 294] predict differences in the preferred geometries of clusters of different metals. However, the result [294] that Pd and Ni clusters are more stable in the icosahedral form, but Ag and Cu in the fcc cuboctahedral configuration cannot be reconciled with experimental findings. The question still remains, whether the tetrahedral growth sequence operates for all metals and the discrepancies have to be attributed to recrystallization processes occurring at different cluster sizes for various metals. The transformation of a particular structure into an energetically more favourable configuration requires the rearrangement of atoms. In very small clusters atoms are very mobile; in large clusters a defect mechanism can explain recrystallization phenomena, but for intermediate size clusters a recrystallization process is difficult to formulate. It is thus conceivable that the icosahedral structure is retained up to fairly large clusters whenever the energy differences between the two structures are large, but that recrystallization to the regular crystal structure occurs already at very small cluster sizes, when the differences are smaller.

2. *Liganded Clusters*

Cluster calculations have predicted packing arrangements of metal atoms in very small particles different to the bulk crystal structure, and particles of such configurations have been identified by high resolution microscopy. However, all those considerations apply only to the "naked" (unliganded) clusters and the assumption that the presence of adsorbants or reactants (ligands) does not affect the cluster geometry is questionable. With the rapid development of the synthesis and characterization of molecular cluster compounds, systems have been produced which by analogies may provide information on the influence of ligands on cluster structures. Analogies of that type have appeared frequently in the recent literature [295–298] and the two entities have been related to each other as follows

$$\text{Very small particle} + n\text{-Ligands} \rightleftharpoons \text{Molecular Clusters}$$

A wide range of metal framework structures are obtained in molecular clusters and Table 8 lists some typical configurations [101, 102, 299]. An

Table 8. Configuration of the metal skeleton in some molecular clusters

Configuration	Cluster compounds
triangle	$Os_3(CO)_{12}$, $Rh_3(CO)_{12}$, $Fe_3(CO)_{12}$
tetrahedron	$Ir_4(CO)_{12}$, $Rh_4(CO)_{12}$, $Co_4(CO)_{12}$
butterfly*	$Ru_4(CO)_{11}(C_8H_{10})$, $Co_4(CO)_{10}(C_6H_{10})$
near planar*	$Re_4(CO)_{16}^{2-}$, $Fe_3Pt_3(CO)_{15}^{2-}$, $Os_6(CO)_{21}$
trigonal bipyramid	$Os_5(CO)_{15}$, $Ni_5(CO)_{12}^{2-}$
spare pyramid	$Fe_5(CO)_{15}C$
oxtahedron	$Os_6(CO)_{18}^{2-}$, $Ru_6H_2(CO)_{18}$, $Rh_6(CO)_{16}$, $Co_6(CO)_{16}$
bicapped tetrahedron*	$Os_6(CO)_{18}$
trigonal prisms*	$Pt_6(CO)_{12}^{2-}$, $Pt_{12}(CO)_{24}^{2-}$, $Pt_{15}(CO)_{30}^{2-}$
capped octahedron	$Rh_7(CO)_{16}^{3-}$, $Os_7(CO)_{21}$
tetracapped octahedron*	$Os_{10}(CO)_{24}C$
close packed cubic (ccp)	$Fe_6Pd_6H(CO)_{24}^{3-}$, $Pt_{38}(CO)_{44}^{2-}$, $Au_{55}(PPh_3)_{12}Cl_6$
hexagonal close packed (hcp)	$Ni_{12}(CO)_{21}H_2^{2-}$, $Rh_{13}(CO)_{25}H_{5-n}^{-n}$, $Pt_{26}(CO)_{32}^{2-}$
body centred cubic (bcc)	$Rh_{14}(CO)_{25}^{4-}$
bcc → hcp, ccp	$Rh_{14}(CO)_{25}H^{3-}$, $Rh_{14}(CO)_{26}^{2-}$, $Rh_{15}(CO)_{27}^{3-}$
icosahedron*	$Rh_{12}(CO)_{27}Sb^{3-}$, $Au_9L_8^{3+}$, $Au_{13}(dppmH)_6^{n+}$
bicapped pentagonal prism*	$Rh_{15}(CO)_{28}C_2^{-}$, $Pt_{19}(CO)_{22}^{4-}$

* less common structures.

examination of Table 8 reveals that most structures considered for very small metal particles also feature in molecular cluster chemistry. Two atoms form a linear geometry; three a triangular array, which is equilateral for highly symmetric clusters ($Ru_3(CO)_{12}$, $Os_3(CO)_{12}$) or more or less distorted for clusters of lower symmetry; four atoms form a tetrahedral framework, either perfect ($Ir_4(CO)_{12}$) or distorted, sometimes so that butterfly type or pseudo square planar arrangements are produced. Square pyramids and trigonal bipyramids feature for clusters with a nuclearity of five. The octahedral structure is most common for $N = 6$, but the bicapped tetrahedral and trigonal prism and antiprism configurations have been observed. The close packed polyhedral structures — pentagonal bipyramid, icosahedron — which were claimed to be of superior stability in naked clusters, do not appear to feature very strongly among cluster compounds. On the contrary molecular clusters seem to favour structures derived by the octahedral growth sequence. However, some compact polytetrahedral structures are known — $[Rh_{12}(CO_{27})\,Sb]^{3-}$ forms an icosahedron framework with antimony as a central atom [300] and $Au_{11}(PPh_3)\,X_3$, $[Au_{13}(dppmH_6)]^{n+}$ also adopt icosahedral symmetry [301]. An interesting structure of five fold symmetry has been reported for $[Pt_{19}(CO)_{22}]^{4-}$ [101]. This cluster adopts a bicapped pentagonal prism configuration with a stacking sequence 1—5—1—5—1—5—1 which is stabilized by bridging CO groups. A similar structure has been recently observed when nickel bromide was reduced in an organic solvent with magnesium metal [302]. Other large molecular metal clusters ($N = 10$ to $N = 55$) adopt hexagonal packing, cubic close packing and even body centred cubic fragments. Among those are the largest molecular clusters prepared up to date, which contain numbers of atoms expected in very small metal particles — namely the hexagonal stacked $[Pt_{26}(CO)_{32}]^{2-}$ cluster, and the cubic close packed $[Pt_{38}(CO)_{44}]^{2-}$ and $Au_{55}(PPh_3)_{12}Cl_6$ clusters. An interesting relationship exists between the rhodium clusters with $N = 13$, 14 and 15 (*c.f.* Figure 11) which represent an

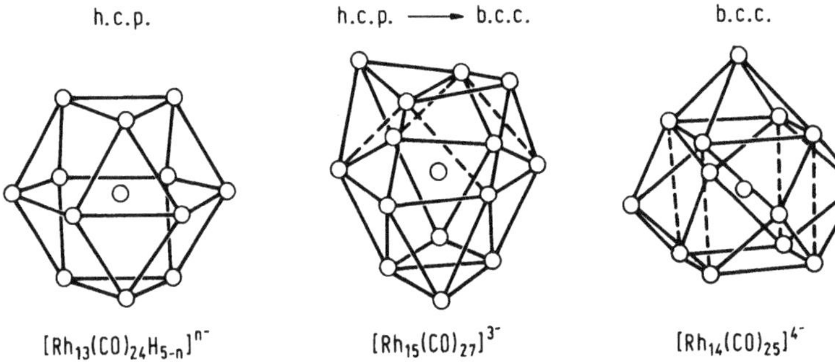

Figure 11. Structural relationship between high nuclearity rhodium carbonyl structures. (Reproduced with permission from ref. [297])

analogy to the transformation between hcp and bcc lattices [303]. The regular twinned cuboctahedron formed in the three rhodium clusters $[Rh_{13}(CO)_{24}H_3]^{2-}$, $[Rh_{13}(CO)_{24}H_2]^{3-}$ and $[Rh_{13}(CO)_{24}H_4]^{-}$ corresponds to a hcp lattice fragment, the $[Rh_{14}CO_{25}]^{4-}$ cluster is a fragment of a body centred cubic lattice; $[Rh_{15}(CO)_{27}]^{3-}$ is an intermediate between both configurations related to both structures with reconstruction at the surface. The reorganization of the metallic skeleton is quite common with molecular clusters and occurs with remarkable ease and the following reactions demonstrate some of those transformations [101]

$$Os_6(CO)_{18} + 3\,I^- \rightleftharpoons [Os_6(CO)_{18}]^{2-} + I_3^-$$

bicapped tetrahedron   octahedron

$$[Fe_4(CO)_{13}]^{2-} + H^+ \underset{DMSO}{\overset{THF}{\rightleftharpoons}} Fe_4(CO)_{13}H$$

tetrahedron   "butterfly"

$$[Rh_{14}(CO)_{25}]^{4-} + H^+ \underset{DMSO,\ Na_2CO_3}{\overset{CH_3CN}{\rightleftharpoons}} [Rh_{14}(CO)_{25}H]^{3-}$$

bcc   bcc-ccp

$$\underset{\text{octahedron}}{[Rh_6(CO)_{13}C]^{2-}} + 2\,CO \underset{333\,K,\,N_2}{\overset{298\,K,\,1\,at}{\rightleftharpoons}} \underset{\text{trigonal prism}}{[Rh_6(CO)_{15}C]^{2-}}$$

The important features which can be deducted from cluster chemistry may be summarized as follows:

(i) the most important building units in molecular clusters are triangular metal arrays, with the three metal atoms bonded not only by pairwise interactions along the triangle edges but with three centre bonding within the triangle itself contributing significantly to the metal bonding scheme. It has to be pointed out that in structure calculations of "naked" clusters such interactions were completely neglected.

(ii) Large nuclearity metal clusters ($N > 10$) commonly adopt structures which are fragments of fcc, hcp and bcc crystal structures or intermediates between those. Clusters of five-fold symmetry are rarer, but some examples of pentagonal prisms — $[Pt_{19}(CO)_{22}]^{4-}$, $[Rh_{15}(CO)_{28}C_2]^-$ — and icosahedral symmetry — $[Rh_{12}Sb(CO)_{27}]^{3-}$, $Au_9L_8^{3+}$, $[Au_{13}(dppmH)_6]^{n+}$] — have been prepared.

(iii) The immense mobility of metal atoms within a molecular cluster has been demonstrated by the ease of deformation of the metal skeleton whenever ligands or electric charges on the clusters are changed. It seems therefore reasonable to assume that metal atoms in small metal particles exhibit a similar high freedom of arrangements in response to different surroundings.

3. Supported Metal Particles

When metal particles are grown on a substrate the metal-substrate interaction may play a significant role in determining the particle structure. The problem has been extensively discussed in nucleation and growth of thin films and the field has been reviewed by Venables and Price [304]. In the following example (Figure 12) we look at the stability of a five atom cluster

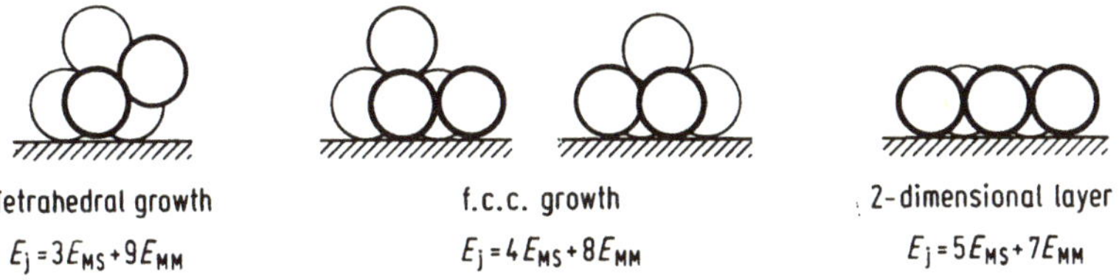

Figure 12. Possible structures of a 5-atom cluster on a substrate. E_j, total energy of the system; E_{MS}, metal-substrate interaction; E_{MM}, metal-metal interaction

on a substrate. The binding energy (E_j) of the cluster is calculated on the basis of summing nearest neighbour bonds: E_{MM} relates to the metal—metal bond and E_{MS} to the metal—substrate bond. Three regimes can be distinguished:

(i) $E_{MM} \gg E_{MS}$

In this case the substrate would not influence the cluster morphology.

(ii) E_{MM} is comparable to E_{MS}

If the interactions between metal atoms and between metal and substrate are equal all four cluster shapes are energetically equivalent and clusters of different morphology coexist. Popescu [305] used an energy minimization calculation of a 13-atom cluster on a substrate and included a potential V_s which simulated the metal substrate interaction. The results clearly demonstrated that with increasing metal-support interactions the normal fcc growth becomes energetically favourable compared to tetrahedral growth.

(iii) $E_{MS} \gg E_{MM}$

A much stronger metal-support interaction would favour the growth of two-dimensional layers. This regime is more common with metals deposited on metal substrates, but has recently become the subject of extensive discussions in catalysis with the discovery of strong metal support interactions (SMSI) in catalysts utilising reducible oxides like TiO_2, Nb_2O_5, Ta_2O_5 and V_2O_5 as supports [306].

Values for the adhesion of metals on oxides under reducing conditions are generally less than 40 kJ mol^{-1}, which implies no more than Van der Waals interactions [413], whereas metal—metal bond energies are in the range 150–250 kJ mol^{-1}. This leads to a strong preference to form metal—metal bonds rather than metal—substrate bonds and near spherical microclusters of a configuration similar to clusters grown in free space result. And indeed metal particles of pentagonal or icosahedral shapes are observed in large numbers, when fcc metals like Ag, Au, Ni, Pd, Pt are condensed onto alkali halides, mica, MoS_2 or MgO [285, 291, 293, 307–310]. In "real" supported metal catalysts, prepared by contacting the substrate with a metal salt solution, followed by removal of the solvent and transformation of the metal compound into the zero valent form, the situation proves much more complex. Although both "non crystalline" and crystalline metal particles seem to be present, their relative concentrations vary considerably with the metal, substrate and treatment conditions. According to Avery and Sanders [310] who examined Ni, Au, Pd and Pt on silica supports, non fcc particles were only present to the extent of 2%. A similar conclusion was reached by Chen and Schmidt [311] for Pt on silica and on titania and in Pt/graphite catalysts fcc-cuboctahedral particles of platinum were detected exclusively [312]. Highly dispersed rhodium ($\bar{d}_{Rh} < 2.0$ nm) seem to form predominantly icosahedral clusters, when supported on Al_2O_3, SiO_2 and carbon, but normal fcc growth seems to occur on TiO_2 and MgO. Pentagonal silver particles were mainly present in Ag/α-Al_2O_3 catalysts after heat treatment in vacuum or under mild oxidising conditions, but regular fcc crystals form after heat treatment in hydrogen [292(b)]. EXAFS experiments carried out on Pt/Y zeolite catalysts [314] have been interpreted with the presence of small icosahedral platinum particles (< 1.2 nm) and larger fcc platinum particles after heating in vacuum. Upon hydrogen chemisorption the Pt-Pt distances relax and all particles assume cubic symmetry. Those findings agree with EXAFS studies by Via *et al.* [219], who reported for highly dispersed osmium, iridium and platinum on silica in hydrogen interatomic distances similar to the bulk values.

The structural relationship between palladium crystallites and γ-Al_2O_3 was studied by microdiffraction techniques in a scanning transmission electron microscope [315]. In catalysts prepared from palladium nitrate the resulting metal particles were randomly oriented on the support. If however, palladium acetylacetonate was used as a metal precursor compound an epitaxial relationship between the metal and the substrate developed. Chemoepitaxy was previously reported in nickel catalysts prepared from nickel antigorite [316]. In this system the orientation of the formed nickel platelets on the silica sheets depended strongly on the reduction conditions.

High resolution microscopy and X-ray methods like EXAFS have been successfully utilised to analyse the structure of extremely small metal particles commonly present in industrial catalysts, but the data are still scarce and the discrepancies between various studies point to the need to obtain a better understanding of the relationship between cluster structure, type of substrate, and treatment conditions.

D. Structure of Multimetallic Catalysts

1. Particle Structure

Possible microstructures which may exist in a catalyst containing a mixture of two metals (A and B) are outlined in Figure 13. Four different cases can be distinguished:

(i) separate particles of A and B (Figure 13(a)),
(ii) particles of varying compositions (Figure 13(c)) "inhomogeneous" catalyst,
(iii) single phase alloy particles (Figure 13(d)) of uniform composition,
(iv) the particle composition is uniform throughout the catalyst — "homogeneous catalyst", but particles are biphasic (Figure 13(b, e–g)).

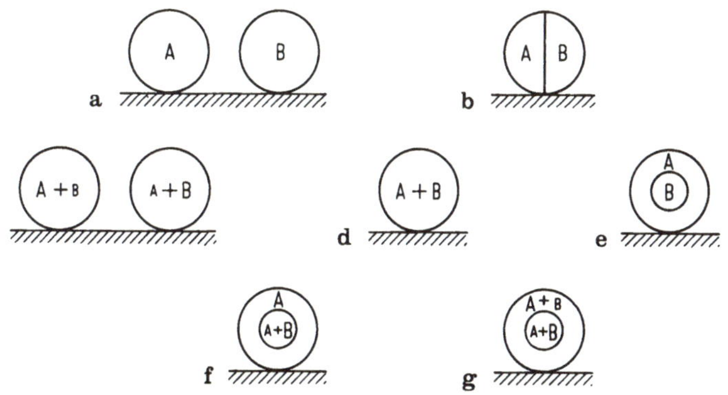

Figure 13. Possible microstructures of metal particles in a bimetallic catalyst

If two miscible metals are codeposited on a substrate the formation of clusters containing both metals might be expected on pure statistical grounds, but the pioneering work by Sinfelt [125, 126] on dispersed Ru—Cu, Os—Cu and Pt—Ir demonstrated that bimetallic clusters form also for metal combinations which are essentially immiscible in the bulk. Those findings questioned the applicability of bulk phase diagrams to microclusters. Ollis [317] proposed on thermodynamic grounds that small metal particles (<10 nm) can remain in one single phase well below the critical temperature of the bulk system. Hoffman [318] challenges the validity of Ollis' hypothesis, and suggests that the higher stability of small particles is caused by a suppression of the nucleation and growth mode of phase transformations. Anderson [319] discusses both theories in his review and arrives at the conclusion that only reliable experimental data can clarify the situation. To identify the microstructure of highly dispersed multimetallic catalysts is by no means a simple task. X-ray diffraction, the traditional method for studying alloy formation, ceases to be feasible for particles <3.0 nm and even for moderately larger particles the method is questionable as Sinfelt [207] recently demonstrated. In the Pt/Ir case the addition of two broad Lorentzian shaped lines at positions corresponding to compositions of 30% Pt/70% Ir and 70% Pt/30% Ir yielded a symmetric XRD profile, "characteristic" of uniform alloy particles. Mössbauer spectroscopy is applicable to highly dispersed metal phases and has been successfully used to determine alloy formation and internal particle structures in systems containing Fe, Sn and Au [254]. X-ray emission in a high resolution scanning electron microscope able to analyse individual particles of 2.0 nm in size, seems to develop into an extremely valuable facility to study alloy formation and catalyst homogeneity [243, 320]. Blanchard *et al.* [243] used the technique to identify separate Pt and Ru particles and bimetallic Pt—Ru particles of different compositions in Pt—Ru catalysts prepared by successive impregnation or by co-impregnation methods. EXAFS spectroscopy permits the determination of the internal structure, provided that catalyst inhomogeneities can be excluded. Up to date only few bimetallic systems have been studied by this technique, the group VIII–Ib combinations Ru—Cu, [222], Os—Cu on SiO_2 [223], Ru—Au on MgO [321] and the group VIII–VIII combinations Pt—Ir [207] and Ir—Rh [322].

Reliable data on internal structures of bimetallic particles are still only available for the few systems listed above and it would be premature to make a definite decision whether the predictions by Ollis and Hoffmann are supported by experimental data. It is however, worthwhile noting that single phase alloy particles have only been observed for systems completely miscible in the bulk like Pd—Au [323], Pt—Fe ([324, 325], Fe—Pd [326], whereas combinations with wide miscibility gaps or complete immiscibility in the bulk form biphasic particles, typical examples are Ru—Cu and Os—Cu dispersed on SiO_2 [222, 223] which formed clusters where a copper layer envelops essentially pure ruthenium or osmium cores, and supported bimetallic Pt—Ir clusters [207] which consisted of a platinum-rich outer layer surrounding an iridium-rich core.

2. Surface Composition

Since it first became apparent [280] that surface composition of an alloy did not necessarily correspond with that of the bulk, considerable progress has been made in developing theoretical models to predict and calculate surface segregation phenomena [327–332], and in measuring surface compositions with Auger spectroscopy, work function measurements, and chemisorptive titrations. The most extensive review on surface segregation models, supported by experimental data has been presented recently by Kelley and Ponec [330]. All surface segregation models are based on establishing an energy relationship between a segregated surface and solute in the bulk. Gibbs [333] originally related the surface concentration of an alloy to its bulk concentration by the equation

$$X_s - X_b = -\frac{a}{RT}\frac{\mathrm{d}\sigma}{\mathrm{d}a}$$

where a represents the activity of the solute and σ the surface free energy of the alloy. This *rigorous thermodynamic approach* has limited usefulness, for the simple reason that the necessary thermodynamic quantities are extremely difficult to measure. Thus *quasichemical* theories have been developed to calculate the free energy change for the interchange process of a solute atom in the bulk with a solvent atom on the surface. To create a surface we can imagine two concepts: (i) a surface forms by breaking bonds between two atomic planes in a crystal. In this case the surface free energy consists of the energy to break the bonds and the entropy gained by the surface atoms, and (ii) an existing surface is increased in which case the energy is given by the surface tension. For many systems bond strength and surface tension differ for different elements, and surface segregation of one component occurs. In general, the energy of the system is reduced if weakest bonds are broken; this is achieved if the component with lower sublimation energy segregates to the surface. Those models have been classified as *broken bond theories* [280, 327–329].

The alternative approach is pursued in the *elastic strain theories* [331] which argue that whenever the atoms which constitute an alloy system differ in size, significant elastic strain exists in such a system, which can only be reduced by moving one atom species to the surface. The predictive power of both theories in isolation is limited as its failures indicates [334, 335], but the model can be improved by combining the two theories and defining the surface segregation driving force as the total surface energy from mixing energy and strain energy contributions [328, 330, 336]. Exact atomic calculations [337, 338] have been carried out where the total potential energy of various solid solutions of atoms in bulk and surface configurations is obtained through atomic relaxation consistent with long range pairwise interactions between the atoms. Such a scheme combines all segregation driving force contributions due to interaction strength differences and atom size differences.

The basic assumption in surface segregation calculations that the surface energies of the constituents in an alloy are the same as those in the pure solid has been recently attacked by Shek *et al.* [339] who claimed that such an assumption is only justified in the absence of significant electronic interactions. For the Pt—Cu alloy, UPS experiments indicated that the two metals interact electronically with the effect that the magnitude of copper segregation to the surface is reduced.

The segregation theories, as discussed, modelled the ultrahigh vacuum situation, leaving vacant surface valencies unsaturated. During a catalytic reaction, reactants are chemisorbed and take up those "dangling" bonds. Chemisorption enthalpies on transition metals have been found to be of the order of 100–400 kJ mol^{-1} [340] and thus the chemisorption contribution cannot be neglected in surface composition calculations, especially if the adsorption enthalpies differ markedly for each of the alloy constituents. Adsorbing H_2, CO and O_2 on a Cu—Ni alloy surface decreases the surface enrichment in Cu found for clean surfaces and may even induce enrichment in Ni [341] — *c.f.* Figure 14. Tomanek *et al.* [341] carried out extensive

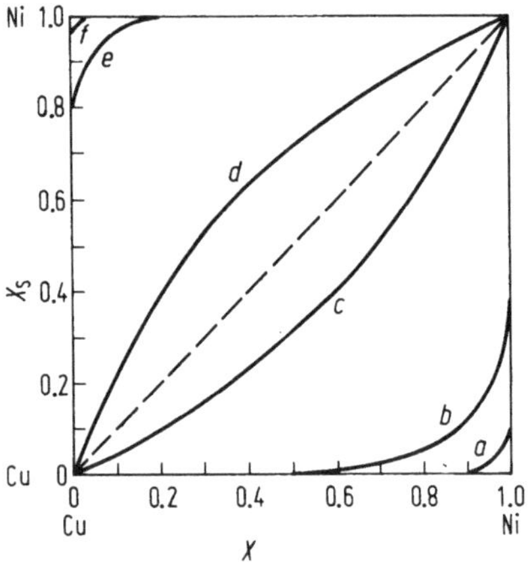

Figure 14. Surface segregation calculated for clean and covered surfaces of a Ni—Cu alloy. *a* Clean surface at $T = 600$ K; (b–f) covered surfaces: *b* H_2 ($\theta = 1$) at 300 K; *c* CO ($\theta = 0.5$) at 600 K; *d* O_2 ($\theta = 0.5$) at 600 K; *e* CO ($\theta = 1$) at 600 K; *f* O_2 ($\theta = 1$) at 600 K. (Reproduced with permission from ref. [341])

calculations on chemisorption induced surface segregations and the data for catalytically important bimetallic systems are collected in Table 9.

Table 9 demonstrates a prominent influence of adsorbates on the surface composition of alloys and those effects have to be considered when the surface composition is determined by selective chemisorption. If the surface composition data are related to reaction results, it would be advantageous to measure the surface composition in the presence of reactants.

Surface segregation in bimetallic microclusters appears to be similar to segregation in bulk alloys as the following experimental data (Table 10) confirm, but three additional factors have to be included in surface segregation models of small metal particles: (i) conservation of mass, (ii) increasing numbers of low coordination surface sites, which are preferred segregation sites, and (iii) metal support interactions.

Table 9. Heat of segregation in alloy systems[a]

Alloy A—B	Segregating element (exp.)	Q_{seg}[b]/ kJ mol^{-1}	Q_{seg}^{chem}(H)[c]/ kJ mol^{-1}	Q_{seg}^{chem}(O)[c]/ kJ mol^{-1}	Q_{seg}^{chem}(CO)[c]/ kJ mol^{-1}
Ag—Pd	Ag	+ 4.7		+ 5,4	−75.6
Ag—Rh	Ag	+103		+59.0	−62.0
Cu—Ni	Cu	+ 35.5	+10.5	−44.5	−27.5
Fe—Pt	none	− 11.1	+17.9	+38.9	+21.9
Ni—Pt	none	+ 0.8	+25.8	+60.8	− 7.2
Pd—Pt	Pd	+ 27.7	+52.7	+85.7	+44.7
Pt—Au	Au	− 54.6			+41.4
Pt—Cu	Cu	− 48.6	−48.6	−28.6	+22.4
Pt—Ir	Pt	+ 32.7	+ 7.7	−37.3	+27.7
Pt—Rh	Pt	+ 10.1	+ 6.1	−39.9	−48.9

[a] Ref. [341]
[b, c] Calculated with the equations:

$$Q_{seg} = \frac{1}{2}[f\,\Delta H_{sol}(A, B) - g(\gamma_A - \gamma_B)\,V_A^{2/3}]$$

f = 0.71 — deviation from average crystal plane due to atomic relaxation
$\Delta H_{sol}(A, B)$ — heat of solution of A in B
$g = 4.0 \times 10^8$
$\gamma_{A(B)}$ — surface energy of A(B)
V_A — molar volume of A

$$Q_{seg}^{chem} = Q_{seg} + (E_A - E_B)\,\theta$$

$E_{A(B)}$ — chemisorption energy
θ — adsorbate coverage

Mass balance criteria dictate that the magnitude of surface segregation phenomena decreases as the dispersion increases, and finally surface segregation effects have to disappear for catalysts with dispersion values of one. In contrast site segregation to low coordination surface sites — steps, kinks, corners — will still occur, even if the surface composition has equalled the bulk composition [342, 343] and as a result the relative population of differently coordinated sites on the particle surface will change markedly. The influence of metal-support interactions on surface segregation is little enriched in ruthenium. This may have been caused by a specific effect understood, but Bassi *et al.* [321] reported recently that contrary to the predictions of surface enrichment theories, Ru—Au particles on MgO were of magnesia on the particle nucleation and growth, but further experimental evidence is required to clarify the influence of support effects in surface segregation.

Group VIII-Ib combinations feature prominently in studies concerned with elucidating reaction mechanisms and identifying active sites [2, 3, 6, 272]. Surface segregation has been reported for most of those systems with the exceptions of Pd—Au [323] and Pd—Ag [344]. Since Group Ib metals are considerably less active for a large number of reactions, a Group VIII-Ib alloy surface can be visualised in terms of active ensembles of VIII metal

Table 10. Surface enrichment in supported bimetallic particles

Metal Combination	Bulk Alloy		Support bimetallic particles			Reference
	Solubility[a]	Segregating element	Support	Segregating element	Characterization Method	
Ni—Cu	MG	Cu	SiO_2	None	Chemisorption, Magnetic measurements	361
			Y-zeolite	Cu	Ferromagnetic resonance	362
			SiO_2	Cu	Chemisorption, Magnetic measurements	363
Ru—Cu	IM	Cu	SiO_2, Al_2O_3	Cu	XPS, EXAFS	364, 222
Ru—Au	IM	Au	SiO_2	None	Chemisorption, XPS	365
			MgO	Ru	Chemisorption, XPS	321
Pd—Au	M	None	SiO_2	None	^{197}Au Mössbauer	323
Pd—Ag	M	Ag	SiO_2	Ag	Reaction data	366
Pd—Ni	M	Pd	SiO_2	Pd	AES, XPS	367
Pt—Cu	OA	Cu	SiO_2	Cu	IR of chemisorbed CO	368
			Al_2O_3	Cu	XPS	369
Pt—Au	MG	Au	SiO_2	Au	Chemisorption	162
			SiO_2	None, Pt	Chemisorption, XPS	370
Pt—Fe	OA	None	Graphite	None	Mössbauer	324
			SiO_2	None	Mössbauer	256
Pt—Ni	M	None	Graphite	None	Reaction data	371
Pt—Pd	M	Pd	SiO_2	None	IR of chemisorbed CO	372
			SiO_2	None	Reaction data,	373
Pt—Rh	MG	Pt	SiO_2	None	XPS	360
Pt—Ir	MG	Pt	SiO_2, Al_2O_3	Pt	Reaction data, Oxidation — TPR	
			SiO_2, Al_2O_3	Pt	EXAFS	207
Pt—Ru	M	—	SiO_2	None	Chemisorption	243
				None	IR of adsorbed CO, NO	374
				Pt	O_2—CO titration	375, 376

[a] M, completely miscible; IM, immiscible; OA, ordered alloy phases; MG, miscibility gap-according to Hansen, M.: Constitution of Binary Alloys. New York, McGraw-Hill, 1958.

within an inactive matrix of Ib metal [272, 344]. Thus for reactions requiring large ensembles (*e.g.* hydrocarbon hydrogenolysis reactions) alloying will have a substantial effect, which is possibly reinforced by the tendency of Ib metals, due to their lower surface energy, to occupy highly active low coordination sites like step, kink and corner sites. However, one has to keep in mind that during reactions, the surface composition of the bimetallic particles may change considerably due to the preferential chemisorption of reactants on the active Group VIII component.

Bimetallic systems containing two active constituents (Group VIII-VIII combination, etc.) have gained significant commercial importance in the chemical and petroleum industry. Such systems offer improved activities, selectivities and longer lifetimes. In catalytic reforming Pt—Re, Pt—Sn and Pt—Ir on γ-alumina have in part replaced the traditional monometallic Pt-alumina catalysts [345]. The question of alloying and of reducibility of Re and Sn in commercial Pt—Re and Pt—Sn catalysts has been extensively considered [346–353] and it appears that the state of those catalysts is extremely sensitive to preparation conditions. Catalysts calcined in air at 773 K prior to reduction in hydrogen, a procedure commonly used in the preparation of commercial catalysts, seem to exhibit less interaction of Re or Sn with platinum [351–353] and confirm the idea that in commercial Pt—Re and Pt—Sn reforming catalysts alloy formation does not occur [346, 349, 350].

Pt—Ir/γ-Al_2O_3 gives improved octane fuel and exhibits lower coking rates [345]. To be effective the two metals have to form bimetallic clusters [126, 207]. The microstructure of such a particle can be visualised as a cherry type model with a platinum-rich surface region surrounding an iridium-rich core [207]. Mild oxidation treatment (<600 K) leads to increased amounts of iridium to the surface of the particles [357], whereas treatment in oxygen at higher temperatures destroys the bimetallic cluster — IrO_2 starts to grow out of a bimetallic particle (Figure 15(b)) and eventually complete separation occurs (Figure 15). After subsequent reduction smaller pure

Figure 15. Pt—Ir particles on SiO_2 heated in flowing 1% O_2 in He at $T > 700$ K. Some IrO_2 crystals are still attached to original metal particles **a**, others are completely separated platinum or platinum-rich alloy particles and extremely large Ir particles are present on the alumina support [345, 357]. Chemical segregation in oxidising atmospheres was also reported for Pt—Pd [358] and Pt—Rh [359, 360]. The surface of Pt—Pd is expected to be enriched in Pd, and oxidation would further enhance Pd surface enrichment and, similar to the Pt—Ir case, complete separation into PdO and Pt particles eventually occurs. Pt—Rh bimetallic particles according to surface segregation theories enriched in Pt, should upon oxygen treatment become enriched in Rh [359]. After high temperature oxidation (<670 K) Rh_2O_3 was observed by electron microscopy to grow as a skin over the Pt—Rh particles [359], which upon reduction would result in crystallites exposing predominantly Rh atoms [360]. Pt—Fe bimetallics have been studied extensively [324, 325, 3]. In reducing atmospheres surface compositions were essentially the same as bulk compositions. Exposure to oxygen at temperatures >570 K lead to segregation of Fe to the surface [324], and after mild reduction treatments a cluster surface enriched in Fe could be preserved [3].

The presented examples show that significant progress has been made over the past decade both theoretically and experimentally regarding alloy formation and surface segregation in bi- and multimetallic microclusters. The effects of reactive gases on the surface composition are in principle understood and some aspects of catalyst preparation methods on its microstructure have been highlighted, but reliable experimental data are still only available for few systems, the question of support effects has hardly been raised at all, and more information about the active entity during a catalytic reaction is urgently required.

E. Surface Structures and Active Ensembles

Surface atoms differ from bulk atoms in that they have an incomplete set of neighbours, *e.g.* a bulk atom in a fcc or hcp crystal is surrounded by 12 nearest neighbours, whereas atoms on the surface are bound to 3 to 11 neighbours, and on the surface of a bcc crystal, characterized by a bulk coordination of 8, atoms are bonded to 1 to 7 nearest neighbours. A representation of an imperfect close packed surface is shown in Figure 16. The coordination unsaturation of surface atoms increases in the order: terrace atoms < step atoms < kink atoms < adatoms. Those dangling valencies are taken up by adsorbates and reactants and one can imagine that atoms in different surroundings display distinct adsorption and reaction properties. "Active sites" may consist of a single atom (on top site) or of an ensemble of atoms (various hollow sites with *n*-number of metal atoms), and a selection of such sites is presented in Figure 17. The number of possible "active" ensembles increases dramatically when the surface is composed of two metals (A and B). Figure 18 demonstrates possible variations for a two-dimensional close packed ensemble of 7 atoms where a central atom (A) is surrounded by a ring of 6 atoms of various compositions. The particular types of sites predominantly present within a two-dimensional array of atoms A and B are

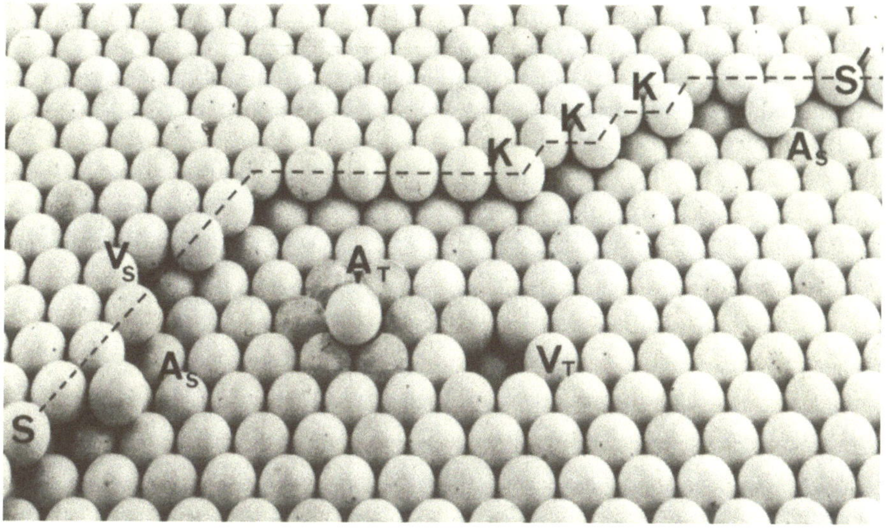

Figure 16. Ball model of close packed imperfect surface: S-step; K-kink; A_T-terrace adatom; A_S-step adatom; V_T-terrace vacancy; V_S-step vacancy

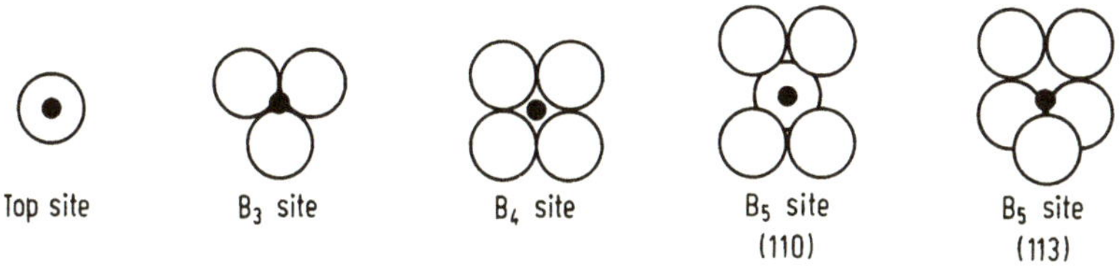

Figure 17. Examples of adsorption sites: black dot represents adsorbate

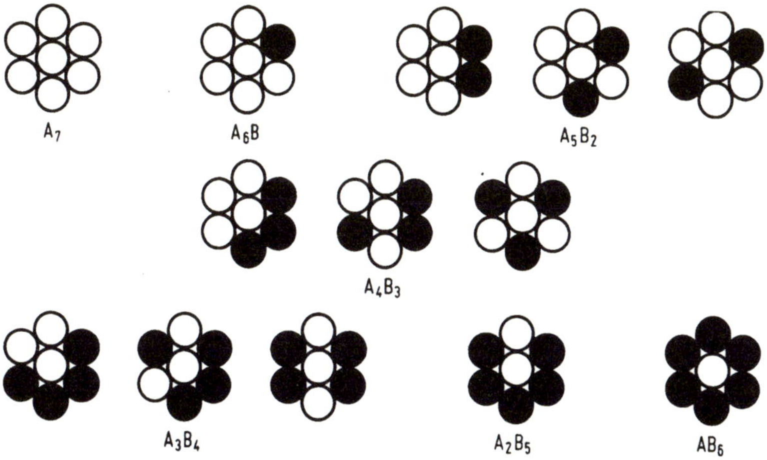

Figure 18. Possible environment of a central atom of species A in a closed packed AB two-dimensional array

determined by the interactions AA, BB and AB, and accordingly three distinct cases can be identified.
(i) The total energy of the system is lowest when AB interactions are minimised — a single boundary develops between two essentially pure A and B surfaces.
(ii) AA and BB interactions are preferred over AB interactions, but to a lesser extent than in the former case — cluster formation occurs.
(iii) AA, BB and AB interactions are equal — A and B atoms are randomly distributed in the surface.

Those models can be quantified by a computer simulation which minimises

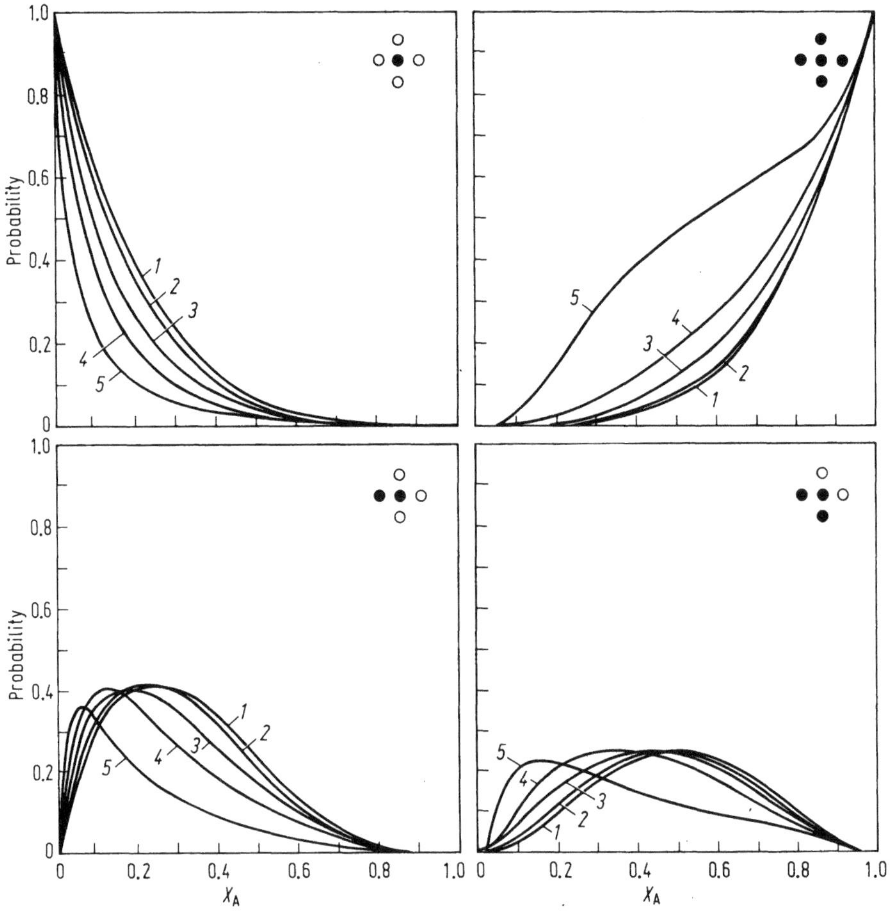

Figure 19. Probabilities of occurrence of defined ensembles (right hand corner of each graph) in a square lattice as functions of surface composition X_A. Curves 1 to 5 correspond to Φ values of 0.02, 0.1, 0.3, 0.5 and 0.8 respectively. $\Phi = \varepsilon/kT$; ε represents interchange energy. Ref. [377, 472]

the energy of the two dimensional array of specified composition and interchange energy [472], which is defined as:

$$\left[\varepsilon = E_{AB} - \frac{E_{AA} + E_{BB}}{2}\right]$$

Some results of those calculations are displayed in Figures 19 and 20, which demonstrates the marked influence of clustering on the statistics of various sites.

When a metal crystal is divided into increasingly smaller crystallites, the number of surface atoms of high valence unsaturation increases. The statistics of surface atoms and surface sites on polyhedra derived from fcc/hcp and bcc

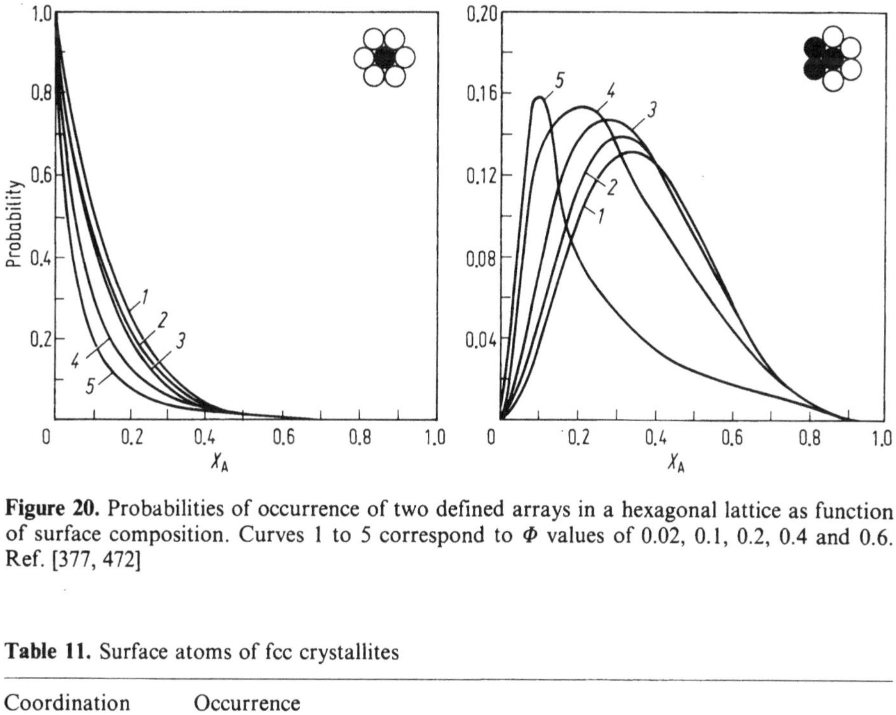

Figure 20. Probabilities of occurrence of two defined arrays in a hexagonal lattice as function of surface composition. Curves 1 to 5 correspond to Φ values of 0.02, 0.1, 0.2, 0.4 and 0.6. Ref. [377, 472]

Table 11. Surface atoms of fcc crystallites

Coordination Number	Occurrence
3	Adatom on (111) (Vertex of tetrahedron, or cube or rhombic dodecohedron)
4	Adatom on (100) (Vertex of octahedron and rhombic dodecohedron)
5	Edge of cube or rhombic dodecahedron
6	Vertex of cuboctahedron, edge of tetrahedron
7	(110) plane, edge of octahedron and cuboctahedron
8	(100) plane
9	(111) plane
10	(311) plane, steps at (111) and (100) planes
11	(110) plane, steps at (111) plane

crystal structures have been treated thoroughly by Van Hardefeld and Hartog [378]. The fcc polyhedra — octahedron, cuboctahedron, rhombic dodecahedron and tetrahedron — expose a number of coordinative unsaturated atoms and the most common ones are listed in Table 11.

A plot of the relative properties of corner, edge and face atoms for a complete octahedron and cuboctahedron versus particle size (Figure 21(a), (b)) demonstrates two important points:

(a) low coordination surface atoms — corner atoms, edge atoms (i ≦ 7) — are present in significant proportions only in crystallites of sizes $d_{rel} < 15$ or $\bar{d} < 4.0$ nm. Larger crystals consist almost exclusively of face atoms (i ≧ 8).

(b) Corner atoms predominate on the surface of extremely small crystallites, whereas in the size region around $d_{rel} = 5$, a considerable fraction of surface atoms are edge atoms.

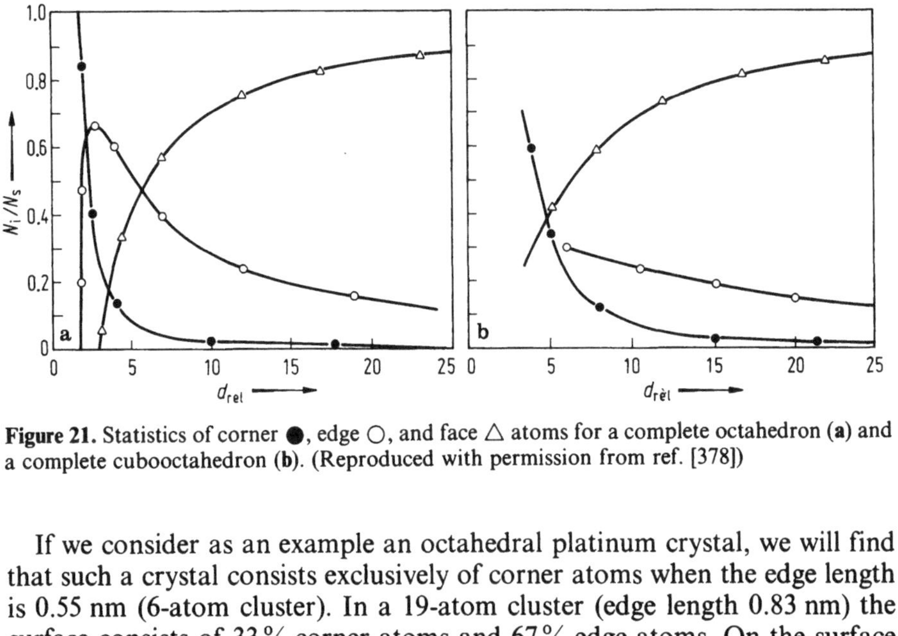

Figure 21. Statistics of corner ●, edge ○, and face △ atoms for a complete octahedron (**a**) and a complete cubooctahedron (**b**). (Reproduced with permission from ref. [378])

If we consider as an example an octahedral platinum crystal, we will find that such a crystal consists exclusively of corner atoms when the edge length is 0.55 nm (6-atom cluster). In a 19-atom cluster (edge length 0.83 nm) the surface consists of 33% corner atoms and 67% edge atoms. On the surface of a 44-atom cluster (edge length 1.11 nm) 16% are corner atoms, 63% edge atoms and 21% face atoms and a 85-atom cluster (edge length 1.38 nm) exposes 9% corner atoms, 54% edge atoms and 37% face atoms. Thus, if the highest catalytic activity resides with corner atoms, the specific activity will decrease steeply with increasing particle size, whereas if corner and edge atoms are equally active a slower decrease will result. A series of adsorption processes and catalytic reactions require defined ensembles of surface atoms as active sites [271], *e.g.* the isomerization of neopentane has been associated with triangular (111) facets [379]. Foger and Anderson [380] correlated the activity for neopentane isomerisation with the number of such sites present in catalysts of different particle sizes and obtained a reasonable relationship (Figure 22). The sites were calculated utilising the statics which were developed by Van Hardefeld and Hartog [378] for the simple ensembles B_2, B_3, B_4 and

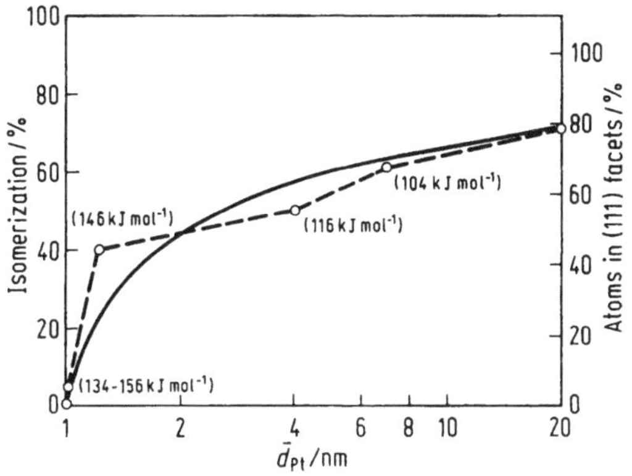

Figure 22. Correlation between neo-pentane isomerisation and surface atoms in (111) facets. The full line represents the percentage of surface atoms in (111) facets for cubo-octahedral crystallites. (Reproduced with permission from ref. [379])

B_5 on the surface of a complete octahedron and cuboctahedron and for two incomplete structures, the fcc octahedron max B_5, and the fcc cuboctahedron max B_5. Van Hardefeld and Hartog further point out that the geometric configuration may not be sufficient to define a particular site, but that in addition the coordination numbers of the atoms constituting the site may have to be considered. An octahedron for example exposes four different B_3 sites, consisting of a combination of corner and edge atoms, edge and face atoms, or pure face atoms. Which ones are in the majority is a critical function of the cluster size and of the type of polyhedron.

The models proposed by Van Hardefeld and Hartog have serious limitations: firstly ideal polyhedra with complete shells hardly exist in real catalysts, and secondly the two incomplete structures proposed as models have not been observed experimentally [312]. More recently Perez *et al.* [381] concluded from electron micrographs of Pt on graphite catalysts that cuboctahedral crystallites with incomplete shells are mainly present in the catalyst and some of those structures were used for model calculations of surface sites. A total of 18 possible sites have been considered and their numbers calculated as a function of particle size. All sites with a coordination higher than B_2 show pronounced oscillation of population versus particle size; thus for a given size a particular type of site will predominate. Similar calculations for other structures (*e.g.* icosahedral clusters) have been promised by the same authors.

The question how far such rigid hard sphere models can be correlated with reaction parameters is difficult to assess, but one has to keep in mind that high nuclearity molecular clusters exhibited a high atom mobility when the surrounding ligands were changed, and evidence has been forthcoming that the atoms in small metal particles change their configuration in a similar manner when exposed to reactive gases [382].

F. Electronic Properties

The electronic factor in chemisorption and catalysis has been a subject of interest for many years [383]. Chemisorption and reactivity were attributed to *d*-band holes, which according to the rigid band theory were thought to be filled by alloying a Group VIII metal with a group Ib metal, an idea which sparked off an enormous interest in bimetallics as catalysts. In the past decade the questions of the electronic structure of small metal particles and its dependence on particle size, electronic properties of surface atoms in various co-ordinations, and the effect of alloying on the electronic structures of the individual components have been an area of active research, and attempts were made to correlate such properties with particle size effects in chemisorption and catalysis [399]. Electronic structures (density of states — DOS — curves) can be calculated [384, 385] adopting approaches such as the semi-empirical extended-Hückel (EH) method, the complete neglect-of-differential-overlap (CNDO) method, or the self-consistent-field-Xα scattered wave (SCF-Xα-SW) method. Experimentally density of states curves can be reliably measured with UPS and XPS, and X-ray absorption spectroscopy (XAS) supplies data about the amount of *d*-character in vacancies near the Fermi level [386].

A early model by Kubo [387] based on a simple Sommerfeld-free-particle-in-a-box picture predicted that the spacing of energy levels is inversely proportional to the cluster size ($\delta = 4E_F/3N$, where E_F is the Fermi energy measured from the bottom of the conduction band and N is the atom number in the cluster). In an infinite crystal the energy levels are quasi continuous and the gap between the highest occupied and the lowest unoccupied energy level (δ) is negligible compared to kT, but this is no longer so in a finite cluster, and applying Kubo's model a deviation from "bulk-like" electronic properties should occur when clusters become <500 atoms in size. Over the past decade density of states curves have been calculated for a great number of transition metal and noble metal clusters using modern molecular orbital methods, but considerable controversy still surrounds the question of the cluster size above which "bulk-like" behaviour is observed. Messmer [388] concluded from an SCF-Xα-SW calculation that 13-atom Cu, Ni, Pd and Pt clusters exhibited all the main features of the bulk band structure, *e.g.* overlap of the *d*-band by the *s,p*-band, a sharp peak in density of states around the Fermi level and a valence band width comparable to the bulk value. However, the ionization potential converged much more slowly and was found to be still approximately 2 eV greater than the corresponding average bulk work function. Similarly, the calculated DOS for a 25-atom aluminium cluster seemed to agree with experimental results on clean aluminium [389]. Other calculations exhibited much slower convergences to bulk-like behaviour. Semi-empirical extended Hückel calculations resulted in a *d*-band width for a 39-atom Ag cluster of only 80% of the bulk value [390] and ionization potential, binding energy and energy gap values differed markedly from bulk values for a 55-atom Ag cluster [391]. The electronic properties of Au, Ru, Rh and Pd clusters up to 79 atoms in size were examined by Baetzold [392] with

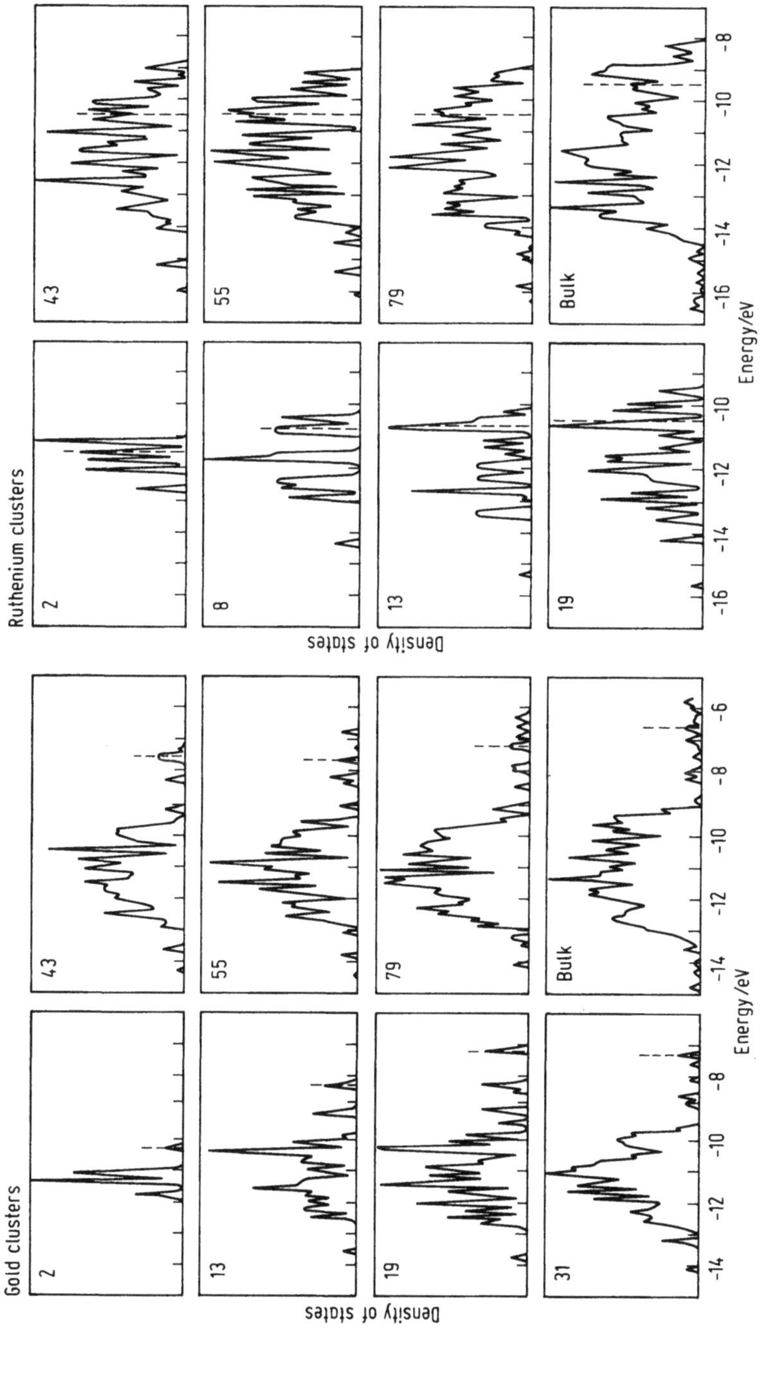

Figure 23. Density of states plots for fcc spherical gold and ruthenium clusters. The dotted line denotes the HOMO for each sized cluster. (Reproduced with permission from ref. [392])

the extended Hückel theory. Density of state plots shown in Figures 23(a) and 23(b) indicate a rather slow convergence toward the bulk DOS. Certain gaps in the d-portion of the DOS only begin to fill up in clusters containing more than 31 atoms and some portion of energy space remain unoccupied even at 79 atoms. The fraction of bulk d-band width calculated for Ru, Pd, Ag and Au clusters converges toward the bulk value, but even in the largest clusters the band width only constitutes 86% of the bulk value. The dependence of the d-band width on the cluster size was found to be the same for all four metals (the band width is proportional to the square root of the average coordination number), but if the gap between the highest occupied molecular orbital (HOMO) and the lowest unoccupied molecular orbital (LUMO) is used as a measure for metallic properties the convergence strongly depended on the type of metal. Pd_{79} and Rh_{79} exhibited bulk-like conduction properties, but not Ag_{79}, Au_{79} and Rh_{79}. Melius [393] calculated the electronic properties of Ni_{13} to Ni_{87} clusters and claimed that from the band width a 87-atom nickel cluster is essentially bulk-like; the ionization potential converged in a similar manner, but electron affinity values differed significantly from the bulk value for a 87-atom cluster. Local densities of states (LDS) calculations by Cyrot-Lackmann [394] on icosahedral and cub-octahedral Ni and Pt clusters containing 13 to 2089 atoms showed that d-band features approached those of the bulk only for large cluster sizes (several hundred of atoms).

Experimental band width measurements [395] on metal films deposited on carbon substrates are shown in Figure 24. In all cases the bulk band width is not achieved until coverage of $>2\times10^{15}$ atoms cm^{-2} are reached. The average particle size determined for this coverage by TEM was approximately 1.5 nm, and such a cluster would contain several hundred atoms. From UV-photoemission spectroscopy, used to measure DOS of Pd-clusters on carbon, an estimate of 140 atoms was given for bulk convergence [396] and Pd deposited on silica films showed metallic properties — a valence band width of ~ 5 eV with emission intensity at the system E_F — at particle sizes

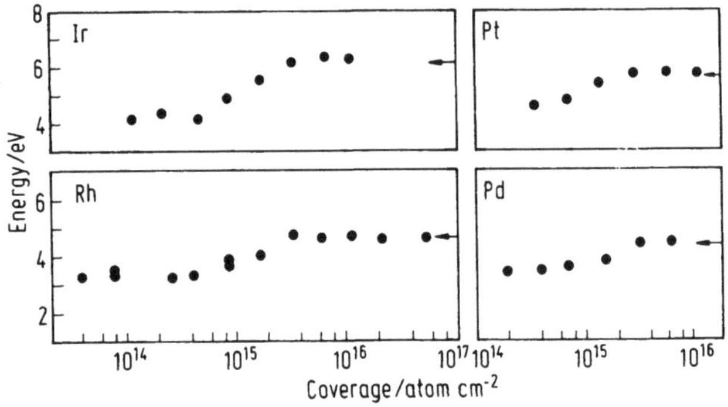

Figure 24. Measured valence band width (full width at half maximum) as function of deposited metal. (Reproduced with permission from ref. [395])

2–3 nm [397] and ESCA spectra looked bulk-like for metal clusters 100 to 200 atoms in size [398].

The question about the number of atoms required for a metallic cluster to exhibit bulk electronic properties seems to be too general, since the various criteria used to characterize bulk-like electronic structure converge at different rates with particle size, and furthermore, seem to depend on the theoretical approaches used to calculate DOS. Thus a larger number of reliable experimental data would be required against which theoretical predictions could be tested.

The *rigid band theory* of alloys so popular in the fifties to explain catalytic and chemisorptive properties of alloy systems, was discredited when measurements of density of states curves were made possible by the development of photoelectron spectroscopy in the late sixties. UPS and optical reflectance spectra of Cu—Ni alloys [400] agreed much better with calculated density of states curves [401] using the *coherent potential approximation* (CPA) as developed by Soven [402] which postulates that the elemental components preserve a good deal of their individuality in alloys, so that the atoms of both components are in principle distinguishable. The value of the coherent potential approximation was confirmed in further theoretical and experimental work [403, 404] and in one review on the theory of electronic structure of substitutional binary alloys, Ehrenreich and Schwartz [405] showed that results obtained by the Coherent Potential Approximation and the Average T-matrix Approximation [406] fitted experimental values so extremely well that detailed features of the Ni density of states could be reproduced.

Yu *et al.* [407] noted that UPS in the energy range $7 < h\nu < 21.2$ eV probes both bulk and surface electronic structure. By tuning through the escape depth curve, surface and bulk electronic structures can be studied essentially independently. The surface UPS spectra for Cu—Ni alloys of different composition and surface conditions consisted of two Ni and Cu *d*-peaks about 2 eV apart. The magnitudes of the peaks correlated to the surface composition, but neither the shape nor the energy position of these peaks was found to be sensitive to changes in bulk composition, surface crystallinity and local environment, which was attributed to the strong localization of the surface electronic structure of Cu—Ni alloys. Later work by Ling *et al.* [408] confirmed those findings.

The idea that surface atoms in an alloy surface retain much of their atomic character, led to the view that chemisorption is mainly influenced by local site effects — ensemble effects — and that ligand effects are of secondary importance [155]. Burch [409] in a review article on the importance of electronic ligand effects in metal alloy catalysis challenged the validity of neglecting electronic changes in interpreting the chemisorption and reaction behaviour on alloy surfaces and points out that although the electronic properties of a metal atom in an alloy are mainly affected by the local potential, some changes due to alloying are occurring at a local level, and the magnitude of those electronic changes is strongly dependent on the type of metals constituting an alloy.

6. Metal-Support Interactions

The unambiguous identification of a metal's modification by a support-metal interaction may prove rather complex, because a wide range of other phenomena may give rise to a "support effect", such as various particle size effects, poisoning of the metal by the presence of support impurities [410], diffusion of the metal into the support or encapsulation of metal particles by the support [411], and spillover and reverse spillover effects [412]. In this context a metal-support interaction effect is defined as *a direct influence* of the support on the chemisorption and catalytic properties of the metal phase either by stabilising *unusual metal particle structures*, by changing the *electronic properties due to electron transfer processes between the metal particles and the support*, or *chemical bonding* — compound formation — between metal and support. Unless reduced at extremely high temperatures metals interact only weakly with non-reducible carriers such as alumina, silica, magnesia, zirconia. Values of 8 to 30 kJ mol^{-1} are commonly measured for the work of adhesion between a metal film and a refractory oxide [413] which implies Van der Waals bonding only. Thus the support is expected to exert little influence on the catalytic and adsorptive properties of metal particles, especially large ones. But the metal-support interactions, although weak, may force very small particles to adopt structures different to those observed with metal particles formed in "free space" (*c.f.* section 5.C.3).

Metal particles inside zeolite channels [414] or supported on reducible oxides like titania and reduced at high temperatures [306(a)] exhibit strongly modified properties caused by support effects. Over the past five years those aspects have attracted considerable interest. Bond [415] has pointed out the diverse nature of support effects, and suggested a classification of metal-support interactions according to their strength into

(i) weak metal-support interactions (WMSI) as discussed above,
(ii) medium metal-support interactions (MMSI) as found with metal particles in zeolites,
(iii) strong metal-support interactions (SMSI) for metals on certain reducible oxides after high temperature reduction treatment.

A. The MMSI State

In the early sixties Rabo *et al.* [416] reported that hydrocarbon reactions on Pt/CaY catalysts were unaffected by the introduction of 10 ppm of thiophen — in general a strong poison for platinum — to the feed. Chukin *et al.* [417] found a similar resistance to sulfur poisoning for Pd/$NaNH_4$Y catalysts and Gallezot *et al.* [418] observed that sulfur poisoned Pt/Y zeolite catalysts are more readily regenerated than other supported platinum catalysts, while ammonia acted as stronger poison for zeolite supported platinum. Enhanced activities were reported with platinum and palladium supported on Y zeolites for hydrogenolysis reactions [174, 380, 419, 420] and to a lesser extent for some hydrogenation reactions [174, 420, 421]. Foger and Anderson [380] demonstrated the effects of cation charge on the catalytic properties of platinum clusters inside Y-zeolite supercages. To exclude other influences (particle size effects) the cations exchange was carried out after the platinum clusters were formed. Pt/NaY and Pt/SiO_2 were found to be of similar activity for the hydrogenolysis of neopentane, if however, the Na^+ ions were exchanged for the divalent ions Ca^{2+} and Mg^{2+} a three to four fold increase in activity was measured, and a further activity increase was achieved by exchanging the triply charged ion La^{3+}. Similarly the hydrogenolysis rate of *n*-butane was strongly affected by the charge of the zeolite cation [419].

The structure of platinum particles in zeolites determined by the RED method and from radial distribution functions derived from EXAFS measurements [218, 382, 422] proved to be very similar to that of platinum clusters comparable in size on silica and alumina [219], but electronically they differed markedly. Positive shifts up to 1.2 eV were measured for XPS core level peaks (*3d*, *4f*) for palladium and platinum microcluster encaged in Y-zeolite [380, 423, 424]. The (L_{III}) X-ray absorption edge of platinum in CaY, CeY and HY is shifted to higher energy values by 0.3, 0.5 and 0.6 eV and the L_{III} peak area was increased by factors 1.4, 1.5 and 1.6 compared to bulk platinum [213] and the IR-stretching frequency of NO adsorbed on Pt/Y-zeolite was found to be shifted to higher values than when NO was adsorbed on Pt/Al_2O_3 [418]. Those findings imply that metal particles inside zeolite channels are electrophilic (electron deficient) due to electron transfer from the metal to acceptor sites on the support surface. The electrophilic character increases with increasing acidity of the zeolite and with the presence of multivalent cations. Electrophilic platinum is expected to exhibit a lower affinity to electron acceptor molecules like sulfur and oxygen [422], and form stronger bonds to donor molecules like ammonia [418]. The enhanced activity has been interpreted either by assuming a higher affinity between the metal and hydrocarbon molecules [419] or a more qualitative interpretation argues that the catalytic behaviour of electrophilic platinum and palladium should compare closer to the behaviour of iridium and rhodium, two metals of significant higher hydrogenolysis activity.

B. The SMSI State

The term, initially introduced by Tauster *et al.* [306] to interpret the diminished H_2 and CO chemisorption values on titania-supported platinum groups metals reduced above 700 K, is now generally used to explain the modification of the chemisorption and reaction properties of supported metals caused by exposure to hydrogen at high temperatures, which can be partially or fully reversed by oxidation treatment. Supports which are claimed to form strong interactions with metals include *reducible oxides* such as TiO_2, Nb_2O_5,

Table 12. Selected chemisorption and reaction studies on catalysts modified by SMSI

A. Metals on Reducible Oxides

Catalyst Metal	Support	Adsorbants reactants	Effects of SMSI	References
Ru, Rh, Pd, Os, Ir, Pt	TiO_2, V_2O_3, Nb_2O_5, Ta_2O_5	H_2, CO	Reduced uptake	306(a), 425
Pt	TiO_2, Ti_2O_3, TiO	H_2	Zero uptake	426
Pt, Ir, Rh	TiO_2	H_2, O_2	H_2 uptake reduced. O_2 chemisorption normal	427–429
Rh, Pt, Ni	TiO_2	H_2, $O_2—H_2$, $H_2—O_2$	H_2 uptake and $H_2—O_2$ titration values low $O_2—H_2$ titration values high	430
Ph, Pd, Ru, Ni, Pt, Ir, Fe, CO	TiO_2	H_2, CO	Reduced uptake	431
Ni, Rh	TiO_2	N_2	Increased uptake	432, 433
Fe, Co, Ni, Ru, Rh, Pd, Os, Ir, Pt	TiO_2	ethane	SMSI effect decreases in the order $Fe \sim Pt < Pd < Co < Ir < Ni < Rh < Os < Ru$	434
Pt, Ir, Rh	TiO_2	*n*-butane benzene cyclohexane ethene, styrene	Strongly suppressed activities	429
Ir	TiO_2	*n*-butane neo-pentane	Hydrogenolysis rates lowered by several magnitudes	435
Pt	TiO_2	*n*-hexane, benzene	Suppressed activities	436
Rh	TiO_2	ethane *n*-butane cyclohexane	Reactivity decrease	437, 439
Pt	TiO_2	2-methylpentane 3-methylpentane	Decreased rates, bond shift mechanism dominates over cyclic mechanism	438
Ni	TiO_2	CO-hydrogenation	Increased activity selectivity shifted towards higher molecular weight products	440–442
Pd	TiO_2	CO-hydrogenation	Increased activity	443
Fe, Co, Ni, Ru, Rh, Pd, Ir, Pt	TiO_2	CO-hydrogenation	Increased activity for Ni, Pd, Pt, Rh, Ir, little effect on Ru and Co	431

Table 12. (continued)

Catalyst Metal	Support	Adsorbants reactants	Effects of SMSI	References
Ni	TiO_2 (100)	CO-hydrogenation	Increased activity	444, 445
Ru	TiO_2	CO-hydrogenation	High olefin production	446
Rh	TiO_2	CO-hydrogenation to alcohol	No effect	447
Rh, Pt	TiO_2	NO—CO	Increased activity	448, 449
B. Metals on Non-Ruducible Oxides				
Pt	Al_2O_3	H_2, O_2, O_2—H_2, H_2—O_2	Decreased H_2 chemisorption unchanged O_2 chemisorption	450
Pt	SiO_2	H_2	Strongly decreased	451
Pt	Al_2O_3	H_2	Reduced chemisorption platinum occlusion and sulfur poisoning excluded	452
Ni	SiO_2	benzene/H_2	Low activity	453
Pt	Al_2O_3	*n*-hexane *n*-pentane	Low activity	454, 450
Pt	SiO_2	benzene/H_2 ethane	Low activity	451
Ni	SiO_2	benzene/H_2 ethane	Low activity	455
		CO/H_2	Not affected	

Ta_2O_5, V_2O_3 and *non-reducible oxides* such as Al_2O_3, SiO_2 and MgO. A selection of systems for which support effects have been reported after high temperature reductions and the type of modifications observed have been collected in Table 12, which shows the following general features of metals in the SMSI state:

(i) greatly diminished H_2 and CO adsorption,
(ii) unchanged O_2 chemisorption,
(iii) enhanced N_2 chemisorption as demonstrated for Ni/TiO_2, Rh/TiO_2,
(iv) decreased activities for hydrocarbon hydrogenolysis, isomerisation and hydrogenation reactions,
(v) enhanced activities for reactions involving CO (Fischer-Tropsch synthesis, methanation, NO—CO reaction).

The extent to which chemisorption and reaction properties are modified depended critically on the *severity of the reduction treatment*, the *type of support* — the temperatures necessary to induce SMSI increases in the order TiO $\sim$ Ti_2O_3 < TiO_2 < Al_2O_3 < SiO_2 —; *the metal* involved — in ethane hydrogenolysis the SMSI effect increased in the order Ru < Rh < Os < Ir < Ni < Pd < Pt, Fe and for CO hydrogenation Ru and Co appeared unaffected, while Ni, Pd, Rh, Ir and Pt show a SMSI effect —; and the *metal dispersion* — the SMSI effect is more pronounced for highly dispersed catalysts [435, 437, 439], but the stability of the SMSI state towards oxygen treatment decreases with increasing dispersion. The SMSI state in highly dispersed Ir/TiO_2 catalysts ($D > 0.8$) was completely destroyed by oxygen treatment at 473 K, while for catalysts of low dispersion more severe oxidation treatments were required [435]. Mechanistic concepts which have been proposed to account for SMSI effects include *strongly adsorbed hydrogen* [454], *intermetallic compound formation* [450, 456] and *electronic changes* due to electron interchange processes between metal particles and the support [457] possibly accompanied by *structural changes* [306(b), 458], Cairns *et al.* [456] who studied 200 nm thick films of platinum and rhodium deposited on Al_2O_3 and TiO_2 by nuclear back scattering spectrometry [459] after treatment in hydrogen between 1200 K and 1300 K and after subsequent oxidation at similar temperatures, interpreted their results with intermetallic compound formation (*e.g.* Pt_xAl_y). Exposure to oxygen resulted in the destruction of the alloy and the formation of either small oxide globules (partial reactivation) or complete coating of the metal with the support (irreversible change). Hydrogen-oxygen consumption values on Pt/Al_2O_3 were explained by a similar model [452], and magnetic measurements on Ni/SiO_2 reduced above 800 K indicated the formation of a Ni—Si alloy [453]. For metals supported on TiO_2 or related compounds, intermetallic compound formation seems unlikely, since the SMSI state can be induced in those systems at temperatures far below those expected for the formation of compounds like Pt_3Ti, Ir_3Ti, Ni_3Ti. Furthermore, metallic titanium has never been observed [248, 466], and recent results on Ni—Ti alloys containing up to 25% Ti [463] demonstrated their ability to chemisorb hydrogen and carbon monoxide. The oxides for which Tauster and Fung [425] observed SMSI effects (TiO_2, Nb_2O_5, Ta_2O_5, V_2O_3) have in common that defect states (Ti^{3+}, V^{2+}, Ta^{2-4+}, Nb^{2-4+}) are created upon H_2 reduction at elevated temperatures [248, 460] and this partial reduction is catalysed by the presence of Groups VIII metals due to spillover phenomena [461]. The SMSI effect was attributed to a charge transfer from the partially reduced carrier to the metal resulting in "electron rich" metal particles, which have been identified from negative shifts of core level metal peaks in XPS spectra of SMSI catalysts [248, 462–466]. Horsley [457] suggested from a molecular orbital calculation that the electron transfer occurs between individual surface Ti^{3+} centres and the metal particles, a mechanism recently critized for the following reasons:

(i) ESR experiments [467, 468] proved that surface Ti^{3+} centres are present in significant concentration in non-SMSI catalysts;

(ii) support effects were also absent in Rh/Y zeolite catalysts containing Ti^{3+} as charge compensating cation [429];

(iii) SMSI properties were observed with platinum on titanium oxides, which contained Ti^{3+} centres in the bulk but the topmost layers were fully oxidised [426];

(iv) recent experiments [469] with Ir on supports consisting of alumina particles surrounded by titania layers, showed SMSI properties only when several layers of titania envelop the alumina.

Therefore several groups [426–429] favour electron transfer from the bulk of the support to the metal as the key for SMSI behaviour. Chen and White [426] who correlated the SMSI behaviour with the electronic properties of supports (*c.f.* Table 13) proposed that materials which show an SMSI

Table 13. Correlation of SMSI behaviour with electronic properties of supports[a]

Support material	Conduction type	Electrical conductivity Ω^{-1} cm^{-1}	SMSI
TiO	metal	10^{-1}	Yes
Ti_2O_3	n	10^{-3}	Yes
TiO_2 (unred.)	n	10^{-11}	No
Nb_2O_5	n	10^{-1} (473 K)	Yes
V_2O_3	metal	10^{+3}	Yes
HfO_2	p	10^{-5} (673 K)	No
ZrO_2	p	10^{-5} (673 K)	No
Sc_2O_3	p	10^{-7} (723 K)	No
MgO	n	10^{-12}	No
SiO_2	insulator	10^{-12}	No
Al_2O_3	insulator	10^{-12}	No
SiC	metal	10^{-3}	Yes

[a] Ref. [426]

effect are characterized by high electrical conductivity and a work function lower than the supported metal. When metal particles are brought in contact with such oxides the resulting flow of electrons toward the metal causes a negative charge on the metal particles, which reduces their ability to adsorb hydrogen or carbon monoxide. Bulk conductivity models for electron transfer between a carrier and a metal are certainly not new ideas and go back many years, when Schwab [470] and Solymosi [471] tried to manipulate catalytic behaviour of metals by changing the carriers semi-conducting properties, but the value of this early work is questionable because of the insufficient characterization of the systems, due to the unavailability of sensitive analysis methods.

A strong interaction between a metal particle and its substrate is expected to affect its morphology, as discussed in detail in Section 5.C.3, and such changes have been observed by Baker *et al.* [306(b), 458], who found that platinum particles in the SMSI state are present as flat pillbox structures whereas before high temperature reduction and after oxidation they assume hemispherical shapes.

In conclusion we suggest that SMSI effects on *reducible oxides* (n-type semi-conductors) and on *non-reducible oxides* (non-conducting oxides) are of different origin: in the former case electron transfer from the support to the metal leads to negative charged metal particles (SMSI state) while in the latter case support and metal atoms form intermetallic compounds.

Abbreviations

TPD — Temperature programmed desorption
TPR — Temperature programmed reduction
XRD — X-ray diffraction
SAXS — Small angle X-ray scattering
EXAFS — Extended X-ray absorption fine structure spectroscopy
TEM — Transmission electron microscopy
SEM — Scanning electron microscopy
STEM — Scanning transmission electron microscopy
CAEM — Controlled atmosphere electron microscopy
XPS — X-ray photoelectron spectroscopy
UPS — UV photoelectron spectroscopy
AES — Auger electron spectroscopy
IR — Infrared spectroscopy

References

1. Anderson, J. R.: Structure of Metallic Catalysts. London, New York: Academic Press 1975
2. Sinfelt, J. H., Cusumano, J. A.: Advanced Materials in Catalysis, Burton, J. J., Garten, R. L. (eds.). London, New York: Academic Press 1977, p. 1
3. Burton, J. J., Garten, R. L.: ibid., p. 33
4. Dowden, D. A.: In: Catalysis (Specialist Periodical Report) The Chemical Society, London 1978, Vol. 2, p. 1
5. Acres, G. J. K., Bird, A. J., Jenkins, J. W., King, F.: In: Catalysis (Specialist Periodical Report), The Chemical Society, London 1981, Vol. 4, p. 1
6. Moss, R. L.: ibid., p. 31
7. Iler, R. K.: Chemistry of Silica. New York: John Wiley and Sons 1979
8. Okkerse, C.: In: Physical and Chemical Aspects of Adsorbents and Catalysts. London, New York: Academic Press 1970, p. 213
9. Lippens, B. C., Steggerda, J. J.: ibid., p. 171
10. Wefers, K., Bell, G. M.: Alcoa Technical Paper No. 19, 1972
11. Kotanigawa, T., Yamamoto, M., Utiyama, M., Hattori, H., Tanabe, K.: Appl. Catal. **1**, 185 (1981)
12. Kirschner, H.: Ber. dt. Ceram. Gesellschaft **43**, 479 (1966)
13. Leonard, A. J., Semaille, P. N., Fripiat, J. J.: Proc. Brit. Ceram. Soc. **1969**, 103
14. Russel, A. S., Cochran, C. N.: Ind. Eng. Chem. **42**, 1336 (1950)
15. Ryland, L. B., Tamele, M. W., Wilson, J. N.: In: Catalysis. Emmett, P. H. (ed.). New York: Reinhold 1960, Vol. 7, p. 1
16. Leonard, A. J., Ratnasamy, P., Declerck, F. D., Fripiat, J. J.: Disc. Farad. Soc. **52**, 98 (1971)
17. Andreu, P., Martin, G., Noller, H.: J. Catal. **21**, 255 (1971)
18. Hamamoto, Y., Mitsutani, A.: Kogyo Kagaku Zasshi **67**, 1227 (1964)
19. Thomas, C. L., Hickey, J., Stecher, G.: Ing. Eng. Chem. **42**, 866 (1950)
20. Swift, H. E.: In: Advanced Materials in Catalysis. Burton, J. J., Garten, R. L. (eds.). London, New York: Academic Press 1977, p. 209

21. Barrer, R. M.: Brit. Chem. Eng. May 1959, p. 1
22. Breck, D. W.: Zeolite Molecular Sieves. New York: J. Wiley 1974
23. Faith, W. L., Keyes, D. B., Clark, R. L.: Industrial Chemicals. New York: Wiley 1965, 3rd edition
24. Barksdale, J. (ed.): Titanium, Its Occurrence, Chemistry and Technology. New York: Ronald Press 1951
25. Jones, P., Hockey, J. A.: Trans. Farad. Soc. **67**, 2669 (1971)
26. Krebs, H.: Grundzüge der Anorganischen Kristallchemie. Stuttgart: Enke 1968
27. Clark, R. J. H.: The Chemistry of Titanium and Vanadium. Amsterdam, London, New York: Elsevier 1968
28. McQuillan, A. D., McQuillan, M. K.: Titanium. London: Butterworths 1956
29. Cornet, D., Burwell, R. L., Jr.: J. Am. Chem. Soc. **90**, 2489 (1968)
30. Burwell, R. L., Jr., Haller, G. L., Taylor, K. C., Read, J. F.: Adv. Catal. **20**, 1 (1969)
31. Burwell, R. L., Jr., Taylor, K. C., Haller, G. L.: J. Phys. Chem. **71**, 4580 (1967)
32. Carruthers, J. D., Sing, K. S. W.: Chem. Ind. (London) **1967**, 1919
33. Kim, G., Krieger, K. A.: In: Actes du Deuxieme Congres de Catalyse. Paris: Editions Technip 1961, p. 687
34. Zecchina, A., Coluccia, S., Gungliminotti, E., Ghiotto, G.: J. Phys. Chem. **75**, 2774 1971
35. Poole, C. P., MacIver, D. S.: Adv. Catal. **17**, 223 (1967)
36. Anderson, P. J., Horlock, R. F.: Trans. Farad. Soc. **58**, 1993 (1962)
37. Horlock, R. F., Morgan, P. L., Anderson, P. J.: Trans. Farad. Soc. **59**, 721 (1963)
38. Anderson, P. J., Morgan, P. L.: Trans. Farad. Soc. **60**, 930 (1964)
39. De Vleesschauwer, W. F. N. M.: In: Physical and Chemical Aspects of Adsorbents and Catalysts. Linsen, B. G. (ed.). London, New York: Academic Press 1970, p. 265
40. Walker, P. L. (ed.): Chemistry and Physics of Carbon. New York: Marcel Dekker 1965
41. Van der Plas, Th.: In: Physical and Chemical Aspects of Adsorbents and Catalysts. Linsen, B. G. (ed.). London, New York: Academic Press 1970, p. 425
42. Medalia, A. I., Rivin, D.: In: Characterisation of Powder Surfaces. Parfitt, G. D., Sing, K. S. W. (eds.). London, New York: Academic Press 1976, p. 279
43. Trimm, D. L.: In: Catalysis (Specialist Periodical Report). The Chemical Society, London 1981, vol. 4, p. 210
44. Trimm, D. L., Cooper, B. J.: J. Chem. Soc. 1970, 477
45. Hucke, E. E.: U.S. Pat. 3,859,421 (1975)
46. Moreno-Castilla, C., Mahajan, O. P., Walker, P. L., Jung, H. J., Vannice, M. A.: Carbon **18**, 271 (1980)
47. DeLuca, J. P., Campbell, L. E.: In: Advanced Materials in Catalysis. Burton, J. J., Garten, R. L. (eds.). London, New York: Academic Press 1977, p. 293
48. Parks, G. A., DeBruyn, P. L.: J. Phys. Chem. **66**, 967 (1962)
49. Parks, G. A.: Chem. Rev. **65**, 177 (1965)
50. Jiratova, K.: Appl. Catal. **1**, 165 (1981)
51. Tanabe, K.: In: Metal Oxides and Complex Oxides. Tanabe, K., Seiyama, T., Huecki, K. (eds.). Tokyo: Kodansha Scientific 1978, p. 407
52. Tanabe, K.: Solid Acids and Bases. Tokyo: Kodansha and New York, London: Academic Press 1970
53. Tanabe, K.: In: Catalysis. Science and Technology. Anderson, J. R., Boudart, M. (eds.). Berlin, Heidelberg, New York: Springer 1981, vol. 2, p. 231
54. Happel, J., Hnatow, M., Bajars, L.: In: Base Metal Oxide Catalysts. New York, Basel: Marcel Dekker 1977
55. Benesi, H. A., Winquist, B. H. C.: Adv. Catal. **27**, 97 (1977)
56. Knözinger, H.: Adv. Catal. **25**, 184 (1976)
57. Rosynek, M. P.: Catal. Rev. — Sci. Eng. **16**, 111 (1977)
58. Stone, F. S.: Adv. Catal. **13**, 1 (1962)
59. Maxwell, I. E.: Adv. Catal. **31**, 2 (1982)
60. Che, M., Tench, A. J.: Adv. Catal. **31**, 77 (1982)
61. Boehm, H. P., Knözinger, H.: In: Catalysis. Science and Technology. Anderson, J. R., Boudart, M. (eds.). Berlin, Heidelberg, New York: Springer 1983, Vol. 4, p. 40

62. Krylov, O. V.: Catalysis by Non Metals. New York, London: Academic Press 1970
63. Jacobs, P. A.: Carboniogenic Activity of Zeolites. Amsterdam, Oxford, New York: Elsevier 1977
64. First International Symposium on the Preparation of Catalysts. Brussels 1975. Proceedings. Delmon, B., Jacobs, P. A., Poncelet, G. (eds.). Amsterdam: Elsevier 1976
65. Second International Symposium on the Preparation of Catalysts. Louvain-la-Neuve 1978. Proceedings. Delmon, B., Grange, P., Jacobs, P. A., Poncelet, G. (eds.). Amsterdam: Elsevier 1979
66. Third International Symposium on the Preparation of Catalysts, Louvain-la-Neuve 1982
67. Baker, R. T. K., Prestridge, E. B., Garten, R. L.: J. Catal. **59**, 293 (1979)
68. Ruckenstein, E., Chu, Y. F.: J. Catal. **59**, 109, 254 (1979)
69. Berebi, G., Bernusset, Ph.: In: Preparation of Catalysts. Delmon, B., Jacobs, P. A., Poncelet, G. (eds.). Amsterdam: Elsevier 1976, p. 13
70. Brunelle, J. P.: Pure Applied Chem. **59**, 1211 (1978)
71. Kennedy, D. C.: Chem. Eng. **87**, 106 (1980)
72. D'Aniello, M. J.: J. Catal. **69**, 9 (1981)
73. Santecesaria, E., Carra, S., Adami, I.: Ind. Eng. Chem. Prod. Res. Dev. **16**, 41 (1977)
74. Parkash, S., Chakrabartty, S. K., Koanigawa, T.: Fuel Process. Technol. **6**, 177 (1982)
75. Summers, J. C., Ausen, S. A.: J. Catal. **52**, 455 (1978)
76. Wei, J., Becker, E. R.: Advan. Chem. Ser. **143**, 116 (1975)
77. Becker, E. R., Wei, J.: J. Catal. **46**, 365, 372 (1977)
78. Summers, J. C., Hegedus, L. L.: J. Catal. **51**, 185 (1978)
79. Kunimori, K., Nakajama, I., Uchijima, T.: Chem. Lett. (Japan), 1982, 1165
80. Maatman, R. W., Prater, C. D.: Ind. Eng. Chem. **49**, 253 (1957)
81. Komiyama, M., Merrill, R. P., Harnsberger, H. F.: J. Catal. **63**, 35 (1980)
82. Neimark, A. V., Kheifets, L. I., Fenelov, V. B.: Ind. Eng. Chem. Proc. Res. Dev. **20**, 439 (1981)
83. Weisz, P. B.: J. Chem. Soc. Farad. I **63**, 1801 (1967)
84. Shyr, Y. S., Ernst, W. R.: J. Catal. **63**, 425 (1980)
85. Kulkarni, S. S., Mauze, G. R., Schwarz, J. A.: J. Cata. **69**, 445 (1981)
86. Becker, E. R., Nuttall, T. A.: In: Preparation of Catalysts II. Delmon, B., Grange, P., Jacobs, P. A., Poncelet, G. (eds.). Amsterdam: Elsevier 1979, p. 159
87. Castro, A. A., Scelza, O. A., Benvenuto, E. R., Baronetti, G. T., De Miguel, S. R., Parera, J. M.: In: Proc. 3rd Int. Symp. on the Prep. of Catalysts. Louvain-la-Neuve 1982. Paper A3
88. Lee, S. Y., Aris, R.: ibid., paper A2.1
89. Hegedus, L. L., Chou, T. S., Summers, J. C., Potter, N. M.: In: Preparation of Catalysts II. Delmon, B., Grange, P., Jacobs, P. A., Poncelet, G. (eds.). Amsterdam: Elsevier 1979, p. 171
90. Fenelov, V. B., Neimark, A. V., Kheifets, L. I., Samakhov, A. A.: ibid., p. 233
91. Kotter, M., Riekert, L.: ibid., p. 51
92. Anderson, J. R., Mainwaring, D. E.: J. Catal. **35**, 162 (1974)
93. Anderson, J. R., Elmes, P. S., Howe, R. F., Mainwaring, D. E.: J. Catal. **50**, 508 (1977)
94. Yermakov, Yu. I., Kuznetsov, B. N., Zakharov, V. A.: Catalysis by Supported Complexes. Amsterdam: Elsevier 1981
95. Ugo, R., Psaro, R., Zanderighi, G. M., Basset, J. M., Theolier, A., Smith, A. K.: In: Fundamental Research in Homogeneous Catalysis 3. Tsutsui, M. (ed.). New York: Plenum Press 1979, p. 579
96. Yermakov, Yu. I., Kuznetsov, B. N.: J. Molec. Catal. **9**, 13 (1980)
97. Martino, G.: In: Growth and Properties of Metal Clusters. Bourdon, J. (ed.). Amsterdam: Elsevier 1980, p. 399
98. Ichikawa, M.: J. Chem. Soc. Chem. Comm. **1976**, 11
99. Ichikawa, M.: Bull. Chem. Soc. Japan **51**, 2268 (1978)
100. Ichikawa, M.: J. Chem. Soc. Chem. Comm. **1976**, 26
101. Chini, P.: J. Organomet. Chem. **200**, 37 (1980)

102. Comprehensive Organometallic Chemistry. Wilkinson, G., Stone, F. G. A., Abel, E. W. (eds.). Oxford, New York: Pergamon Press 1982
103. Anderson, S. L. T., Watters, K. K., Howe, R. F.: J. Catal. **69**, 212 (1981)
104. Smith, A. K., Hughes, F., Theolier, A., Basset, J. M., Ugo, R., Zanderighi, G. M., Bilhou, J. L., Bilhou-Bougnol, V., Graydon, F.: Inorg. Chem. **18**, 3104 (1979)
105. Simpson, A. F., Whyman, R.: J. Organomet. Chem. **213**, 157 (1981)
106. Zahraa, O., Garin, F., Maire, G.: Farad. Discuss. **72**, 45 (1982)
107. Lazutkin, A. M., Latzutkina, A. I., Yermakov, Yu. I.: React. Kinet. Catal. Lett. **8**, 81 (1978)
108. Yermakov, Yu. I.: Catal. Rev. — Sci. Eng. **13**, 77 (1976)
109. Yermakov, Yu. I., Kuznetsov, B. N., Zakharov, V. A.: Catalysis by Supported Complexes. Amsterdam: Elsevier 1981, pp. 337–415
110. Highfield, J. G., Bossi, A., Stone, F. S.: Proc. 3rd Int. Symp. on the Prep. of Catalysts. Louvain-la-Neuve 1982. Paper B5.1
111. Bruce, L., Mathews, J. F.: Appl. Catal. **4**, 353 (1982)
112. Kruissink, E. C., Pelt, H. L., Ross, J. R. H., Van Reijen, L. L.: Appl. Catal. **1**, 23 (1981)
113. Alzamora, L. E., Ross, J. R. H., Kruissink, E. C., Van Reijen, L. L.: J. Chem. Soc. Farad. Trans. I **77**, 665 (1981)
114. Doesburg, E. B. M., Orr, S., Ross, J. R. H., Van Reijen, L. L.: J. Chem. Soc. Chem. Comm. **1977**, 734
115. Klabunde, K. J., Ralston, D. H., Zoellner, R. W., Hattori, H., Tanaka, Y.: J. Catal. **55**, 213 (1978)
116. Klabunde, K. J., Davis, S. C., Hattori, H., Tanaka, Y.: J. Catal. **54**, 254 (1978)
117. Matsuo, K., Klabunde, K. J.: J. Catal. **73**, 216 (1982)
118. Ralston, D. H., Klabunde, K. J.: Appl. Catal. **3**, 13 (1982)
119. Tanaka, K., Watters, K. L., Howe, R. F.: J. Catal. **75**, 23 (1982)
120. Wu, W., Hercules, D. M.: J. Phys. Chem. **83**, 2003 (1979)
121. Yao, H. C., Japan, S., Shelef, M.: J. Catal. **50**, 407 (1977)
122. Kruissink, E. C., Alzamora, L. E., Orr, S., Doesburg, E. B. M., Van Reijen, L. L., Ross, J. R. H., Van Veen, G.: In: Preparation of Catalysts II. Delmon, B., Grange, P., Jacobs, P. A., Poncelet, G. (eds.). Amsterdam: Elsevier 1979, p. 143
123. Dorling, T. A., Lynch, B. W. J., Moss, R. L.: J. Catal. **20**, 190 (1971)
124. Bournonville, J. P., Franck, J. P., Martino, G.: Proc. 3rd Int. Symp. on the Prep. of Catalysts. Louvain-la-Neuve 1982. Paper A6.1
125. Sinfelt, J. H.: J. Catal. **29**, 308 (1973)
126. Sinfelt, J. H.: U.S. Pat. 3,953,368 (1976)
127. Isaacs, B. H., Petersen, E. E.: J. Catal. **77**, 43 (1982)
128. Penchev, V., Neynska, Ya., Kanazirev, V.: In: Preparation of Catalysts II. Delmon, B., Grange, P., Jacobs, P. A., Poncelet, G. (eds.). Amsterdam: Elsevier 1979, p. 497
129. Boldyrev, V. V., Bulens, M., Delmon, B.: The Control of the Reactivity of Solids. Amsterdam: Elsevier 1979
130. Haber, J.: J. Less Common Met. **54**, 243 (1977)
131. Delmon, B.: Introduction a la Cinetique Heterogene. Paris: Edition Technip 1969
132. Coenen, J. W. E.: In: Preparation of Catalysts II. Delmon, B., Grange, P., Jacobs, P. A., Poncelet, G. (eds.). Amsterdam: Elsevier 1979, p. 89
133. Robertson, S. D., McNicol, B. D., De Baas, J. H., Kloet, S. C., Jenkins, J. W.: J. Catal. **37**, 424 (1975)
134. Hurst, N. W., Gentry, S. J., Jones, A., McNicol, B. D.: Catal. Rev. — Sci. Eng. **24**, 233 (1982)
135. Gentry, S. J., Hurst, N. W., Jones, A.: J. Chem. Soc. Trans. Farad. I, **75**, 1688 (1979)
136. Condurier, G., Decamp, T., Praliaud, H.: J. Chem. Soc. Farad. Trans. I, **78**, 2661 (1982)
137. Kulifay, S. M.: J. Amer. Chem. Soc. **83**, 4916 (1961)
138. McKee, D. W., Norton, F. J.: J. Phys. Chem. **68**, 481 (1964)
139. De Jong, K. P., Geus, J. W.: Proc. 3rd Int. Symp. on the Prep. of Catalysts. Louvain-la-Neuve 1982. Paper A9.1
140. LePage, J. F., Miguel, J.: In: Preparation of Catalysts I. Delmon, B., Jacobs, P. A., Poncelet, G. (eds.). Amsterdam: Elsevier 1976, p. 39

141. Gregg, S. J., Sing, K. S. W.: Adsorption, Surface area and Porosity. London, New York: Academic Press 1967, 1982
142. Sing, K. S. W.: In: Characterisation of Catalysts. Thomas, J. M., Lambert, R. M. (eds.). Chichester, New York, Brisbane, Toronto: Wiley & Sons 1980, p. 12
143. Sing, K. S. W.: In: Characterisation of Powder Surfaces. Parfitt, G. D., Sing, K. S. W. (eds.). Academic Press 1976, p. 1
144. Broekhoff, J. C. F.: In: Preparation of Catalysts II. Delmon, B., Grange, P., Jacobs, P. A., Poncelet, G. (eds.). Amsterdam: Elsevier 1979, p. 663
145. Lecloux, A. J.: In: Catalysis, Science and Technology. Anderson, J. R., Boudart, M. (eds.). Berlin, Heidelberg, New York: Springer-Verlag 1981, vol. 2, p. 171
146. Purdy, G. R., Anderson, R. B.: In: Experimental Methods in Catalytic Research. Anderson, R. B., Dawson, P. T. (eds.). New York: Academic Press 1976, vol. 2, p. 92
147. Brunauer, S., Emmett, P. H., Teller, E.: J. Amer. Chem. Soc. **60**, 309 (1938)
148. Broekhoff, J. C. P., De Boer, J. H.: J. Catal. **9**, 8, 15 (1967). **10**, 153, 368, 377, 391 (1968)
149. Brunauer, S., Mikhail, R. S., Bodor, E. E.: J. Coll. Interface Sci. **24**, 451 (1967)
150. Lecloux, A., Pirard, J. P.: J. Coll. Interface Sci. **70**, 265 (1979)
151. Lippens, B. C., De Boer, J. H.: J. Catal. **4**, 319 (1965)
152. Sing, K. S. W.: Chem. Ind. (London), **1968**, 1520
153. Gregg, S. J., Langford, J. F.: Trans. Farad. Soc., **65**, 1394 (1969)
154. Boudart, M.: Adv. Catal. **20**, 153 (1969)
155. Sachtler, W. M. H., van Santen, R. A.: Adv. Catal. **26**, 69 (1977)
156. Myasaki, E.: J. Catal. **65**, 84 (1980)
157. Anderson, J. R.: Structure of Metallic Catalysts. London, New York: Academic Press 1975, p. 328
158. Moss, R. L.: In: Experimental Methods in Catalytic Research. Anderson, R. B., Dawson, P. T. (eds.). New York: Academic Press 1976, vol. 2, p. 43
159. Rasser, J. C.: Platinum-Iridium Reforming Catalysts. Delft: Delft Univ. Press 1977
160. Gomez, R., Fuentes, S., Fernandez de Valle, F. J., Campero, A., Ferreira, J. M.: J. Catal. **38**, 47 (1975)
161. Stefan, J. J., Ponec, V., Sachtler, W. M. H.: Surf. Sci. **47**, 403 (1975)
162. Anderson, J. R., Foger, K., Breakspere, R. J.: J. Catal. **57**, 458 (1979)
163. Sinfelt, J. H., Lam, Y. L., Cusumano, J. A., Barnett, A. E.: J. Catal. **42**, 227 (1976)
164. Foger, K., Anderson, J. R.: J. Catal. **64**, 448 (1980)
165. Corro, G., Maubert, A., Gomez, R.: Nouveau Journal de Chimie **3**, 755 (1979)
166. Aben, P. C., van der Eijk, H., Oelderik, J. M.: In: Proc. 5th Int. Congr. Catal. Miami 1972. Hightower, J. W. (ed.). Amsterdam: North Holland 1973, p. 717
167. Boer, H., Boersma, W. J., Wagstaff, N.: Rev. Sci. Instr. **53**, 349 (1982)
168. Popova, N. M., Babenkova, L. V.: React. Kinet. Catal. Lett. **11**, 187 (1979)
169. Herz, R. K., Kiela, J. B., Marin, S. P.: J. Catal. **73**, 66 (1982)
170. Brenner, A., Hucul, D. A.: J. Catal. **56**, 134 (1979)
171. Gorte, R. J.: J. Catal. **75**, 164 (1982)
172. Scholten, J. J. F.: In: Preparation of Catalysts II. Delmon, B., Grange, P., Jacobs, P. A., Poncelet, G. (eds.). Amsterdam: Elsevier 1979, p. 685
173. Candy, J. P., Fouilloux, P., Renouprez, A. J.: J. Chem. Soc. Farad. Trans. I, **76**, 616 (1980)
174. Dalla Betta, R. A., Boudart, M.: In: Proc. 5th Intern. Congr. Catal. Miami 1972. Hightower, J. W. (ed.). Amsterdam: North-Holland 1973, p. 1329
175. Kummori, K., Uchijima, T., Yamada, M., Matsumoto, H., Hattori, T., Murakami, Y.: Appl. Catal. **4**, 67 (1982)
176. Wanke, S. E., Dougharty, N. A.: J. Catal. **24**, 367 (1972)
177. Yates, J. C., Murrell, L. L., Prestridge, E. B.: In: Growth and Properties of Metal Clusters. Bourdon, J. (ed.). Amsterdam: Elsevier 1980, p. 137
178. Yates, J. C., Murrell, L. L., Prestridge, E. B.: J. Catal. **57**, 41 (1979)
179. Primet, M.: J. Chem. Soc. Farad. Trans. I, **74**, 2570 (1978)
180. Dalla Betta, R. A.: J. Catal. **34**, 57 (1974)
181. Taylor, K. C.: J. Catal. **38**, 299 (1975)
182. Goodwin, J. G.: J. Catal. **68**, 227 (1981)

183. Dalla Betta, R. A.: J. Phys. Chem. **79**, 251 (1975)
184. Brooks, C. S.: J. Coll. Interface Sci. **34**, 419 (1970)
185. Vannice, M. A.: J. Catal. **37**, 449 (1975)
186. McVicker, G. B., Baker, R. T. K., Garten, R. L., Kugler, E. L.: J. Catal. **65**, 207 (1980)
187. Krishnamurthy, S., Landolt, G. R., Schoenagel, H. J.: J. Catal. **78**, 319 (1982)
188. Batholomew, C. H., Parnell, R. B.: J. Catal. **65**, 390 (1980)
189. Parnell, R. B., Chung, K. S., Batholomew, C. H.: J. Catal. **46**, 340 (1977)
190. Topsoe, H., Topsoe, N., Bohlbro, H.: In: Proc. 7th Int. Congr. Catalysis, Tokyo 1980. Tokyo: Kadansha. Amsterdam: Elsevier 1981
191. Jung, H. J., Vannice, M. A., Mulay, L. N., Stanfield, R. M., Delgass, W. N.: J. Catal. **76**, 208 (1982)
192. Boudart, M., Delbouille, A., Dumesic, J. A., Khammouma, S., Topsoe, H.: J. Catal. **37**, 486 (1975)
193. Scholten, J. J. F., Konvalinka, J. A., Beekman, F. W.: J. Catal. **28**, 209 (1973)
194. Fukushima, T., Galvagno, S., Parravano, G.: J. Catal. **57**, 177 (1979)
195. Klug, H. P., Alexander, L. E.: X-ray Diffraction Procedures for Polycrystalline and Amorphous Materials. New York: Wiley 1974
196. Le Bail, A., Louer, D.: J. Appl. Cryst. **11**, 50 (1978)
197. Sashital, S. R., Cohen, J. B., Burwell, R. L., Butt, J. B.: J. Catal. **50**, 479 (1977)
198. Jaeger, H.: private communication
199. Fagherazzi, G., Cocco, G., Schiffini, L., Enzo, S., Benedetti, A., Passerini, R., Tauszik, G. R.: Chim. Ind. (Milan) **60**, 892 (1978)
200. Ganesan, P., Kno, H. K., Saavedra, A., De Angelis, R. J.: J. Catal. **52**, 310 (1978)
201. Gerold, V.: J. Appl. Cryst. **11**, 376 (1978)
202. Kolomiichuk, V. N.: React. Kinet. Catal. Lett. **20**, 123 (1982)
203. Whyte, F. E.: Catal. Rev. **8**, 117 (1973)
204. Goodisman, J., Brumberger, H., Cupelo, R.: J. Appl. Cryst. **14**, 305 (1981)
205. Gallezot, P., Bergeret, G.: In: Stud. Surf. Sci. Catal. 12: Metal Microstructures in Zeolites. Jacobs, P. A., Jaeger, N. I., Jiru, P., Schulz-Ekloff, G. (eds.). Amsterdam: Elsevier 1982, p. 167
206. Moss, R. L., Whalley, L.: Adv. Catal. **22**, 115 (1972)
207. Silfelt, J. H., Via, G. H., Lytle, F. W.: J. Chem. Phys. **76**, 2779 (1982)
208. Greegor, R. B., Lytle, F. W.: J. Catal. **63**, 476 (1980)
209. Lytle, F. W., Via, G. H., Sinfelt, J. H.: Synchrotron Radiat. Res., 1980, 401
210. Joyner, R.: In: Characterisation of Catalysts. Thomas, J. M., Lambert, R. M. (eds.). Chichester, New York, Brisbane, Toronto: John Wiley & Sons 1980, p. 237
211. Pettifer, R. F.: ibid., p. 264
212. Lytle, F. W., Wei, P. S. T., Greegor, R. B., Via, G. H.: J. Chem. Phys. **70**, 4849 (1979)
213. Gallezot, P., Weber, R., Dalla Betta, R. A., Boudart, M.: Z. Naturforsch. A. **34**, 40 (1979)
214. Knopp, G. S., Chen, H., Klippert, T. E.: Rev. Sci. Instrum. **49**, 1658 (1978)
215. Cohen, G. G., Fisher, D. A., Colbert, J., Shevchik, N. J.: Rev. Sci. Instrum. **51**, 3 (1980)
216. Khalid, S., Emrich, R., Dujari, R., Shultz, J., Katzer, J. R.: Rev. Sci. Instrum. **53**, 22 (1982)
217. Joyner, R. W.: J. Chem. Soc., Farad. Trans. I, **76**, 357 (1980)
218. Moraweck, B., Clugnet, G., Renouprez, A. J.: Surf. Sci. **81**, L631 (1979)
219. Lytle, F. W., Via, G. H., Sinfelt, J. H.: J. Chem. Phys. **67**, 3831 (1977)
220. Friedman, R. M., Freeman, J. J., Lytle, F. W.: J. Catal. **55**, 10 (1978)
221. Greegor, R. B., Lytle, F. W., Chin, R. L., Hercules, D. M.: J. Phys. Chem. **85**, 1232 (1981)
222. Sinfelt, J. H., Via, G. H., Lytle, F. W.: J. Chem. Phys. **72**, 4832 (1980)
223. Sinfelt, J. H., Via, G. H., Lytle, F. W., Greegor, R. B.: J. Chem. Phys. **75**, 5527 (1981)
224. Edington, J. W.: Practical Microscopy in Materials Science. (4 volumes). London: Macmillan 1975
225. Fryer, J. R.: The Chemical Applications of Transmission Electron Microscopy. London, New York: Academic Press 1979
226. Goldstein, J. I., Yakowitz, H. (eds.): Practical Scanning Microscopy. New York, London: Plenum Press 1975

227. Howie, A.: In: Characterisation of Catalysts. Thomas, J. M., Lambert, R. M. (eds.). Chichester, New York, Brisbane, Toronto: John Wiley & Sons 1980, p. 89
228. Baird, T.: In: Catalysis (Specialist Periodical Report). The Chemical Society, London 1982, vol. 5, p. 172
229. Sprys, J. W., Bartosiewicz, L., McCune, R., Plummer, H. K.: J. Catal. **39**, 91 (1975)
230. Dalmai-Imelik, G., Leclercq, C., Mutin, I.: J. Microscopie **20**, 123 (1974)
231. Millward, G. R.: J. Catal. **64**, 381 (1980)
232. Flynn, P. C., Wanke, S. E., Turner, P. S.: J. Catal. **33**, 233 (1974)
233. Treacy, M. M. J., Howie, A.: J. Catal. **63**, 265 (1980)
234. Freeman, L. A., Howie, A., Treacy, M. M. J.: J. Microscopy **111**, 165 (1977)
235. Zenith, J., Contreras, J. L., Dominguez, J. M., Yacaman, M. J.: J. Micros. Spectros. Electron **5**, 291 (1980)
237. Treacy, M. M. J., Howie, A., Wilson, C. J.: Phil. Mag. A. **38**, 569 (1978)
238. Baker, R. T. K., Prestridge, E. B., Garten, R. L.: J. Catal. **56**, 390 (1979)
239. Crewe, A. V., Langmore, J. P., Isaacson, M. S.: In: Physical Aspects of Electron Microscopy and Microbeam Analysis. Siegel, B. M., Beaman, D. R. (eds.). New York: Wiley 1975, p. 47
240. Howie, A.: Ultramicroscopy **8**, 163 (1982)
241. Cowley, J. M., Spence, J. C. H.: Ultramicroscopy **6**, 359 (1981)
242. Lynch, J. P., Lesage, E., Dexpert, H., Freund, E.: Inst. Phys. Conf. Ser. No. 61, Chapter 2, p. 67
243. Blanchard, G., Charcosset, H., Dexpert, H., Freund, E., Leclercq, C., Martino, G.: J. Catal. **70**, 168 (1981)
244. Poppa, H., Heinemann, K.: Optik **56**, 183 (1980)
245. Baker, R. T. K.: Catal. Rev. — Sci. Eng. **19**, 161 (1979)
246. Edmonds, T.: In: Characterisation of Catalysts. Thomas, J. M., Lambert, R. M. (eds.). Chichester, New York, Brisbane, Toronto: John Wiley & Sons 1980, p. 30
247. Hercules, D. M., Klein, J. C.: Appl. Electron. Spectrosc. Chem. Anal. **63**, 147 (1982)
248. Sexton, B. A., Hughes, A. E., Foger, K.: J. Catal. **77**, 85 (1982)
249. Chien, S. H., Shelimov, B. N., Resasco, D. E., Lee, E. H., Haller, G. L.: J. Catal. **77**, 301 (1982)
250. Fung, S. C.: J. Catal. **76**, 225 (1982)
251. Topsoe, H., Dumesic, J. A., Morup, S.: In: Applications of Mössbauer Spectroscopy. Cohen, R. L. (ed.). New York: Academic Press 1980
252. Delgass, W. N., Haller, G. L., Kellerman, R., Lunsford, J. H.: Spectroscopy in Heterogeneous Catalysis. New York: Academic Press 1979, p. 132
253. Garten, R. L., Sinfelt, J. H.: J. Catal. **62**, 127 (1980)
254. Bussiere, P.: Revue Phys. Appl. **16**, 477 (1981)
255. Jones, W.: In: Characterisation of Catalysts. Thomas, J. M., Lambert, R. M. (eds.). Chichester, New York, Brisbane, Toronto: John Wiley & Sons 1980, p. 114
256. Garten, R. L.: In: Mössbauer Effect Methodology. Gruvermann, I. J. (ed.). New York, London: Plenum Press 1976, vol. 10, p. 69
257. Raupp, G. F., Delgass, W. N.: J. Catal. **58**, 337, 348, 361 (1979)
258. Becaud, R., Bussiere, P., Figueras, F.: J. Catal. **69**, 399 (1981)
259. Lee, J. B.: J. Catal. **68**, 27 (1981)
260. Eischens, R. P., Pliskin, W. A.: Advan. Catal. **10**, 1 (1958)
261. Howe, R. F.: J. Catal. **50**, 196 (1977)
262. Griffiths, P. R.: Chemical Infrared Fourier Transform Spectroscopy. New York: John Wiley 1975
263. Moon, S. H., Windawi, H., Katzer, J. R.: Ind. Eng. Fundam. **20**, 396 (1981)
264. Little, L. H.: Infrared Spectra of Adsorbed Species. New York: Academic Press 1966
265. Hair, M. L.: Infrared Spectroscopy in Surface Chemistry. New York: Dekker 1967
266. Sheppard, N., Nguyen, T. T.: In: Advances in Infrared and Raman Spectroscopy. Clark, R. J. H., Hester, R. E. (eds.). London: Heyden & Sons 1978, vol. 5, p. 67
267. Angell, C. L.: In: Fourier Transform Infrared Spectroscopy. Ferraro, J. R., Basile, L. J. (eds.). New York: Academic Press 1982, p. 1
268. Selwood, P. W.: Chemisorption and Magnetisation. New York: Academic Press 1975

269. Wright, C. J.: In: Characterisation of Catalysts. Thomas, J. M., Lambert, R. M. (eds.). Chichester, New York, Brisbane, Toronto: John Wiley & Sons 1980, p. 169
270. Cairns, J. A.: ibid., p. 185
271. Gault, F. G.: Adv. Catal. **30**, 1 (1981)
272. Clarke, J. K. A., Creaner, A. C. M.: Ind. Eng. Chem. Prod. Res. Dev. **20**, 574 (1981)
273. Davis, S. C., Klabunde, K. J.: Chem. Rev. **82**, 153 (1982)
274. Somorjai, G. A.: Chemistry in two Dimensions. Ithaca: Cornell University Press 1981
275. Boudart, M.: Adv. Catal. **20**, 153 (1969)
276. Katzer, J. R., Manogue, W. H.: J. Catal. **32**, 166 (1974)
277. Ponec, V., Sachtler, W. M. H.: J. Catal. **24**, 250 (1972)
278. Hegedus, L. L., McCabe, R. W.: Catal. Rev. — Sci. Eng. **23**, 377 (1981)
279. Batholomew, C. H., Agrawal, P. K., Katzer, J. R.: Adv. Catal. **31**, 135 (1982)
280. Sachtler, W. M. H.: Catal. Rev. — Sci. Eng. **14**, 193 (1975)
281. Ponec, V.: In: The Chemical Physics of Solid Surfaces and Heterogeneous Catalysis. King, D. A., Woodruff, D. P. (eds.). Amsterdam: Elsevier 1982, p. 365
282. Hoare, M. R.: Advances in Chemical Physics. Prigogine, I., Price, S. A. (eds.). New York: Wiley & Sons 1979, vol. 40, p. 68
283. Hoare, M. R., Pal, P.: Adv. Phys. **20**, 161 (1971)
284. Burton, J. J.: J. Chem. Phys. **52**, 345 (1970)
285. Allpress, J. G., Sanders, J. V.: Aust. J. Phys. **23**, 23 (1970)
286. Fripiat, J. G., Chow, K. T., Boudart, M., Diamond, J. B., Johnson, K. H.: J. Molec. Catal. **1**, 59 (1975)
287. Gordon, M. B., Cyrot-Lackmann, F., Desjonqueres, M. C.: Surf. Sci. **80**, 59 (1979)
288. Kimoto, K.: Nippon Kessho Seicho Gakkaishi **6**, 122 (Eng.) (1979)
289. Kimoto, K., Nishida, I.: J. Phys. Soc. Japan **22**, 940 (1967)
290. Solliard, C.: Surf. Sci. **106**, 58 (1981)
291. Renou, A., Gillet, M.: Surf. Sci. **106**, 27 (1981)
292. a. Marks, L. D., Howie, A.: Nature **282**, 196 (1979)
b. Marks, L. D., Smith, D. J.: J. Crystal Growth **54**, 425 (1981)
293. Gillet, M.: J. Crystal Growth **36**, 239 (1976)
294. Baetzold, R. C.: J. Phys. Chem. **80**, 1504 (1976)
295. Muetterties, E. L.: C. & E. N., Aug. 1982, 28
296. Basset, J. M., Ugo, R.: In: Aspects of Homogeneous Catalysis 3. Tsutsui, M. (ed.). New York: Plenum Press 1979, p. 579
297. Chini, P.: Gasz. Chim. Ital. **109**, 225 (1979)
298. Jackson, S. D., Wells, P. B., Whyman, R., Worthington, P.: In: Catalysis (Specialist Periodical Report). The Chemical Society, London 1981, vol. 4, p. 75
299. Muetterties, E.: J. Organomet. Chem. **200**, 177 (1980)
300. Vidal, J. L., Troup, J. N.: J. Organomet. Chem. **213**, 351 (1981)
301. Steggerda, J. J., Bour, J. J., van der Velden, J. W. A.: Recueil **101**, 164 (1982)
302. Brieu, M., Gillet, M.: Thin Solid Films **100**, 53 (1983)
303. a. Martinengo, S., Ciani, G., Sironi, A., Chini, P.: J. Amer. Chem. Soc. **100**, 7096 (1978)
b. Ciani, G., Sironi, A., Martinengo, S.: J. Chem. Soc. Dalton Trans. **1982**, 1099
304. Venables, J. A., Price, G. L.: In: Epitaxial Growth. Part B. Matthews, J. W. (ed.). New York: Academic Press, p. 1
305. Popescu, M.: Proc. Int. Symp. on Heterogeneous Catalysis, 4th, 1979, part 1, p. 79
306. a. Tauster, S. J., Fung, S. C., Garten, R. L.: J. Amer. Chem. Soc. **100**, 170 (1978)
b. Baker, R. T. K., Prestridge, E. B., Garten, R. L.: J. Catal. **59**, 293 (1979)
307. Ino, S.: J. Phys. Soc. Japan **27**, 941 (1969)
308. Ino, S., Ogawa, S.: J. Phys. Soc. Japan **22**, 1365 (1967)
309. Heinemann, K., Yacaman, M. J., Yang, C. Y., Poppa, H.: J. Crystal Growth **47**, 177 (1979)
310. Avery, N. R., Sanders, J. V.: J. Catal. **18**, 129 (1970)
311. Chen, M., Schmidt, L. D.: J. Catal. **55**, 348 (1978)
312. Yacaman, M. J., Dominguez, J. M.: J. Catal. **64**, 213 (1980)
313. Yacaman, M. J., Fuentes, S., Dominguez, J. M.: Surf. Sci. **106**, 472 (1981)
314. Moraweck, B., Renouprez, A. J.: Surf. Sci. **106**, 35 (1981)

315. Dexpert, H., Freund, E., Lesage, E., Lynch, J. P.: In: Metal-Support and Metal-Additive Effects in Catalysis. Imelik, B., Naccache, C., Condurier, G., Praliaud, H., Meriaudeau, P., Gallezot, P., Martin, G. A., Vedrine, J. C. (eds.). Amsterdam: Elsevier 1982, p. 53
316. Dalmai-Imelik, G., Leclercq, C., Maubert-Muguet, A.: J. Solid State Chem. **16**, 129 (1976)
317. Ollis, D. F.: J. Catal. **23**, 131 (1971)
318. Hoffmann, D. W.: J. Catal. **27**, 374 (1972)
319. Anderson, J. R.: Structure of Metallic Catalysts. London, New York: Academic Press 1975, pp. 263–266
320. Dexpert, H., Freund, E., Lynch, J.: Quant. Microanal. High Spat. Resolut. Proc. Congr. 1981, p. 101
321. Bassi, I. W., Garbassi, F., Vlaic, G., Marzi, A., Tauszik, G. R., Cocco, G., Galvagno, S., Parravano, G.: J. Catal. **64**, 405 (1980)
322. Meitzner, G., Via, G. H., Lytle, F. W., Sinfelt, J. H.: J. Chem. Phys. **78**, 2533 (1983)
323. Lam, Y. L., Boudart, M.: J. Catal. **50**, 530 (1977)
324. Bartholomew, C. H., Boudart, M.: J. Catal. **29**, 278 (1973)
325. Vannice, M. A., Garten, R. L.: J. Molec. Catal. **1**, 201 (1975/76)
326. Garten, R. L.: J. Catal. **43**, 18 (1976)
327. Williams, F. L., Nason, D.: Surface Sci. **45**, 377 (1974)
328. Overbury, S. H., Bertrand, P. A., Somorjai, G. A.: Chem. Rev. **75**, 547 (1975)
329. Miedema, A. R.: Z. Metall. **69**, 455 (1979)
330. Kelley, M. J., Ponec, V.: Progress in Surface Science II (1981)
331. McLean, D.: Grain Boundaries in Metals. Oxford: Clarendon Press 1957
332. Seah, M. P.: J. Catal. **57**, 450 (1979)
333. Gibbs, J. W.: Collected Works. New York: Longmans 1928
334. Burton, J. J., Machlin, E. S.: Phys. Rev. Lett. **37**, 1433 (1976)
335. Tsai, N. H., Pound, G. M., Abraham, F. F.: J. Catal. **50**, 200 (1977)
336. Wynblatt, P., Ku, R. G.: Surface Sci. **65**, 511 (1977)
337. Abraham, F. F., Tsai, N. H., Pound, G. M.: Surface Sci. **83**, 406 (1979)
338. Abraham, F. F., Brundle, G. R.: J. Vac. Sci. Technol. **18**, 506 (1981)
339. Shek, M. L., Stefan, P. M., Weissman-Wenocur, D. L., Pate, B. B., Lindau, I., Spicer, W. E.: Surface Sci. **115**, L86 (1982)
340. Ertl, G.: In: The Nature of the Surface Chemical Bond. Rhodin, T., Ertl, G. (eds.). Amsterdam: North Holland 1980, p. 313
341. Tomanek, D., Mukherjee, S., Kumar, V., Bennemann, K. H.: Surface Sci. **114**, 11 (1982)
342. Burton, J. J., Hyman, E., Fedak, D. G.: J. Catal. **37**, 106 (1975)
343. Ng, Y. S., McLane, S. B., Tsong, T. T.: J. Vac. Sci. Technol. **17**, 154 (1980)
344. Moss, R. L.: In: Catalysis (Specialist Periodical Report). The Chemical Society, London 1977, vol. 1, p. 37
345. Sinfelt, J. H.: In: Catalysis. Science and Technology. Anderson, J. R., Boudart, M. (eds.). Berlin, Heidelberg, New York: Springer 1983, vol. 1, p. 257
346. Johnson, M. F. L., Leroy, V. M.: J. Catal. **35**, 434 (1974)
347. Webb, A. N.: J. Catal. **39**, 485 (1975)
348. Bolivar, C., Charcosset, H., Frety, R., Primet, M., Tournayan, L., Bertineau, C., Leclercq, G., Maurel, R.: J. Catal. **45**, 163 (1976)
349. Bertolacini, R. J., Pellet, R. J.: In: Deactivation of Catalysts. Delmon, B., Froment, G. F. (eds.). Amsterdam: Elsevier 1980, p. 73
350. Short, D. R., Khalid, S. M., Katzer, J. R., Kelley, M. J.: J. Catal. **72**, 288 (1981)
351. McNicol, B. D.: J. Catal. **46**, 438 (1977)
352. Charcosset, H., Frety, R., Leclercq, G., Mendes, E., Primet, M., Tournayan, L.: J. Catal. **56**, 468 (1978)
353. Dautzenberg, F. M., Helle, J. N., Biloun, P., Sachtler, W. M. H.: J. Catal. **63**, 119 (1980)
354. Burch, R.: J. Catal. **71**, 348 (1981)
355. Bacaud, R., Bussiere, P., Figueras, F.: J. Catal. **69**, 399 (1981)
356. Yermakov, Yu. I., Kuznetsov, B. N., Zakharov, V. A.: Catalysis by Supported Complexes. Amsterdam: Elsevier 1981, p. 342
357. Foger, K., Jaeger, H. K.: J. Catal. **70**, 53 (1981)

358. Chen, M., Schmidt, L. D.: J. Catal. **56**, 198 (1979)
359. Chen, M., Wang, T., Schmidt, L. D.: J. Catal. **60**, 356 (1979)
360. Wang, T., Schmidt, L. D.: J. Catal. **71**, 411 (1981)
361. Dalmon, J. A.: J. Catal. **60**, 325 (1979)
362. Maskos, Z., van Hooff, J. H. C.: J. Catal. **66**, 73 (1980)
363. Cale, F. S., Richardson, J. T.: J. Catal. **79**, 378 (1983)
364. Helms, C. R., Sinfelt, J. H.: Surface Sci. **72**, 229 (1978)
365. Galvagno, S., Schwank, J., Parravano, G., Garbassi, F., Marzi, A., Tauszik, G. R.: J. Catal. **69**, 283 (1981)
366. Moss, R. L., Pope, P., Davis, B. J.: J. Catal. **61**, 57 (1980)
367. Mintsa-Eya, V., Hilaire, L., Touroude, R., Gault, F. G., Moraweck, B., Renouprez, A.: J. Catal. **76**, 169 (1982)
368. Toolenaar, F. J. C. M., Reinalda, D., Ponec, V.: J. Catal. **64**, 110 (1980)
369. Ping-Chau Liao, Fleisch, T. H., Wolf, E. E.: J. Catal. **75**, 396 (1982)
370. Galvagno, S., Parravano, G.: J. Catal. **57**, 272 (1979)
371. Dominguez, M. E., Vazquez, A. S., Renouprez, A. J., Yacaman, M. J.: J. Catal. **75**, 101 (1982)
372. Grill, C. M., McLaughlin, M. L., Stevenson, J. M., Gonzalez, R. D.: J. Catal. **69**, 454 (1981)
373. Karpinski, Z., Koscielski, T.: J. Catal. **63**, 313 (1980)
374. Ramamoorthy, P., Gonzalez, R. D.: J. Catal. **58**, 188 (1979)
375. Miura, H., Gonzalez, R. D.: J. Catal. **74**, 216 (1982)
376. Miura, H., Gonzalez, R. D.: Ind. Eng. Chem. Prod. Res. Dev. **21**, 274 (1982)
377. Harding, M. P.: Ph. D. Thesis. University of Melbourne 1980
378. Van Hardefeld, R., Hartog, F.: Surface Sci. **15**, 189 (1969)
379. Anderson, J. R., Avery, N. R.: J. Catal. **5**, 446 (1966)
380. Foger, K., Anderson, J. R.: J. Catal. **54**, 318 (1978)
381. Perez, O. L., Romeu, D., Yacaman, M. J.: Appl. Surface Sci. **13**, 402 (1982)
382. Gallezot, P.: Surface Sci. **106**, 459 (1981)
383. Dowden, D. A.: In: Proc. 5th Int. Congr. Catal. Miami 1972. Hightower, J. W. (ed.). Amsterdam: North-Holland 1973, p. 621
384. Baetzold, R. C., Hamilton, J. F.: Prog. Solid State Chem. **15**, 1 (1983)
385. Messmer, R. P.: In: Semiempirical Methods of Electronic Structure Calculations. Segal, G. A. (ed.). New York: Plenum 1977, vol. 8, chapter 6
386. Mason, M. G.: Phys. Rev. B. 27, 748 (1983)
387. Kubo, R.: J. Phys. Soc. Japan **17**, 975 (1962)
388. Messmer, R. P., Knudson, S. K., Johnson, K. H., Diamond, J. B., Wang, C. Y.: Phys. Rev. B. **13**, 1396 (1976)
389. Salahub, D. R., Messmer, R. P.: Phys. Rev. B. **16**, 2526 (1977)
390. Baetzold, R. C.: J. Chem. Phys. **68**, 555 (1978)
391. Baetzold, R. C., Mack, R. E.: J. Chem. Phys. **62**, 1513 (1975)
392. Baetzold, R. C.: Inorg. Chem. **20**, 118 (1981)
393. Melius, C. F., Upton, T. H., Goddard, W. A.: Solid State Commun. **1978**, 28
394. Gordon, M. B., Cyrot-Lackmann, F., Desjonqueres, M. C.: Surface Sci. **68**, 359 (1977)
395. Baetzold, R. C., Mason, M. G., Hamilton, J. F.: J. Chem. Phys. **72**, 366 (1980)
396. Unwin, R., Bradshaw, A. M.: Chem. Phys. Lett. **58**, 58 (1977)
397. Takasu, Y., Unwin, R., Tesche, B., Bradshaw, A. M., Grunze, M.: Surface Sci. **77**, 219 (1978)
398. Mason, M. G., Gerenser, L. J., Lee, S. T.: Phys. Rev. Lett. **39**, 288 (1977)
399. Cyrot-Lackmann, F.: In: Quantum Theory of Chemical Reactions. Daudel, R., Pullman, A., Salem, L., Veillard, A. (eds.). Dordrecht, Boston, London: Reidel Publishing Company 1982, p. 35
400. Seib, D. H., Spicer, W. E.: Phys. Rev. B. **2**, 1676, 1694 (1970)
401. Stock, G. M., Williams, R. W., Faulkner, J. S.: Phys. Rev. B. **4**, 4390 (1971)
402. Soven, P.: Phys. Rev. **156**, 809 (1967); Phys. Rev. **158**, 1136 (1969)
403. Bansil, A., Ehrenreich, H., Schwartz, L., Watson, R. E.: Phys. Rev. B. **9**, 445 (1974)
404. Hufner, S., Wertheim, G. K., Wernick, J. H.: Phys. Rev. B. **8**, 4511 (1973)

405. Ehrenreich, H., Schwartz, L. M.: Solid State Phys. **31**, 149 (1976)
406. Korringa, J.: J. Phys. Chem. Solids **7**, 252 (1978)
Beeby, J. L.: Proc. Roy. Soc. London, Ser. A. **302**, 113 (1967)
407. Yu. K. Y., Helms, C. R., Spicer, W. E., Chyne, P. W.: Phys. Rev. B. **15**, 1629 (1977)
408. Ling, D. T., Miller, J. N., Stefan, P. M., Spicer, W. E.: Phys. Rev. B. **21**, 1417 (1980)
409. Burch, R.: Acc. Chem. Res. **15**, 24 (1982)
410. Fuentes, S., Figueras, F.: J. Chem. Soc. Farad. Trans. I, **74**, 174 (1978)
411. Tatarchuk, B. J., Dumesic, J. A.: J. Catal. **70**, 308, 323, 335 (1981)
412. Dowden, D. A.: In: Catalysis (Specialist Periodical Report). The Chemical Society, London 1980, vol. 3, p. 136
413. Anderson, J. R.: Structure of Metallic Catalysts. London, New York: Academic Press 1975, p. 275
414. Gallezot, P.: Catal. Rev. — Sci. Eng. **20**, 121 (1979)
415. Bond, G. C.: In: Metal Support and Metal Additive Effects in Catalysis. Imelik, B., Naccache, C., Condurier, G., Praliaud, H., Meriandeau, P., Gallezot, P., Martin, G. A., Vedrine, J. C. (eds.). Amsterdam: Elsevier 1982, p. 1
416. Rabo, J. A., Schomaker, V., Pickert, P. E.: In: Proc. 3rd Int. Congr. Catalysis. Amsterdam 1964. Amsterdam: North-Holland 1965, p. 1264
417. Chukin, M. V., Landau, M. V., Kruglikov, V. Ya., Agievskii, D. A., Smirnov, B. V., Belozerov, A. L., Asreiva, V. D., Goncharova, N. V., Radchenko, E. D., Konovalchikov, O. D., Agafonov, A. V.: in: Proc. 6th Int. Congr. Catalysis. London 1976. The Chemical Society, London 1977, p. 668
418. Gallezot, P., Datka, J., Massadier, J., Primet, M., Imelik, B.: ibid., p. 696
419. Tri, T. M., Massadier, J., Gallezot, P., Imelik, B.: In: Proc. 7th Int. Congr. Catalysis. Tokyo 1980. Tokyo: Kadansha. Amsterdam: Elsevier 1981, p. 266
420. Naccache, C., Kaufmann, N., Dufaux, M., Bandiera, J., Imelik, B.: In: Molecular Sieves II. Katzer, J. R. (ed.). Amer. Chem. Soc. 1977, p. 538
421. Figueras, F., Gomez, R., Primet, M.: Adv. Chem. Ser. **121**, 480 (1973)
422. Weber, R. S., Boudart, M., Gallezot, P.: In: Growth and Properties of Metal Clusters. Bourdon, J. (ed.). Amsterdam: Elsevier 1980, p. 415
423. Vedrine, J. C., Dufaux, M., Naccache, C., Imelik, B.: J. Chem. Soc. Farad. Trans. I, **74**, 440 (1978)
424. Antoshin, G. V., Shpiro, E. S., Tkachenko, O. P., Nikishenko, S. B., Ryashentseva, M. A., Avaev, V. I., Minachev. Kh. M., Zelinsky, N. D.: In: Proc. 7th Int. Congr. Catalysis. Tokyo 1980. Tokyo: Kadansha. Amsterdam: Elsevier 1981, p. 302
425. Tauster, S. J., Fung, S. C.: J. Catal. **55**, 29 (1978)
426. Chen, B. H., White, J. M.: J. Phys. Chem. **86**, 3534 (1982)
427. Ellestad, O. H., Naccache, C.: In: Perspectives in Catalysis. Larsson, R. (ed.). Lund: C.W.K. Gleerup 1981, p. 95
428. Hermann, J. M., Disdier, J., Pichat, P.: In: Metal Support and Metal Additive Effects. Imelik, B., et al. (eds.). Amsterdam: Elsevier 1982, p. 27
429. Meriaudeau, P., Ellestad, O. H., Dufaux, M., Naccache, C.: J. Catal. **75**, 243 (1982)
430. Duprez, D., Miloudi, A.: ibid., p. 179
431. Vannice, M. A.: J. Catal. **74**, 199 (1982)
432. Burch, R., Flambard, A. R.: J. Chem. Soc. Chem. Commun. **1981**, 965
433. Resasco, D., Haller, G. L.: J. Chem. Soc. Chem. Commun. **1980**, 1150
434. Ko, E. T., Garten, R. L.: J. Catal. **68**, 233 (1981)
435. Foger, K.: J. Catal. **78**, 406 (1982)
436. Meriandeau, P., Dutel, J. F., Dufaux, M., Naccache, C.: In: Metal Support and Metal Additive Effects in Catalysis. Imelik, B., et al. (eds.). Amsterdam: Elsevier 1982, p. 95
437. Resasco, D. E., Haller, G. L.: ibid., p. 105
438. Dauscher, A., Garin, F., Luck, F., Maire, G.: ibid., p. 113
439. Ko, E. I., Winston, S., Woo, C.: J. Chem. Soc. Chem. Commun. **1982**, 740
440. Vannice, M. A., Garten, R. L.: J. Catal. **56**, 236 (1979)
441. Bartholomew, C. H., Parnell, R. B., Butler, J. L.: J. Catal. **65**, 335 (1980)
442. Vannice, M. A., Garten, R. L.: J. Catal. **66**, 242 (1980)
443. Wang, S. Y., Moon, S. H., Vannice, M. A.: J. Catal. **71**, 167 (1981)

444. Kao, C. C., Tsai, S. C., Chung, Y. W.: J. Catal. **73**, 136 (1982)
445. Kao, C. C., Tsai, S. C., Chung, Y. W.: In: Metal Support and Metal Additive Effects in Catalysis. Imelik, B., et al. (eds.). Amsterdam: Elsevier 1982, p. 211
446. Vannice, M. A., Garten, R. L.: J. Catal. **63**, 255 (1980)
447. Katzer, J. R., Sleight, A. W., Gajardo, P., Michel, J. B., Gleason, E. F., McMillan, S.: J. Chem. Soc. Farad. Discuss. **72**, 121 (1982)
448. Rives-Arnau, V., Munuera, G.: Appl. Surface Sci. **6**, 122 (1980)
449. Nakamura, R., Yamagami, K., Nishiyama, S., Niiyama, H., Echigoya, E.: Chem. Lett. (Japan) **1981**, 275
450. Otter, G. J., Dautzenberg, F. M.: J. Catal. **53**, 116 (1978)
451. Martin, G. A., Dutartre, R., Dalmon, J. A.: React. Kinet Catal. Lett. **16**, 329 (1981)
452. Kunimori, K., Ikeda, Y., Soma, M., Uchijima, T.: J. Catal. **79**, 85 (1983)
453. Praliaud, H., Martin, G. A.: J. Catal. **72**, 394 (1981)
454. Menon, P. G., Froment, G. F.: Appl. Catal. **1**, 31 (1981)
455. Martin, G. A., Dalmon, J. A.: React. Kinet. Catal. Lett. **16**, 325 (1981)
456. Cairns, J. A., Baglin, J. E. E., Coark, G. J., Ziegler, J. F.: J. Catal. **83**, 301 (1983)
457. Horsley, J. A.: J. Amer. Chem. Soc. **101**, 2870 (1979)
458. Baker, R. T. K.: In: Metal Support and Metal Additive Effects in Catalysis. Imelik, B., et al. (eds.). Amsterdam: Elsevier 1982, p. 37
459. Ziegler, J. F.: In: New Uses of Ion Accelerators. Ziegler, J. F. (ed.). New York: Plenum Press 1975
460. Meriaudeau, P., Che, M., Gravelle, P. C., Teichner, S. J.: Bull. Soc. Chim. Fr. **13** (1973)
461. De Canio, S. J., Apple, T. M., Dybrowski, C. R.: J. Phys. Chem. **87**, 194 (1983)
462. Fung, S. C.: J. Catal. **76**, 225 (1982)
463. Fischer, T. E., Keleman, S. R., Polizzotti, R. S.: J. Catal. **69**, 345 (1981)
464. Bahl, M. K., Tsai, S. C., Chung, Y. W.: Phys. Rev. B. **21**, 1344 (1980)
465. Kao, C. C., Tsai, S. C., Bahl, M. K., Chung, Y. W., Lo, W. J.: Surface Sci. **95**, 1 (1980)
466. Chien, S. H., Shelimov, B. N., Resasco, D. E., Lee, E. H., Haller, G. L.: J. Catal. **77**, 301 (1982)
467. Huizinga, T., Prins, R.: J. Phys. Chem. **85**, 2156 (1981)
468. Conesa, J. C., Soria, J.: J. Phys. Chem. **86**, 1392 (1982)
469. Foger, K.: to be published
470. Schwab, G. M.: Adv. Catal. **27**, 1 (1978)
471. Solymosi, F.: Catal. Rev. **1**, 233 (1967)
472. Anderson, J. R., Harding, M. P.: J. Chem. Research (M), 1984, 0401–0409

Subject Index

Author Index Volume 1–6

Bei Fragen zur Produktsicherheit wenden Sie sich bitte an:
If you have any questions regarding product safety,
please contact:

Walter de Gruyter GmbH
Genthiner Straße 13
10785 Berlin
productsafety@degruyterbrill.com